Adipokines

Editors

Victor R. Preedy
Professor of Nutritional Biochemistry
School of Biomedical & Health Sciences
King's College London
and
Professor of Clinical Biochemistry
King's College Hospital
UK

Ross J. Hunter
Cardiology Research Fellow
St Bartholomew's Hospital
London
UK

CRC Press
Taylor & Francis Group
Boca Raton London New York

CRC Press is an imprint of the
Taylor & Francis Group, an **informa** business

A SCIENCE PUBLISHERS BOOK

CRC Press
Taylor & Francis Group
6000 Broken Sound Parkway NW, Suite 300
Boca Raton, FL 33487-2742

First issued in paperback 2017

© 2011 by Taylor & Francis Group, LLC
CRC Press is an imprint of Taylor & Francis Group, an Informa business

ISBN 13: 978-1-138-11441-8 (pbk)
ISBN 13: 978-1-57808-689-4 (hbk)

Library of Congress Cataloging-in-Publication Data

```
Adipokines / editors, Victor R. Preedy, Ross J. Hunter.
     p. ; cm. --  (Series on modern insights into disease
from
molecules to man)
 Includes bibliographical references and index.
 ISBN 978-1-57808-689-4 (hardcover)
1. Adipose tissues. 2. Fat cells. 3. Cytokines.  I.
Preedy, Victor
R. II. Hunter, Ross, 1977- III. Series: Series on modern in-
sights into
disease from molecules to man.
 [DNLM: 1. Adipokines--physiology.  QU 55.2]
 QP88.15.A325 2011
 612'.015756--dc22
                                              2011004641
```

Preface

The adipokines (also called adipocytokines), are a group of peptides secreted by adipose tissue. They have diverse roles, from the cell to the whole body. For example they have endocrine, autocrine and paracrine functions and they can also influence both transcription and translation. They can be pro-inflammatory and affect virtually all tissue systems either directly or indirectly. Although it is well known that they modulate or influence insulin resistance, endothelial function, cardiovascular disease, diabetes and obesity, they also appear to target other tissues as well. Moreover, they have other properties which are too numerous to document in a single preface. Thus, to understand adipokines in relation to health and disease a comprehensive text is needed. However, many books in health related sciences are written for experts alone. In Adipokines we have attempted to make the text accessible to the novice and expert alike. Each chapter has sections on applications to other areas of health and disease, key facts and definitions or explanations of key terms, genes, chemicals or pathways.

In Adipokines the Editors have invited distinguished scholars to write chapters which together will provide readers with a holistic knowledge-base to enable them to understand adipokines in specific or broad details. Coverage on specific adipokines includes adiponectin, acylation stimulating protein (ASP), adipsin, adrenomedullin, angiotensinogen, apelin, C1q/TNF-related proteins (CTRPs), fasting-induced adipose factor FIAF/Angiopoietin like protein 4, interleukins, leptin, omentin, plasminogen activator inhibitor-1 (PAI-1), resistin and retinol-Binding Protein 4, TNF-α, transforming growth factor, vaspin and visfatin. Other broad area of coverage includes adipokines in relation to thermal stress, neuroendocrine and autonomic function, obesity, nonalcoholic fatty liver disease, endocrine disruptors, the heart, rheumatic disease, allergy, sleep disorders, children and adolescents, immune cell development and innate immunity, hypocaloric diets, energy metabolism, deficiency, fibroblasts and chondrocytes, inflammation and seminal fluid and amniotic fluid.

There are 4 major sections as follows:

[1] Specific adipokines

[2] General cellular aspects

[3] Diseases and conditions

[4] Specific adipokines in disease

The book has a mixture of preclinical and clinical information and is designed for a broad scientific readership such as health scientists, doctors, physiologists, immunologists, biochemists, college and university teachers and lecturers, undergraduates and graduates. The chapters are written either by national or international experts or specialists in their field.

Victor R. Preedy
Ross J. Hunter

Contents

II. General Cellular Aspects

III. Diseases and Conditions

IV. Specific Adipokines in Disease

Section I
Specific Adipokines

CHAPTER 1

Acylation Stimulating Protein (ASP) as an Adipokine

Thea Scantlebury-Manning
Department of Biological and Chemical Sciences
Faculty of Pure and Applied Sciences, Cave Hill Campus
University of the West Indies, Bridgetown, St. Michael, Barbados

ABSTRACT

In recent years, adipose tissue has been recognized as a dynamic organ in that it produces several hormones/proteins (adipokines) in addition to energy storage in the form of triacylglycerides (TG). Acylation stimulating protein (ASP) is an adipokine that has been shown to be produced by adipocytes in a differentiation-dependent manner. All the proteins required for the production of ASP are also expressed and produced by adipocytes. ASP directly regulates glucose uptake and TG storage by increasing glucose transporters (GLUT) on the membrane and via up-regulating diacylglyceride transferase (DGAT) activity, the rate-limiting step of TG synthesis. In addition, dietary fat (CHYLO) has been shown to stimulate ASP production. The identification of C5L2 as the ASP receptor further supported the function of ASP as a hormone in lipid metabolism. Several fat-loaded meal studies in rodents have demonstrated that a loss in ASP or the ASP receptor function results in a reduction in TG clearance from circulation. The aim of this chapter is to give an overview of ASP production and function in lipid metabolism in association with obesity, diabetes and metabolic syndrome.

INTRODUCTION

Adipose tissue can be considered an endocrine organ in that it secretes a variety of factors. These factors, also known as adipokines, can influence energy metabolism. Acylation stimulating protein (ASP) is one such adipokine that is produced by adipose tissue and has been shown to play a role in the regulation of lipid metabolism. The aim of this chapter is to highlight the major functions of ASP in lipid metabolism and in association with obesity, diabetes and metabolic syndrome.

ASP PRODUCTION

ASP (C3adesArg) is a 76 amino acid protein (8932 D) produced via the interaction of complement C3 (ASP precursor), factor B and adipsin (complement factor D) (Hugli 1989). The level of ASP production has been shown to increase with respect to the maturity of the adipocyte (Cianflone and Maslowska 1995). The production of ASP (Fig. 1) is initiated with the formation of a C3bB complex. Adipsin cleaves factor B resulting in the formation of a C3bBb (C3 convertase) and a Ba fragment (Hugli 1989). This C3 convertase (C3bBb) cleaves C3 generating a C3a and a C3b fragment (Lesarve *et al.* 1979). The newly generated C3b fragment is recycled to regenerate the C3 convertase (C3bBb). The carboxy-terminal arginine of C3a fragment is cleaved by carboxypeptidase N to produce C3adesArg (Campbell *et al.* 2002).

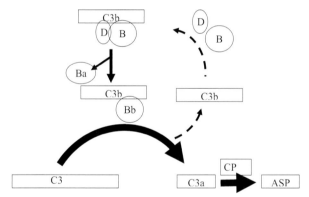

Fig. 1 ASP production. The interaction of complement C3, factor B and adipsin generates C3 convertase (C3bBb), which cleaves C3 into C3a and C3b. Carboxypeptidase (CP) then cleaves terminal arginine from C3a to produce ASP.

ASP RECEPTOR

The C5L2 (GPR77) protein was recently identified as a receptor for ASP (Kalant *et al.* 2003, 2005). Although the signaling pathway of this receptor still remains to be fully elucidated, C5L2 is a potential seven transmembrane G-protein coupled receptor. C5L2 is highly expressed in adipose tissue, muscle and the liver (Kalant *et al.* 2003). Studies performed in 3T3-L1 (murine) and human adipocytes suggested that C5L2 mediated the effects of ASP, since ASP does not bind to the known C3a and C5a receptors (Kalant *et al.* 2003, 2005, Monk *et al.* 2007). Furthermore, recent studies clearly demonstrated a positive correlation between plasma levels of ASP and the expression of C5L2 (Fig. 2) (MacLaren *et al.* 2010). This suggests that varying levels of C5L2 expression may explain the varying degrees of lipid metabolism seen in individuals.

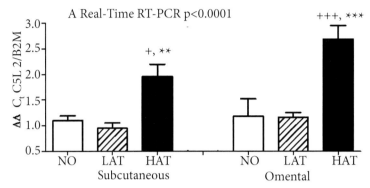

Fig. 2   C5L5 gene expression in adipose tissue is associated with plasma ASP and triacylglyceride levels. This graph shows C5L2 gene expression in subcutaneous (SC) and omental (OM) adipose tissue from control non-obese (NO), LAT (low plasma ASP and TG), and HAT (high plasma ASP and TG) subjects as assessed by ΔΔ real time RT-PCR for C5L2 relative to the housekeeping gene. Statistical analysis indicated HAT had higher levels of C5L2 when compared to LAT and NO in both types of tissues (p<0.001); however, there were no differences between tissues (pNS). With permission from MacLaren *et al.* (2010).

THE INVOLVEMENT OF ASP IN LIPID METABOLISM

Postprandial TG Storage

Adipose tissue is the major site of TG storage. After the ingestion of a meal (postprandial period), TG and cholesterol are packaged into chylomicron particles (CHYLOs). Like other types of lipoproteins, CHYLOs are responsible for transporting lipids in the body via blood. The general overview of postprandial TG storage can be found in Fig. 3. In summary, non-esterified fatty acids (NEFA) are produced as a result of hydrolysed TG from CHYLOs by lipoprotein lipase (LPL). Glycerol-3 phosphate (G-3P) is transported into the adipocyte via glucose transporter 4 (GLUT4). With the transfer of NEFA and G-3P into the cell, TG can be synthesized. Hormone-sensitive lipase (HSL) can hydrolyse the stored TG into NEFA and glycerol, which can be secreted into the blood to support the body's energy requirements.

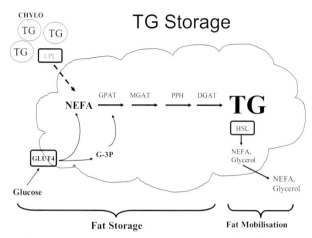

Fig. 3 Postprandial triacylglyceride storage. Lipoprotein lipase (LPL) generates non-esterified fatty acids (NEFA) to be taken up by the cell and esterified onto glycerol-3-phosphate (G3P) to produce TG. TG is synthesized via a series of enzyme reactions. Hormone-sensitive lipase (HSL) hydrolyses the TG into NEFA for release.

ASP and Regulation of TG Storage

After a meal, insulin has a positive effect on TG storage since it simultaneously stimulates LPL activity, decreases HSL activity and increases glucose transport. This ensures that NEFA entering the adipocyte is stored as TG and not released back into the blood. A study in human adipocytes has demonstrated that both ASP and insulin regulate NEFA esterification into TG and inhibit the effects of HSL, as seen in Fig. 4 (Van Harmelen *et al.* 1999). More specifically, ASP has been shown to stimulate the diacylglyerol transferase (DGAT) activity (Yasruel *et al.* 1991), which is the rate-limiting step of TG synthesis, thus demonstrating a direct regulation of ASP on TG storage. A study in human adipocytes has shown that insulin and ASP both increase the translocation of the glucose transporters, GLUT1 and GLUT4, to the cell surface and consequently has a direct positive effect on glucose uptake of the adipocyte (Maslowska *et al.* 1997). It was further demonstrated that CHYLOs (dietary lipoprotein) greatly increased the levels of C3 and ASP from human adipocytes (Maslowska *et al.* 1997) and that ASP levels rise concurrently with TG clearance in the microenvironment of the adipose tissue postprandially (Saleh *et al.* 1998). It can be hypothesized that ASP augmented the TG synthetic capacity of the adipocyte via increasing TG clearance.

A stratified analysis of fasting plasma ASP levels with respect to postprandial TG and NEFA clearances demonstrated that lower plasma ASP levels were associated with an augmented TG clearance in women when compared to men (Cianflone *et al.* 2004), suggesting that women are more sensitive to ASP than

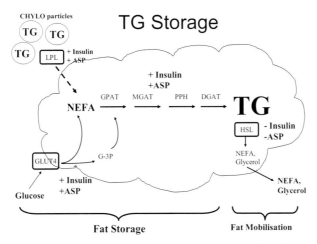

Fig. 4 ASP and insulin regulation of triacylglyceride storage. ASP and insulin positively regulate the uptake of the required components, and the enzymes needed for TG synthesis. ASP and insulin also negatively regulate the release of TG from the tissue.

men. Postprandial TG and ASP levels were measured in Caucasian American (CA) and African American women (AA), as seen in Fig. 5, and BMI correlated positively with ASP levels (Scantlebury-Manning *et al.* 2009). In addition to higher ASP levels, CA women demonstrated slower TG clearance when compared to their AA counterparts, suggesting an ASP-resistant state for CA women and an ASP-sensitive state for AA women. The levels of the receptor and/or response to ASP levels may explain the differences in lipid metabolism with respect to gender and race.

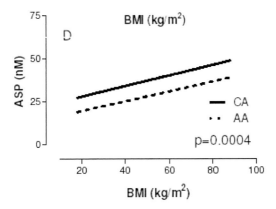

Fig. 5 ASP correlates with BMI. Fasting ASP levels in Caucasian American women (CA) at all BMIs are higher than in African American (AA) women (p=0.0004). ASP and BMI were significantly correlated. With permission from Scantlebury-Manning *et al.* (2009).

Confirmation studies in wild type, obese (ob/ob and db/db), and ASP-deficient (C3 knockout) models demonstrated that the levels of ASP positively correlated with the efficacy of TG and NEFA clearance (Murray *et al.* 1999a, Saleh *et al.* 2001). In addition, the administration of ASP to the above models resulted in an increase in TG and NEFA clearance. Recent studies in C5L2 knockout mice (ASP receptor deficient) (Paglialunga *et al.* 2007) further demonstrated that a lack of the ASP signaling pathway delays TG and NEFA clearance, thus strengthening the role of ASP in lipid metabolism.

Obesity, Diabetes, and Metabolic Syndrome

In the majority of studies (Cianflone *et al.* 2003), ASP levels were positively correlated with adiposity. Non–insulin-dependent diabetes mellitus (NIDDM) or type II diabetes is most often associated with obesity. Recent studies confirmed that serum ASP levels increased in individuals with type I (insulin-dependent diabetes mellitus) and type II (NIDDM) diabetes with a maintained state of hyperinsulinemia and hyperglycemia (Faraj *et al.* 2008). An improvement in glucose homeostasis in NIDDM via treatment resulted in a decrease in ASP production (Tahiri *et al.* 2007), suggesting a complex mechanism whereby ASP is one regulatory factor, among others, responding to the changes in glucose metabolism.

Metabolic syndrome is characterized by a group of metabolic risk factors that are present in the form of hyperglycemia, high blood pressure, elevated TG, central obesity, high LDL levels, and low HDL levels. Several population studies (Zimmet *et al.* 1999, Cianflone *et al.* 2003) have demonstrated that ASP levels change with respect to these metabolic risk factors. Recent studies in children have shown an association of metabolic syndrome with increased levels of ASP with presence of at least one single metabolic syndrome factor (hypertension, central obesity, or hyperglycemia) (Wamba *et al.* 2008, de Lind van Wijngaarden *et al.* 2010). It has been suggested that there is a genetic contribution to and correlation with the variation in circulating levels of ASP with respect to risk factors associated with metabolic syndrome (Martin *et al.* 2004). Moreover, metabolic syndrome is associated with lipodystrophy, NIDDM, coronary heart disease and obesity (Zimmet *et al.* 1999), suggesting that fasting ASP levels may be indicative of the efficiency of TG storage and consequently glucose and lipid metabolism (Table 1).

Summary Points

- ASP is produced by the interaction of C3, factor B and adipsin.
- ASP, C3, factor B and adipsin are produced by adipocytes.
- C5L2 was identified as the ASP receptor and is expressed by adipocytes.
- Dietary fat (CHYLO) increases ASP production.
- ASP increases NEFA and glucose uptake into adipocytes and consequently TG storage.

- Loss in ASP or C5L2 function results in a reduction in TG clearance and glucose uptake and consequently adipose tissue mass.
- ASP is positively correlated with BMI.
- ASP levels are increased in NIDDM, obesity and metabolic syndrome.

Table 1 Key facts about ASP and the role of ASP in lipid metabolism with respect to selected disease stages

- As adipocytes mature, ASP levels increase.
- C5L2 receptor is a G-coupled receptor that is expressed by adipocytes and identified as the ASP receptor.
- ASP stimulates glucose uptake and TG clearance in the postprandial period, while inhibiting TG release.
- ASP regulates TG synthesis via the activation of DGAT, rate-limiting enzyme in TG synthesis.
- Changes in the function of ASP and/or ASP receptor (C5L2) may account for varying metabolic characteristics observed in population studies with respect to gender and race differences.
- Increase in BMI suggests an increase in adipose tissue.
- ASP plasma levels are positively correlated with BMI.
- Obesity is associated with NIDDM and ASP levels are higher in NIDDM individuals.
- Correction of glucose homeostasis reduces ASP levels in NIDDM individuals.
- Symptoms of metabolic syndrome include hyperglycemia, high blood pressure, elevated TG, central obesity, high LDL levels, and low HDL levels.
- Other diseases associated with metabolic syndrome are lipodystrophy, NIDDM, coronary heart disease and obesity.
- ASP levels are higher in individuals diagnosed with metabolic syndrome.

Abbreviations

ASP	:	acylation stimulating protein
C3	:	complement C3
CHYLO	:	chylomicrons
B	:	factor B
D	:	adipsin
DGAT	:	diacylglycerol acyltransferase
GPAT	:	glycerol-3 phosphate acyltransferase
G-3P	:	glycerol-3 phosphate
GLUT	:	glucose transporter
HSL	:	hormone sensitive lipase
LPL	:	lipoprotein lipase
Mg2+	:	magnesium ion
MGAT	:	monoacylglycerol acyltranferase
NEFA	:	non-esterified fatty acids
PPH	:	phosphatidate phosphohydrolase
TG	:	triacylglyceride

Definition of Terms

Postprandial period: The time frame after a meal before the ingestion of another meal. For the studies discussed the postprandial period was 8 h.

References

Campbell, W.D. and E. Lazoura, N. Okada, and H. Okada. 2002. Inactivation of C3a and C5a octapeptides by carboxypeptidase R and carboxypeptidase N. Microbiol. Immunol. 46: 131-134.

Cianflone K. and M. Maslowska. 1995. Differentiation-induces production of ASP in human adipocytes. Eur. J. Clin. Invest. 25: 817-825.

Cianflone K. and Z. Xia, and L.Y. Chen. 2003. Critical review of acylation-stimulating protein physiology in humans and rodents. Biochim. Biophys. Acta. 1609: 127-143.

Cianflone K. and R. Zakarian, C. Couillard, B. Delphanque, J.P. Despres, and A.D. Sniderman. 2004. Fasting acylation stimulating protein is predictive of postprandial triglyceride clearance. J. Lipid Res. 45: 124-131.

de Lind van Wijngaarden R.F. and K. Cianflone, Y. Gao, R.W. Leunissen, and A.C. Hokken-Koelega. 2010. Cardiovascular and Metabolic Risk Profile and Acylation-Stimulating Protein Levels in Children with Prader-Willi Syndrome and Effects of Growth Hormone Treatment. J. Clin. Endocrinol. Metab. Feb 19. [Epub ahead of print].

Faraj, M. and G. Beauregard, A. Tardif, E. Loizon, A. Godbout, K. Cianflone, H. Vidal, and R. Rabasa-Lhoret. 2008. Regulation of leptin, adiponectin and acylation-stimulating protein by hyperinsulinaemia and hyperglycaemia in vivo in healthy lean young men. Diabetes Metab. 34: 334-342.

Hugli, T.E. 1989. Structure and function of C3a anaphylatoxin. Curr. Top. Microbiol. Immunol. 153: 181-208.

Kalant, D. and S.A. Cain, M. Maslowska, A.D. Sniderman, K. Cianflone, and P.N. Monk. 2003. The chemoattractant receptor-like protein C5L2 binds the C3a des-Arg77/acylaton-stimulating protein. J. Biol. Chem. 278: 11123-11129.

Kalant, D. and R. MacLaren, W. Cui, R. Samanta, P.N. Monk, S.A. Lapoorte, and K. Cianflone. 2005. C5L2 is a functional receptor for acylation-stimulating protein. J. Biol. Chem. 280: 23936-23944.

Lesarve, P.H. and T.E. Hugli, A. F. Esser, and H.J. Müller-Eberhard . 1979. The alternative pathway C3/C5 convertase: Chemical basis of factor B activation. J. Immunol. 123: 529-534.

MacLaren, R.E. and W. Cui, H. Hu, S. Simard, and K. Cianflone. 2010. Association of adipocyte genes with ASP expression: a microarray analysis of subcutaneous and omental adipose tissue in morbidly obese subjects. Med. Genomics, Jan 27 (in press).

Martin, L.J. and K. Cianflone, R. Zakarian, G. Nagrani, L. Almasy, D.L. Rainwater, S. Cole, J.E. Hixson, J.W. Blangero, and A.G. Comuzzie. 2004. Bivariate linkage between acylation stimulating protein and BMI and high-density lipoproteins. Obes Res. 12: 669-678.

Maslowska, M. and T. Scantlebury, R. Germinario, and K. Cianflone. 1997. Acute in vitro production of ASP in differentiated adipocytes. J. Lipid Res. 38: 21-31.

Monk, P.N. and A.M. Scola, P. Madala, and D.P. Fairlie. 2007. Function, structure and therapeutic potential of complement C5a receptors. Br. J. Pharmacol. 152: 429-448.

Murray, I. and A.D. Sniderman, and K. Cianflone. 1999. Enhanced triglyceride clearance with intraperitoneal human acylation stimulating protein (ASP) in C57Bl/6 mice. Am. J. Physiol. Endocrinol. Metab. 277: E474-E480.

Paglialunga, S. and P. Schrauwen, C. Roy, E. Moonen-Kornips, H. Lu, M.K. Hesselink, Y. Deshaies, D. Richard, and K. Cianflone. 2007. Reduced adipose tissue triglyceride synthesis and increased muscle fatty acid oxidation in C5L2 knockout mice. J. Endocrinol. 194: 293-304.

Saleh, J. and J.E. Blevins, P.J. Havel, J.A. Barrett, D.W. Gietzen, and K. Cianflone. 2001. Acylation stimulating protein (ASP) acute effects on postprandial lipemia and food intake in rodents. Int. J. Obes. Relat. Metab. Disord. 25: 705-713.

Saleh, J. and L.K. Summers, K. Cianflone, B.A. Fielding, A.D. Sniderman, and K.N. Frayn. 1998. Coordinated release of acylation stimulating protein (ASP) and triacylglycerol clearance by human adipose tissue *in vivo* in the postprandial period. J. Lipid Res. 39: 884-891.

Scantlebury-Manning, T. and J. Bower, K. Cianflone, and H. Barakat. 2009. Racial difference in Acylation Stimulating Protein (ASP) correlates to triglyceride in non-obese and obese African American and Caucasian women. Nutr. Metab. 6: 18-26 (in press).

Tahiri, Y. and F. Karpe, G.D. Tan, and K. Cianflone. 2007. Rosiglitazone decreases postprandial production of acylation stimulating protein in type 2 diabetics. Nutr. Metab. 4: 11.

Van Harmelen, V. and S. Reynisdottir, K. Cianflone, E. Degerman, J. Hoffstedt, K. Nilsell, A. Sniderman, and P. Arner P. 1999. Mechanisms involved in the regulation of free fatty acid release from isolated human fat cells by acylation-stimulating protein and insulin. J. Biol. Chem. 274: 18243-18251.

Wamba, P.C. and J. Mi, X.Y. Zhao, M.X. Zhang, Y. Wen, H. Cheng, D.Q. Hou, and K. Cianflone. 2008. Acylation stimulating protein but not complement C3 associates with metabolic syndrome components in Chinese children and adolescents. Eur. J. Endocrinol. 159: 781-790.

Yasruel, Z. and K. Cianflone, A.D. Sniderman, M. Walsh, and M.A. Rodriquez. 1991. Effect of acylation stimulating protein on the triacylglycerol synthetic pathway of human adipose tissue. Lipids 26: 495-499.

Zimmet, P. and E.J. Boyko, G.R. Collier, and M. de Courten. 1999. Etiology of the metabolic syndrome: potential role of insulin resistance, leptin resistance, and other players. Ann. N.Y. Acad. Sci. 892: 25-44.

2

Adiponectin as an Adipokine

Zhikui Wei[1] and G. William Wong[1,*]

[1]Department of Physiology and Center for Metabolism and
Obesity Research, Johns Hopkins University School of Medicine,
855 North Wolfe Street, Baltimore, MD 21205, USA

ABSTRACT

Adiponectin, a major insulin-sensitizing, multimeric hormone, is secreted by adipocytes and circulates at high levels in plasma. In the peripheral tissues, adiponectin activates AMPK and PPAR-α to enhance glucose uptake and fatty acid oxidation in muscle; in liver, adiponectin synergizes with insulin to suppress hepatic glucose output. Adiponectin also acts in the central nervous system. By activating AMPK in the hypothalamic neurons, adiponectin increases food intake and decreases peripheral energy expenditure. This protein binds to three known receptors—AdipoR1, AdipoR2 and T-cadherin—but the way in which these receptors are coupled to downstream signaling is poorly understood. Along with metabolic functions, adiponectin also has anti-inflammatory and anti-atherogenic properties, making it a truly multi-functional adipokine.

INTRODUCTION

As the largest endocrine organ, adipose tissue secretes many bioactive molecules that circulate in blood, collectively termed *adipokines* (Scherer 2006). Adipokines play important roles in energy homeostasis and the etiology of metabolic diseases. Adiponectin, the most intensely studied adipokine, has insulin-sensitizing, anti-inflammatory and anti-atherogenic properties. Since the discovery of adiponectin (Scherer *et al.* 1995), over 6000 articles on this molecule have been published. The growing body of evidence implicates adiponectin as a major insulin-sensitizing

---

*Corresponding author

adipokine and an important biomarker and therapeutic target for obesity-associated metabolic diseases (Kadowaki *et al.* 2008).

SYNTHESIS AND STRUCTURE

Adiponectin was discovered in a screen to look for gene highly induced upon 3T3-L1 adipocyte differentiation in culture (Scherer *et al.* 1995). Expression of adiponectin is highly specific to adipose tissue. Within adipose tissue, mature adipocytes express and secrete high levels of adiponectin. Under normal physiological conditions, adiponectin circulates at high concentrations (0.5-30 µg/mL) in plasma; these levels are influenced by hormonal signals and metabolic states (Combs *et al.* 2003). For example, testosterone inhibits the secretion of adiponectin, accounting for lower plasma levels of adiponectin in male rodents and humans (Xu *et al.* 2005). Further, circulating adiponectin levels are decreased in humans with obesity, diabetes and/or cardiovascular diseases (Trujillo and Scherer 2005). Indeed, decreased adiponectin levels in disease states provided the first clue to its metabolic function.

Structurally, adiponectin belongs to the C1q/TNF-α superfamily. The protein has a signal peptide, a variable N-terminal region, a collagen domain and a C-terminal globular domain homologous to the immune complement C1q (Fig. 1). Despite little protein sequence identity, the crystal structure of adiponectin, with a three-fold symmetry, strikingly resembles the trimeric structure of TNF-α (Fig. 2) (Shapiro and Scherer 1998). Most adiponectin in plasma exists as full-length protein. A small amount of the cleaved globular head can also be detected, but the function of this form is unknown (Fruebis *et al.* 2001). Further, adiponectin can form homomultimers through disulfide bond formation. The three most abundant multimers of adiponectin are the trimer, hexamer and higher molecular weight (HMW) oligomer of 18 subunits (Tsao *et al.* 2002). Each of the adiponectin isoforms is stable and does not interchange *in vivo*; their proportions in serum change with metabolic and disease states (Pajvani *et al.* 2004).

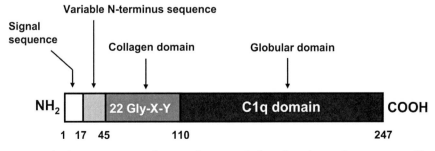

Fig. 1 The domain structures of mouse adiponectin. Indicated are the signal sequence, variable N-terminal region, collagen domain and C-terminal globular domain (Scherer *et al.* 1995).

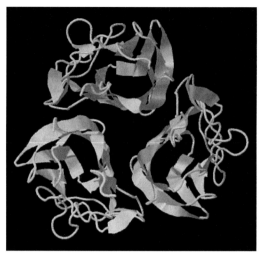

Fig. 2   Structure of adiponectin (Shapiro and Scherer 1998). The homotrimeric crystal structure of Acrp30/adiponectin was downloaded from the PDB protein databank (structure 1C3H).

Color image of this figure appears in the color plate section at the end of the book.

PHYSIOLOGICAL EFFECTS

Plasma levels of adiponectin decrease in conditions of obesity and diabetes. Replenishment of adiponectin by administration of recombinant protein significantly ameliorates insulin resistance in lipoatrophic mouse and diabetic KKAy mouse models (Yamauchi *et al.* 2001). Administration of recombinant adiponectin to normal or obese (ob/ob) mice decreases blood glucose by improving hepatic insulin action (Berg *et al.* 2001). Further, adiponectin enhances fatty acid oxidation and glucose uptake in muscle and induces weight loss in mice fed a high-fat diet (Fruebis *et al.* 2001). A transgenic mouse model with a three-fold elevation of plasma adiponectin levels shows increased lipid clearance, greater lipoprotein lipase activity and improved insulin action in suppressing hepatic glucose output (Combs *et al.* 2004). Impressively, elevating plasma levels of adiponectin in obese and diabetic *ob/ob* mice largely corrects most metabolic dysfunctions through massive expansion of the subcutaneous fat depot (Kim *et al.* 2007). Sequestration of fat in subcutaneous adipose tissue prevents lipotoxicity associated with ectopic lipid deposition in liver and muscle, a major cause of insulin resistance. Several adiponectin knockout mouse models have been generated; all exhibit variable and relatively mild metabolic phenotypes when challenged with a high-fat diet (Yano *et al.* 2008).

Adiponectin also has direct anti-atherogenic and anti-inflammatory effects. Transgenic over-expression of globular adiponectin in atherogenic-prone ApoE-null mice significantly reduces the size of atherosclerotic lesions (Yamauchi *et al.* 2003). In response to vascular injury, adiponectin knockout mice have thicker intima at the vascular wall, indicating a protective role for adiponectin in suppressing neointima formation (Kubota *et al.* 2002, Matsuda *et al.* 2002). Additionally, adiponectin inhibits macrophage uptake of oxidized low-density lipoprotein (LDL) rich in cholesterol, thus reducing the number of foam cells available to form atherosclerotic plaques (Ouchi *et al.* 2001).

As with leptin, adiponectin acts in the central nervous system (CNS) to regulate peripheral energy balance. Adiponectin is present in the cerebrospinal fluid of mice and levels decrease after refeeding. Administration of adiponectin by intracerebroventricular (i.c.v.) injection increases food intake in mice; concomitantly, whole-body oxygen consumption decreases, reflecting a reduction in peripheral energy expenditure (Kubota *et al.* 2007).

MECHANISMS OF ACTION

Adiponectin exerts whole-body insulin-sensitizing effects through multiple direct and indirect mechanisms (Fig. 3). In skeletal muscle, adiponectin activates the conserved kinase AMPK. Activated AMPK phosphorylates and inactivates acetyl coenzyme-A carboxylase, an enzyme that generates malonyl-CoA. Malonyl-CoA is a competitive inhibitor of carnitine palmitoyltransferase 1, which facilitates import of fatty acyl-CoA into mitochondria for β-oxidation. Thus, by activating AMPK, adiponectin enhances fatty acid oxidation in skeletal muscle (Tomas *et al.* 2002, Yamauchi *et al.* 2002). Decreasing circulating free fatty acid improves whole-body insulin action. In addition, adiponectin increases the expression of the nuclear receptor PPAR-α and its endogenous ligand activity. PPAR-α regulates the expression of many important metabolic genes, including fatty-acid transporter CD36, fatty-acid oxidation enzyme, acyl-coenzyme A (acyl-coA) oxidase and uncoupling protein (UCP) 2. Increased expression of these metabolic genes enhances fatty acid oxidation and decreases skeletal triglyceride (TG) accumulation. Adiponectin activation of AMPK also increases glucose uptake into muscle (Tomas *et al.* 2002, Yamauchi *et al.* 2002). In the liver, adiponectin activates AMPK to suppress gluconeogenic gene expression and glucose production (Yamauchi *et al.* 2002). Expression of a dominant-negative AMPK mutant abolishes effects of adiponectin on glucose and fatty acid metabolism. At the level of signal transduction, adiponectin reduces the p70 S6 kinase-mediated serine phosphorylation of insulin receptor substrate 1 that impairs insulin signaling, thereby improving insulin action (Wang *et al.* 2007).

The anti-inflammatory and anti-atherogenic properties of adiponectin stem from an ability to suppress the expression of inflammatory cytokines and adhesion molecules (Ouchi *et al.* 1999, Yamauchi *et al.* 2003). Further, adiponectin suppresses TNF-α-induced NF-κB activation by inhibiting IκB phosphorylation. In macrophages, adiponectin suppresses the expression of scavenger receptor class A1, responsible for LDL uptake, reducing the formation of cholesterol-laden foam cells (Ouchi *et al.* 2001).

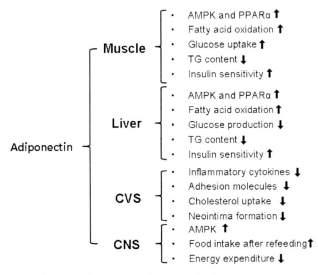

Fig. 3 Function of adiponectin and its mechanisms of action. Adiponectin exerts its biological effects in muscle, liver, cardiovascular system (CVS) and central nervous system (CNS) through the activation of AMPK and up-regulation of PPAR-α nuclear receptor.

The central effect of adiponectin is also mediated by AMPK. AMPK is phosphorylated and activated in the arcuate nucleus of the hypothalamus (ARH) follwing i.c.v. administration of adiponectin, resulting in increased food intake (Kubota *et al.* 2007).

Table I   Key points about adiponectin

- Insulin signaling and actions regulate many metabolic pathways. These include stimulation of glucose uptake into the insulin-sensitive tissues, suppression of endogenous glucose production and promotion of anabolic processes.
- Insulin resistance occurs when insulin-sensitive tissues have diminished response to insulin.
- Adipose tissue-derived adipokines directly or indirectly regulate systemic insulin sensitivity and glucose and fatty acid metabolism. Dysregulation of adipokines contributes to metabolic diseases.
- AMPK is a highly conserved energy gauge that is activated by high AMP/ATP ratio and other stimuli such as cellular stress, hormonal stimulation and pharmacological agents. Activated AMPK turns off anabolic processes that consume ATP and turns on catabolic processes that generate ATP.

CONCLUSION

Adiponectin is a multi-functional adipokine with insulin-sensitizing, anti-inflammatory and anti-atherogenic properties. Circulating adiponectin binds to receptors in both peripheral tissues and the CNS to control whole-body metabolism. Ongoing studies will likely uncover additional physiological processes regulated by adiponectin.

Summary Points

- Adiponectin is a hormone secreted by adipocytes that circulates at high concentrations in plasma.
- Circulating levels of adiponectin are influenced by hormonal signals and metabolic states.
- Adiponectin has insulin-sensitizing, anti-inflammatory and anti-atherogenic properties.
- Adiponectin acts in the CNS and peripheral tissues.
- Adiponectin increases glucose uptake and fatty acid oxidation in muscle.
- Adiponectin suppresses hepatic glucose output.
- Adiponectin is known to act by activating AMPK and up-regulating PPAR-α expression.

Abbreviations

Acyl-CoA	:	acyl-coenzyme A
AMPK	:	AMP activated protein kinase
ApoE	:	apolipoprotein E
CNS	:	central nervous system
i.c.v.	:	intracerebroventricular
LDL	:	low-density lipoprotein
NF	:	nuclear factor
PPAR-α	:	peroxisome proliferator-activated receptor α
TG	:	triglycerides
TNF-α	:	tumor necrosis factor α

Definition of Terms

Adhesion molecules: Cell surface proteins used by cells to interact with other cells or the extracellular matrix. They are central to the activation of immune cells.

Atherosclerosis: A cardiovascular disease in which the artery walls become thicker, narrower and harder through the build-up of fatty molecules like cholesterol.

Hyperlipidemia: A condition of elevated levels of lipids in the bloodstream.

KKAy mouse: Agouti yellow mutant mouse that has an ectopic expression of the agouti protein. These mice have hyperglycemia, hyperinsulinemia, glucose intolerance and obesity by the age of 8 wk and are a useful model of type 2 diabetes.

Lipoprotein: A large assembly of proteins and lipids. Low-density lipoproteins transport cholesterol and triglycerides from the liver to peripheral tissues.

ob/ob *mouse:* A mutant mouse that is leptin-deficient. These mice eat excessively, become morbidly obese and are a useful model of type 2 diabetes.

References

Berg, A.H. and T.P. Combs, X. Du, M. Brownlee, and P.E. Scherer. 2001. The adipocyte-secreted protein Acrp30 enhances hepatic insulin action. Nat. Med. 7: 947-953.

Combs, T.P. *et al.* 2004. A transgenic mouse with a deletion in the collagenous domain of adiponectin displays elevated circulating adiponectin and improved insulin sensitivity. Endocrinology 145: 367-383.

Fruebis, J. and T.S. Tsao, S. Javorschi, D. Ebbets-Reed, M.R. Erickson, F.T. Yen, B.E. Bihain, and H.F. Lodish. 2001. Proteolytic cleavage product of 30-kDa adipocyte complement-related protein increases fatty acid oxidation in muscle and causes weight loss in mice. Proc. Natl. Acad. Sci. USA 98: 2005-2010.

Kadowaki, T. and T. Yamauchi, and N. Kubota. 2008. The physiological and pathophysiological role of adiponectin and adiponectin receptors in the peripheral tissues and CNS. FEBS Lett. 582: 74-80.

Kim, J.Y. and E. van de Wall, M. Laplante, A. Azzara, M.E. Trujillo, S.M. Hofmann, T. Schraw, J.L. Durand, H. Li, G. Li, L.A. Jelicks, M.F. Mehler, D.Y. Hui, Y. Deshaies, G.I. Shulman, G.J. Schwartz, and P.E. Scherer. 2007. Obesity-associated improvements in metabolic profile through expansion of adipose tissue. J. Clin. Invest. 117: 2621-2637.

Kubota, N. and Y. Terauchi, T. Yamauchi, T. Kubota, M. Moroi, J. Matsui, K. Eto, T. Yamashita, J. Kamon, H. Satoh, W. Yano, P. Froguel, R. Nagai, S. Kimura, T. Kadowaki, and T. Noda. 2002. Disruption of adiponectin causes insulin resistance and neointimal formation. J. Biol. Chem. 277: 25863-25866.

Kubota, N. *et al.* 2007. Adiponectin stimulates AMP-activated protein kinase in the hypothalamus and increases food intake. Cell. Metab. 6: 55-68.

Matsuda, M. *et al.* 2002. Role of adiponectin in preventing vascular stenosis. The missing link of adipo-vascular axis. J. Biol. Chem. 277: 37487-37491.

Ouchi, N. and S. Kihara, Y. Arita, K. Maeda, H. Kuriyama, Y. Okamoto, K. Hotta, M. Nishida, M. Takahashi, T. Nakamura, S. Yamashita, T. Funahashi, and Y. Matsuzawa. 1999. Novel modulator for endothelial adhesion molecules: adipocyte-derived plasma protein adiponectin. Circulation 100: 2473-2476.

Ouchi, N. *et al.* 2001. Adipocyte-derived plasma protein, adiponectin, suppresses lipid accumulation and class A scavenger receptor expression in human monocyte-derived macrophages. Circulation 103: 1057-1063.

Pajvani, U.B. *et al.* 2004. Complex distribution, not absolute amount of adiponectin, correlates with thiazolidinedione-mediated improvement in insulin sensitivity. J. Biol. Chem. 279: 12152-12162.

Scherer, P.E. 2006. Adipose tissue: from lipid storage compartment to endocrine organ. Diabetes 55: 1537-1545.

Scherer, P.E. and S. Williams, M. Fogliano, G. Baldini, and H.F. Lodish. 1995. A novel serum protein similar to C1q, produced exclusively in adipocytes. J. Biol. Chem. 270: 26746-26749.

Shapiro, L. and P.E. Scherer. 1998. The crystal structure of a complement-1q family protein suggests an evolutionary link to tumor necrosis factor. Curr. Biol. 8: 335-338.

Tomas, E. and T.S. Tsao, A.K. Saha, H.E. Murrey, C.C. Zhang, S.I. Itani, H.F. Lodish, and N.B. Ruderman. 2002. Enhanced muscle fat oxidation and glucose transport by ACRP30 globular domain: acetyl-CoA carboxylase inhibition and AMP-activated protein kinase activation. Proc. Natl. Acad. Sci. USA 99, 16309-16313.

Trujillo, M.E. and P.E. Scherer. 2005. Adiponectin—journey from an adipocyte secretory protein to biomarker of the metabolic syndrome. J. Intern. Med. 257: 167-175.

Tsao, T.S. and H.E. Murrey, C. Hug, D.H. Lee, and H.F. Lodish. 2002. Oligomerization state-dependent activation of NF-kappa B signaling pathway by adipocyte complement-related protein of 30 kDa (Acrp30). J. Biol. Chem. 277: 29359-29362.

Wang, C. and X. Mao, L. Wang, M. Liu, M.D. Wetzel, K.L. Guan, L.Q. Dong, and F. Liu. 2007. Adiponectin sensitizes insulin signaling by reducing p70 S6 kinase-mediated serine phosphorylation of IRS-1. J. Biol. Chem. 282: 7991-7996.

Xu, A. and K.W. Chan, R.L. Hoo, Y. Wang, K.C. Tan, J. Zhang, B. Chen, M.C. Lam, C. Tse, G.J. Cooper, and K.S. Lam. 2005. Testosterone selectively reduces the high molecular weight form of adiponectin by inhibiting its secretion from adipocytes. J. Biol. Chem. 280: 18073-18080.

Yamauchi, T. *et al.* 2002. Adiponectin stimulates glucose utilization and fatty-acid oxidation by activating AMP-activated protein kinase. Nat. Med. 8: 1288-1295.

Yamauchi, T. *et al.* 2001. The fat-derived hormone adiponectin reverses insulin resistance associated with both lipoatrophy and obesity. Nat. Med. 7: 941-946.

Yano, W. and N. Kubota, S. Itoh, T. Kubota, M. Awazawa, M. Moroi, K. Sugi, I. Takamoto, H. Ogata, K. Tokuyama, T. Noda, Y. Terauchi, K. Ueki, and T. Kadowaki. 2008. Molecular mechanism of moderate insulin resistance in adiponectin-knockout mice. Endocr. J. 55: 515-522.

CHAPTER 3

Adipsin as an Adipokine

Yu-Feng Zhao[1] and Chen Chen[2,*]

[1]Preclinical Experiment Center, Fourth Military Medical University,
Xi'an 710032, China
[2]School of Biomedical Sciences, University of Queensland,
Brisbane 4072, Australia

ABSTRACT

Adipose tissue is accepted as an endocrine organ that secretes several hormones to regulate mainly metabolic balance of the body. Adipsin is one of these hormones from adipose tissue, also named adipokines. Function of adipsin includes not only metabolic regulatory action but also immune system regulation as a complement. Adipsin is a serine protease and exhibits the activity of complement factor D in mice. Human adipsin is structurally identical to complement factor D. Adipsin is highly expressed in adipose tissue both in rodent and in human, and the expression of adipsin is regulated by the differentiation of adipocytes, hormones such as insulin and adrenal glucocorticoid, and other factors such as retinoic acids and tumor necrosis factor-alpha. Current data reveal that the actions of adipsin include the activation of the alternative pathway of complement signaling, formation of another adipokine, acylation-stimulating protein, and synthesis of triglyceride in adipose tissue. Adipsin levels are closely related to obesity, insulin resistance and some immune diseases. The physiological and pathological roles of adipsin in obesity-related diseases remain to be clarified.

INTRODUCTION

Adipsin is a serine protease primarily synthesized by adipose tissue and to a lesser extent by sciatic nerve (Cook *et al.* 1987). It is abundantly secreted by adipose

---

*Corresponding author

tissue, and the blood adipsin level is about 50 μg/ml in healthy adult mice of normal weight (Rosen *et al.* 1989).

In 1983, adipsin was identified and found to be specifically expressed in differentiated adipocytes. Spiegelman's group isolated several clones from cDNA libraries corresponding to mRNA that were highly expressed during adipocyte differentiation (Spiegelman *et al.* 1983). One of these mRNA encoded a protein that was a novel member of the serine protease family at the time and named adipsin (Cook *et al.* 1985). Enzymatic analysis showed that adipsin has the structure of human complement factor D (Rosen *et al.* 1989). Antibody neutralization experiments suggest that adipsin is the murine homology of human complement factor D (Rosen *et al.* 1989). Recombinant adipsin substituted for complement factor D in functional assays (Rosen *et al.* 1989). Cloning of human adipsin cDNA confirmed that adipsin and complement factor D were identical in human (White *et al.* 1992).

SYNTHESIS AND ITS EXPRESSION

The mouse adipsin gene spans 1700 bases and contains five exons that are interrupted by four introns. The characteristic feature of serine protease genes is conserved in the adipsin gene. The exon structure of adipsin is highly homologous to the serine proteases such as trypsin, chymotrypsin, elastase and tissue plasminogen activator (tPA). The alternative splicing of adipsin gene transcript may occur, and two adipsin mRNA species are generated with difference in only three nucleotides, which encode two different signal peptides of adipsin (Min and Spiegelman 1986).

Mouse adipsin is 28,000 Daltons in size, and its sequence shows 61% identity with human adipsin or complement factor D. Human factor D is a single-chain serine protease 25,000 Daltons in size and having 222 amino acid residues. While a proenzyme form of complement factor D of 239 amino acid residues may exist in the body, human factor D presents mainly as an active form of 222 amino acids in blood circulation with a concentration around 1-2 μg/ml (Volanakis and Narayana 1996).

The expression of adipsin is regulated by the level of differentiation of adipocytes, which is regulated by several hormones, immune factors, and other factors. Blood levels of adipsin change in accordance with the expression of adipsin in adipocytes. Adipsin is expressed in only differentiated adipocytes, but not in undifferentiated preadipocytes, indicating that differentiation initiates the expression of adipsin. Adipogenesis results from a sequential induction of transcriptional factors C/EBPbeta, peroxisome proliferator-activated receptor gamma (PPAR-gamma), and C/EBPalpha (Gregoire *et al.* 1998). Retinoic acids inhibited C/EBPbeta-mediated transcription and blocked adipogenesis in experiments (Schwarz *et al.* 1997). Retinoic acids also reduced adipsin gene expression, supporting that adipsin expression was indeed initiated by the differentiation of adipocytes (Antras *et al.* 1991). In the mouse model of insulin deficiency, adipsin mRNA levels in adipose tissue were 2-3 times the level in normal mice (Miner *et al.* 1993). It is therefore suggested that insulin receptor signaling reduces the expression of adipsin. Adrenal glucocorticoid suppresses the expression of adipsin in genetically obese rat and mouse, and exogenous administration of glucocorticoid causes suppression in levels of adipsin mRNA and protein (Spiegelman *et al.* 1989). It is suggested that hyperglucocorticoid in certain obese models may be partially responsible for the lower adipsin levels in these models. Some immune factors also regulate adipsin expression. Tumor necrosis factor alpha (TNF-alpha) inhibits adipsin expression (Min and Spiegelman 1986). In one report, functional gamma-secretase inhibitors (FGSIs) increased adipsin levels, and this effect might be mediated by the inhibition of Hes-1 expression by FGSIs (Searfoss *et al.* 2003). Hes-1 was proposed to be a proximal promoter in mouse adipsin gene. The molecular detail of adipsin gene expression needs more studies into adipsin gene promoters.

PHYSIOLOGICAL ROLE

Adipsin contributes to the activation of alternative complement pathway. Besides secreting adipsin, adipose tissue produces the other two important complements that are involved in the activation of alternative pathway, C3 and complement factor B (Choy *et al.* 1992). All three proteins cooperate to activate the proximal portion of the alternative pathway to generate several complement-related polypeptides. Factor B and activated C3 (C3b) are first combined together. Adipsin/factor D cleaves the arginine-lysine bond in factor B, creating the C3 convergase (C3b.Bb). The C3 convergase leads to the production of C3a from C3 and the release of C3b. C3b covalently attach to target cell surface for opsonization. More importantly, C3b keeps accelerating the formation of more C3 convergase eventually to assemble a pore complex, which interrupts the cell membrane of infected cell to cause cell lysis (Xu *et al.* 2001b). C3a is the anaphylatoxin that causes changes in vascular permeability and vasoconstriction and is chemotactic for leukocytes (Haas and van Strijp 2007, Malmsten and Schmidtchen 2007). Complement is one of the major immunological defense systems in vertebrates. Adipsin catalyzes the critical first step in the alternative complement pathway, which establishes a connection between the alternative pathway of complement and adipose cell biology.

Another important role of adipsin is to mediate the formation of acylation-stimulating protein (ASP). ASP is an important adipokine that is formed in the activation of alternative pathway of complements. As mentioned above, adipsin causes the formation of C3 convergase, which leads to the formation of C3a, a polypeptide that is composed of 77 amino acid residues. The arginine on the carboxyl end of C3a is quickly cleaved by carboxypeptidase, resulting in a polypeptide of 76 amino acid residues (C3a desArg 77). Due to the stimulating

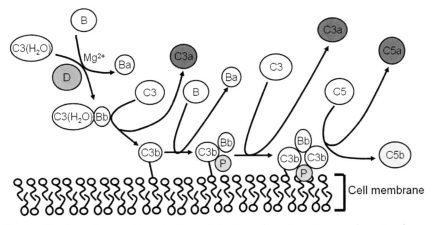

Fig. 1 The alternative complement pathway. Activation of the alternative complement pathway is initiated by the spontaneous hydrolysis of C3. Binding of factor B to $C3(H_2O)$ allows factor D/ adipsin to cleave factor B into Ba and Bb. Bb and $C3(H_2O)$ forms the complex, also known as C3 convertase. This convertase, although only produced in small amounts, can cleave multiple C3 proteins into C3a and C3b. After the creation of C3 convertase, the complement system follows the same path to the classical pathway. Binding of another C3b-fragment to the C3-convertase of the alternative pathway creates a C5-convertase analogous to the classical pathway. (This figure is modified from a file in Wikimedia Commons. Wikimedia Commons declares that permission is granted to copy, distribute and/or modify this document under the terms of the GNU Free Documentation License.)

Color image of this figure appears in the color plate section at the end of the book.

action of C3a desArg 77 on acylation, it is named ASP (Baldo *et al.* 1993). ASP is a potent stimulator of the esterification of fatty acids to form intracellular triglyceride in adipocytes (Cianflone 1997). ASP increases membrane transport of glucose by accelerating the translocation of glucose transporters from intracellular vesicles to the plasma membrane (Tao *et al.* 1997). Moreover, ASP increases the activity of diacylglycerol acyltransferase, which controls the rate limiting step in the synthesis of triglyceride (Yasruel *et al.* 1991). Both effects markedly increase the rate of triglyceride synthesis. Therefore, adipsin takes part in the regulation of triglyceride synthesis in adipocytes by promoting ASP production. Adipsin and ASP form the adipsin-ASP system, which is important in the regulation of adipocyte function (Sniderman and Cianflone 1994).

ROLE IN BODY DISEASES

It was reported that circulating levels of adipsin were significantly decreased in the genetic obesity model, db/db mice and ob/ob mice (Flier *et al.* 1987). The acquired obesity model made from newborn mice also showed reduction in adipsin levels. However, the obesity rat model induced by high energy overfeeding showed little

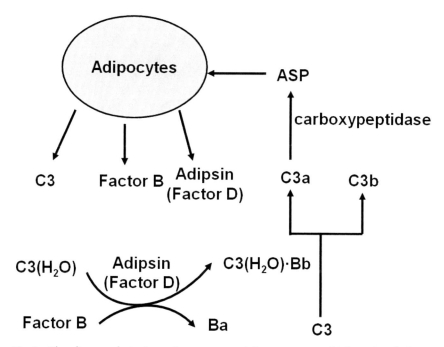

Fig. 2 The adipocyte-derived complement system. Adipocytes secrete C3, factor B and adipsin (factor D). All three proteins cooperate to activate the proximal portion of the alternative pathway to generate several complement-related polypeptides. ASP is created after cleavage of the arginine on the carboxyl end of C3a by carboxypeptidase. (Figure created by Yu-Feng Zhao and Chen Chen.)

Table I The key points of adipsin

- Adipsin is the adipocyte-secreted serine protease that has the activity of complement factor D.
- Adipsin is highly expressed in differentiated adipocytes and its expression is regulated by many factors including adipogenesis, insulin, adrenal glucocorticoid, retinoic acid and tumor necrosis factor alpha.
- Adipsin cooperates with C3 and factor B to activate alternative complement pathway. It cleaves the arginine-lysine bond in factor B to create C3 convergase, which is the critical step in the activation of alternative complement pathway.
- Adipsin mediates the formation of acylation-stimulating protein, which is a derivative of C3 and has strong stimulatory effects on triglyceride synthesis in adipocytes. Adipsin and ASP form the adipsin-ASP system, which is important in the regulation of adipocyte function.
- Adipsin levels in blood are altered after obesity, but no apparent abnormality in developmental biology and body weight was observed in the complement factor D-deficient mice. The role of adipsin in the etiology of obesity remains largely unknown.
- Adipsin levels were significantly increased in patients with immune diseases, which indicates that adipsin may take part in immune damage in some diseases.
- Adipsin, as one of the adipokines, connects adipose tissue to the function of immune system. The role of adipocyte-derived complement system in the regulation of energy metabolism and immunological action needs to be thoroughly studied.

change in adipsin expression in adipose tissues (Flier 1987). The defect of adipsin expression in genetic obesity model was later confirmed as a secondary feature to the onset of obesity (Dugail *et al.* 1990). In human, plasma adipsin levels were significantly high in obese group (Maslowska *et al.* 1999). Regression analysis demonstrated that adipsin was positively related to body mass index (BMI) in both obese and non-obese populations (Maslowska 1999). It is indicated that adipsin levels change in different pattern along with the progression of obesity. In the beginning of obesity, adipsin expression increases and accelerates the deposit of triglyceride, accompanying the increase in ASP. In the later stage of severe obesity, the decrease in adipsin may cause adipose tissue to negatively control fat deposit. In the complement factor D-deficient mice that were generated by gene targeting, no apparent abnormality in developmental biology and body weight was observed (Xu *et al.* 2001a). Therefore, the role of adipsin in the etiology of obesity remains largely unknown.

The involvement of adipsin/factor D in immune diseases, on the other hand, is well established. Recent data show that adipsin levels were significantly increased in patients with immune diseases such as seasonal allergic rhinitis and multiple sclerosis (Ciprandi *et al.* 2009, Hietaharju *et al.* 2009). Complement factor D deficiency may protect mice from gastrointestinal ischemia/reperfusion injury, suggesting that adipsin/factor D and alternative complement pathway play an important role in local tissue injury (Stahl *et al.* 2003). Elimination of adipsin/factor D also protects photoreceptors from light-induced damage in mice (Rohrer *et al.* 2007). These results indicate that adipsin/factor D may take part in the immune damage in some diseases and that inhibition of adipsin/factor D may be effective in the treatment.

In conclusion, adipose tissue is connected to the function of immune system. Adipsin, as one of the adipokines, plays an important role in the activation of alternative complement pathway and is involved in triglyceride synthesis via promoting ASP formation. Further investigation is warranted to clarify physiological and pathological roles of adipsin.

Summary Points

- Adipsin is a serine protease found in rodents and identical to complement factor D in human.
- Adipsin is primarily synthesized and secreted by differentiated adipocytes as an adipokine.
- Adipsin catalyses the formation of C3 convergase, the initial proteolytic step in the alternative complement pathway.
- Adipsin promotes the formation of ASP and composes adipsin-ASP system to stimulate triglyceride synthesis in adipocytes.

- Adipsin expression is regulated by hormones such as glucocorticoid and insulin.
- Adipsin levels are linked to obesity and some immune diseases.

Abbreviations

ASP	:	acylation-stimulating protein
BMI	:	body mass index
C/EBP	:	CCAAT/enhancer-binding protein
PPAR-gamma	:	peroxisome proliferator-activated receptor gamma
TNF-alpha	:	tumor necrosis factor alpha
tPA	:	tissue plasminogen activator

Definition of Terms

Acylation-stimulating protein: A derivative of C3 that increases triglyceride synthesis by stimulating fatty acids incorporation into adipose triglycerides.

Adipogenesis: The well-controlled process of cell differentiation by which preadipocytes become adipocytes.

Adipsin: A serine protease secreted by adipocytes into the bloodstream and has the same activity as human complement factor D.

Alternative complement pathway: One of the complement pathways to opsonize and kill pathogens, initiated by the spontaneous hydrolysis of C3 and activated by subsequent involvement of factor B and factor D.

Complement factor D: A serine protease involved in the alternative complement pathway of the complement system where it cleaves factor B.

Acknowledgment

Work in authors' laboratories is supported by Australian NHMRC, the University of Queensland, Australia, and the Fourth Military Medical University, China.

References

Antras, J. *et al.* 1991. Adipsin gene expression in 3T3-F442A adipocytes is posttranscriptionally down-regulated by retinoic acid. J. Biol. Chem. 266: 1157-1161.

Baldo, A. *et al.* 1993. The adipsin-acylation stimulating protein system and regulation of intracellular triglyceride synthesis. J. Clin. Invest. 92: 1543-1547.

Choy, L.N. *et al.* 1992. Adipsin and an endogenous pathway of complement from adipose cells. J. Biol. Chem. 267: 12736-12741.

Cianflone, K. 1997. Acylation stimulating protein and the adipocyte. J. Endocrinol. 155: 203-206.

Ciprandi, G. *et al.* 2009. Serum adipsin levels in patients with seasonal allergic rhinitis: preliminary data. Int. Immunopharmacol. 9: 1460-1463.

Cook, K.S. *et al.* 1985. A developmentally regulated mRNA from 3T3 adipocytes encodes a novel serine protease homologue. Proc. Natl. Acad. Sci. USA 82: 6480-6484.

Cook, K.S. *et al.* 1987. Adipsin: a circulating serine protease homolog secreted by adipose tissue and sciatic nerve. Science 237: 402-405.

Dugail, I. *et al.* 1990. Impairment of adipsin expression is secondary to the onset of obesity in db/db mice. J. Biol. Chem. 265: 1831-1833.

Flier, J.S. *et al.* 1987. Severely impaired adipsin expression in genetic and acquired obesity. Science 237: 405-408.

Gregoire, F.M. *et al.* 1998. Understanding adipocyte differentiation. Physiol. Rev. 78: 783-809.

Haas, P.J. and J. van Strijp. 2007. Anaphylatoxins: their role in bacterial infection and inflammation. Immunol. Res. 37: 161-175.

Hietaharju, A. *et al.* 2009. Elevated cerebrospinal fluid adiponectin and adipsin levels in patients with multiple sclerosis: a Finnish co-twin study. Eur. J. Neurol.

Malmsten, M. and A. Schmidtchen. 2007. Antimicrobial C3a—biology, biophysics, and evolution. Adv. Exp. Med. Biol. 598: 141-158.

Maslowska, M. *et al.* 1999. Plasma acylation stimulating protein, adipsin and lipids in non-obese and obese populations. Eur. J. Clin. Invest. 29: 679-686.

Min, H.Y. and B.M. Spiegelman. 1986. Adipsin, the adipocyte serine protease: gene structure and control of expression by tumor necrosis factor. Nucleic Acids Res. 14: 8879-8892.

Miner, J.L. *et al.* 1993. Adipsin expression and growth in rats as influenced by insulin and somatotropin. Physiol. Behav. 54: 207-212.

Rohrer, B. *et al.* 2007. Eliminating complement factor D reduces photoreceptor susceptibility to light-induced damage. Invest. Ophthalmol. Vis. Sci. 48: 5282-5289.

Rosen, B.S. *et al.* 1989. Adipsin and complement factor D activity: an immune-related defect in obesity. Science 244: 1483-1487.

Schwarz, E.J. *et al.* 1997. Retinoic acid blocks adipogenesis by inhibiting C/EBPbeta-mediated transcription. Mol. Cell. Biol. 17: 1552-1561.

Searfoss, G.H. *et al.* 2003. Adipsin, a biomarker of gastrointestinal toxicity mediated by a functional gamma-secretase inhibitor. J. Biol. Chem. 278: 46107-46116.

Sniderman, A.D. and K. Cianflone. 1994. The adipsin-ASP pathway and regulation of adipocyte function. Ann. Med. 26: 388-393.

Spiegelman, B.M. *et al.* 1983. Molecular cloning of mRNA from 3T3 adipocytes. Regulation of mRNA content for glycerophosphate dehydrogenase and other differentiation-dependent proteins during adipocyte development. J. Biol. Chem. 258: 10083-10089.

Spiegelman, B.M. *et al.* 1989. Adrenal glucocorticoids regulate adipsin gene expression in genetically obese mice. J. Biol. Chem. 264: 1811-1815.

Stahl, G.L. *et al.* 2003. Role for the alternative complement pathway in ischemia/reperfusion injury. Am. J. Pathol. 162: 449-455.

Tao, Y. *et al.* 1997. Acylation-stimulating protein (ASP) regulates glucose transport in the rat L6 muscle cell line. Biochim. Biophys. Acta. 1344: 221-229.

Volanakis, J.E. and S.V. Narayana. 1996. Complement factor D, a novel serine protease. Protein Sci. 5: 553-564.

White, R.T. *et al.* 1992. Human adipsin is identical to complement factor D and is expressed at high levels in adipose tissue. J. Biol. Chem. 267: 9210-9213.

Xu, Y. *et al.* 2001a. Complement activation in factor D-deficient mice. Proc. Natl. Acad. Sci. USA 98: 14577-14582.

Xu, Y. *et al.* 2001b. Structural biology of the alternative pathway convertase. Immunol. Rev. 180: 123-135.

Yasruel, Z. *et al.* 1991. Effect of acylation stimulating protein on the triacylglycerol synthetic pathway of human adipose tissue. Lipids 26: 495-499.

CHAPTER **4**

Adrenomedullin as a Cytokine

Alexis Elias Malavazos[1,*] **and Gianluca Iacobellis**[2]

[1]Department of Medical and Surgical Sciences, University of Milan, IRCCS Policlinico San Donato, 30, Morandi Street, 20097 San Donato Milanese, Italy
[2]Department of Medicine, McMaster University, St. Joseph's Hospital, 50 Charlton Avenue East, Fontbonne Bldg, Hamilton, ON, L8N 4A6, Canada

ABSTRACT

In the last few years, adipose tissue (AT) has been widely investigated and has been revealed to act as a real endocrine organ, secreting various active biomolecules named adipokines, involved in the development of obesity-related diseases.

One of these molecules is adrenomedullin, a cytokine expressed in many different cellular types that has been reported to regulate renal function, neurotransmission, and cellular proliferation and also to suppress insulin secretion and oxidative stress, in addition to its vasoactive and hypotensive properties.

In particular, it has a crucial role in the adipose tissue, where it appears to be related to the lipidic metabolism, pathophysiology of obesity and obesity-related diseases, through insulin secretion, hypoxia and inflammation, which are the main up-regulators of adrenomedullin release.

Adrenomedullin is also related to multiple metabolic factors such as insulin resistance (IR) and diabetes mellitus, C reactive protein, LDL cholesterol, adiponectin and basal microcirculatory perfusion.

Furthermore, it has been recently suggested that adrenomedullin, synthesized by the epicardial fat (EF), which is the visceral heart fat depot along the distribution of the coronary arteries, could also exert a protective action to the heart, through its vasodilator and antioxidative actions.

It is therefore possible to consider adrenomedullin as an interesting target for investigation because of its vascular implications.

---

*Corresponding author

INTRODUCTION

Adrenomedullin is a potent vasoactive and anti-oxidative peptide originally isolated from human pheochromocytoma.

Studies have detected that adrenomedullin is ubiquitously expressed in many cellular types, including vascular endothelial and vascular smooth muscle cells, cardiomyocytes, neurons, glial cells, fibroblasts and adipocytes.

A variety of biological actions of adrenomedullin have been reported: regulation of renal function, neurotransmission, cell proliferation, suppressive actions on insulin secretion and oxidative stress. The complex of calcitonin receptor-like receptor and receptor activity-modifying protein-2 or -3 generate adrenomedullin receptor, a seven trans-membrane domains receptor protein.

ADIPOSE TISSUE AND ADRENOMEDULLIN

AT has recently been considered to play an active endocrine and paracrine role. In fact, the adipocyte is an active organ that secretes a variety of bioactive molecules, named adipokines, which are directly and indirectly related to metabolic and cardiovascular diseases, such as diabetes mellitus, obesity, hypertension and coronary artery disease.

Among these is adrenomedullin, produced by subcutaneous and visceral AT. Several studies on murine cell lines demonstrated that 3T3-L1 and 3T3-F442 A cells synthesize adrenomedullin and that rat AT contains adrenomedullin mRNA. Experimental studies reported that AT obtained from rats exposed to high-fat diet expresses greater amounts of adrenomedullin than that obtained from rats fed with low-fat diet (Fukai *et al.* 2004).

The presence of adrenomedullin mRNA has been also confirmed in human AT. Adrenomedullin peptide is particularly produced by isolated human adipocytes and human mesenchymal stem cell (hMSC)–derived adipocytes when exposed to an inflammatory environment (Harmancey *et al.* 2005, Linscheid *et al.* 2005). It has been reported that adrenomedullin levels in the visceral AT were significantly higher in obese women than in lean ones (Paulmyer-Lacroix *et al.* 2006).

The mechanisms regulating the secretion of adrenomedullin by the AT are still unclear, but the main up-regulators of preproadrenomedullin gene transcription and adrenomedullin release seem to be insulin, inflammation and hypoxia, all closely related to visceral obesity.

Adrenomedullin appears to be related to the lipidic metabolism.

Treatment with adrenomedullin evoked a dose-dependent activation of lipolysis, assessed by measuring glycerol released into medium of hMSC-derived adipocytes (Linscheid *et al.* 2005).

However, adrenomedullin failed to modify non-estereated fatty acid (NEFA) release into the culture medium of 3T3-F442A-derived adipocytes, whereas a beta-adrenergic receptor agonist, isoproterenol, caused a strong increase in NEFA release. Adrenomedullin causes a shift to the right of the concentration–response curve to isoproterenol in lipolysis, suggesting that adrenomedullin may inhibit beta-adrenergic–stimulated lipolysis. This inhibition by adrenomedullin seems to be mediated by nitric oxide (NO), which oxidizes isoproterenol to an inactive compound, isoprenochrome. The β3-adrenergic receptor, predominantly expressed in adipocytes, mediates the catecholamine action on lipolysis (Harmancey *et al.* 2005). Although the mechanism is still uncertain, adrenomedullin may modulate the expression of β3-adrenergic receptors via the cAMP pathway as an autocrine and/or paracrine factor.

INFLAMMATION AND ADRENOMEDULLIN

Loss of endothelial barrier function is a central feature of acute inflammation and contributes essentially to organ dysfunction in severe infection. Several studies tested the potency of adrenomedullin as a therapeutic molecule for the direct stabilization of endothelial barrier function during inflammation and infection. Cumulative evidence demonstrates increased NO release, vasodilatation in different vascular beds, and subsequent hypotension due to adrenomedullin. With respect to hypotension in severe infection and sepsis it is noteworthy that septic mice over-expressing adrenomedullin in their vasculature are resistant to endotoxin shock despite hypotension. Adrenomedullin-related reduction of mortality is also observed in models of polymicrobial sepsis, endotoxin-related shock, and hemorrhage. Besides its action on the macro-circulation, intravital microscopy studies demonstrated that adrenomedullin diminishes microcirculatory disturbances under inflammatory conditions. In addition, adrenomedullin attenuates endothelial cell apoptosis in polymicrobial sepsis possibly by increasing antiapoptotic Bcl-2 expression. In concert with the reduction of pro-inflammatory cytokine release, these vascular effects of adrenomedullin may improve circulation during severe infection (Temmersfeld-Wollbrück *et al.* 2007).

Furthermore, inflammatory cytokines like tumor necrosis factor α and lipopolysaccharide positively regulate adrenomedullin expression in a time- and dose-dependent manner (Takahashi *et al.* 2005). The induction of adrenomedullin expression by those pro-inflammatory factors might indicate the possible protective role of adrenomedullin against the inflammation-induced dysfunction of adipocytes.

OBESITY AND ADRENOMEDULLIN

Vila *et al.* recently reported that plasma concentration of an inactive product of post-translational preproadrenomedullin, the midregional proadrenomedullin, significantly correlates with body mass index (BMI). Interestingly, the same authors showed that the preproadrenomedullin significantly decreases in severely obese subjects who underwent Roux-en-Y gastric bypass (Vila *et al.* 2009). These findings suggest that the reduction in body weight and AT inflammation might be the denominators of decreased plasma adrenomedullin.

Nomura *et al.* 2009 measured the biologically active adrenomedullin and its intermediate form in obese and non-obese subjects without overt cardiovascular disease. In both these adrenomedullin forms, plasmatic levels were higher in obese subjects than in non-obese subjects, with significant relationships between BMI and metabolic factors. These results suggest active roles of adrenomedullin in metabolic syndrome (Nomura *et al.* 2009).

Circulating adrenomedullin is elevated in overweight patients with essential hypertension and is decreased by hypocaloric diet (Minami *et al.* 2000). This vasodilator peptide may act against further elevation in blood pressure in obese patients with essential hypertension.

DIABETES AND ADRENOMEDULLIN

Plasma levels of adrenomedullin were elevated in patients with type 2 diabetes mellitus (T2DM) (Hayashi *et al.* 1997). It has been hypothesized that the increased visceral adiposity commonly observed in T2DM patients may play a contributory role as source of circulating adrenomedullin.

However, adrenomedullin seems to act directly on the pancreatic beta-cells, inhibiting insulin secretion. Two experimental models have been used to study adrenomedullin's effects in pancreatic physiology: (1) Analysis of isolated rat islets showed that adrenomedullin inhibits insulin secretion in a dose-dependent manner. The monoclonal antibody MoAb-G6, neutralizing adrenomedullin bioactivity, was able to increase insulin release 5-fold; this effect was reversed by the addition of synthetic adrenomedullin. (2) Oral glucose tolerance tests showed that intravenous injection of adrenomedullin reduces insulin bloodstream levels with a concomitant increase in circulating glucose.

It is therefore plausible that adrenomedullin acts on muscle or liver as a positive regulator in the insulin signaling pathway, such as the phosphatidylinositol 3-kinase/Akt pathway. However, adipocyte differentiation has been associated with a down-regulation of adrenomedullin and up-regulation of resistin, raising the hypothesis that low expression of adrenomedullin and high expression of resistin in adipocytes may be related to the higher occurrence of IR in obese subjects (Li *et al.* 2003).

Studies on adrenomedullin gene knockout mice demonstrated that adrenomedullin deficiency leads to IR in aged mice, possibly due to the decreased intrinsic antioxidant effect of adrenomedullin (Shimosawa *et al.* 2003).

Individuals with T2DM are at risk for vascular injury. Plasma MR-proADM concentration was elevated in subjects with T2DM with preserved renal function. This was further accentuated when nephropathy set in and could be an appropriate physiological response to ongoing vascular injury. Adrenomedullin-mediated vasodilation could be determinant in basal microcirculatory perfusion and might play a role in diabetic vasculopathy pathogenesis (Lim *et al.* 2007).

Adrenomedullin was also higher in patients with diabetic retinopathy and epiretinal membranes. This mediator appears to be involved in the development of angiogenesis and proliferation in the eye. Since retinal hypoxia and proliferation are the main features of vascular and proliferative disease, it is possible to consider adrenomedullin as an interesting target for investigation because of its vascular implications (Er *et al.* 2005).

HEART AND ADRENOMEDULLIN

Several cardiac physiological and pathophysiological roles have been attributed to adrenomedullin. Ischemic and hypoxic conditions seem to stimulate the production and secretion of adrenomedullin. The increase in adrenomedullin circulating levels and AT expression might be induced by cerebral or cardiac ischemia. It has been suggested that adrenomedullin synthesis from the EF could play a protective role on the coronary arteries. Iacobellis *et al.* 2009 demonstrated that chronic and stable coronary artery disease (CAD) was associated with low intracoronaric and systemic adrenomedullin levels and also down-regulated EF adrenomedullin gene and protein expression; in fact, intracoronary adrenomedullin concentrations were significantly lower in subjects with CAD than in those without CAD. It has been suggested that EF adrenomedullin synthesis increases in response to chronic hypoxemia. However, the relationship of adrenomedullin with CAD is partly unclear (Iacobellis *et al.* 2009). Recent studies also showed that adrenomedullin is able to up-regulate M2 muscarinic receptors in cardiomyocytes derived from the murine P19 cell line (Buys *et al.* 2003). Adrenomedullin promotes angiogenesis and may therefore participate in angiogenesis under certain conditions.

Adrenomedullin could exert a protective action on the heart. Circulating adrenomedullin levels have been suggested as promising independent predictors of future cardiovascular events in patients with multiple cardiovascular risk factors (Nishida *et al.* 2008). However, the use of adrenomedullin in clinical practice will need further investigation.

Table 1 Key features of adrenomedullin

- Adrenomedullin appears to be related to the activation of lipolysis modulating the expression of β3-adrenergic receptors.
- The induction of adrenomedullin expression by pro-inflammatory factors might indicate its possible protective role against inflammation-induced adipocyte dysfunction through increased release of NO.
- Adrenomedullin and its vasodilator effect may act against further elevation in blood pressure in obese patients with essential hypertension.
- Adrenomedullin secreted by epicardial fat could play a protective role on the coronary arteries, through its vasoactive action.
- Adrenomedullin could be an interesting target for its vascular implications.

Table 2 Key features of endothelial function

- Modulation of coagulation
- Modulation of endothelium-blood cell interactions
- Modulation of vascular permeability
- Modulation of vascular tone
- Modulation of vascular structure

CONCLUSIONS

Adrenomedullin is produced by subcutaneous and visceral AT, including EF, so it can be considered a new member of the adipokine family. The mechanisms that regulate the secretion of adrenomedullin by the AT are still unclear.

Adrenomedullin may have a cardio-protective role, through its antioxidant and potent vasodilator effects. However, future studies are necessary to clarify the role of adrenomedullin, as secreted by AT, particularly in obesity.

Summary Points

- Adrenomedullin can be considered a new member of the adipokine family and increases in obese individuals.
- Adrenomedullin is important for the stabilization of endothelial barrier function during inflammation.
- Adrenomedullin is implicated in obesity-related diseases, mainly in type 2 diabetes and metabolic syndrome.
- Adrenomedullin acts on muscle or liver as a positive regulator in insulin signaling pathway.
- Adrenomedullin could play a protective role on the heart, through its vasoactive action.

Abbreviations

AT	:	adipose tissue
BMI	:	body mass index
CAD	:	coronary artery disease
EF	:	epicardial fat
hMSC	:	human mesenchymal stem cell
IR	:	insulin resistance
NEFA	:	non-estereated fatty acid
NO	:	nitric oxide
T2DM	:	type 2 diabetes mellitus

Definition of Terms

Adipokines: Soluble autocrine, paracrine or systemic mediators produced by adipocytes.

Adipose tissue: A complex endocrine and immune organ involved in insulin sensitivity, appetite, endocrine functions, inflammation and immunity.

Epicardial fat: Visceral heart fat depot surrounding and buffering coronary arteries against the torsion induced by cardiac contraction, facilitating remodelling, regulating fatty acid homeostasis in the coronary microcirculation and providing fatty acids to cardiac muscle as a local energy source in times of high demand.

Metabolic syndrome: A variety of common disorders, including hyperglycemia, hyperlipidemia, hypertension and visceral obesity, predisposing to the development of cardiovascular disease.

Visceral obesity: Accumulation of adipose tissue inside the abdominal cavity.

References

Buys, S. and F. Smih, A. Pathak, and P. Philip-Couderc. 2003. Adrenomedullin upregulates M2-muscarinic receptors in cardiomyocytes from P19 cell line. Br. J. Pharmacol. 139: 1219-27.

Er, H. and S. Doğanay, E. Ozerol, and M. Yürekli. 2005. Adrenomedullin and leptin levels in diabetic retinopathy and retinal diseases. Ophthalmologica 219: 107-111.

Fukai, N. and T. Yoshimoto, T. Sugiyama, and N. Ozawa. 2004. Concomitant expression of adrenomedullin and its receptor components in rat adipose tissues. Am. J. Physiol. Endocrinol. Metab. 288: E56-E62.

Harmancey, R. and J.M. Senard, A. Pathak, and F. Desmoulin. 2005. The vasoactive peptide adrenomedullin is secreted by adipocytes and inhibits lipolysis through NO-mediated β-adrenergic agonist oxidation. FASEB J. 19: 1045-1047.

Hayashi, M. and T. Shimosawa, M. Isaka, and S. Yamada. 1997. Plasma adrenomedullin in diabetes. Lancet 350: 1449-1450.

Iacobellis, G. and C.R.T. Gioia, M. Di Vito, and L. Petramala. 2009. Epicardial adipose tissue and intracoronary adrenomedullin levels in coronary artery disease. Horm. Metab. Res. 41: 1-6.

Li, Y. and K. Totsune, K. Takeda, and K. Furuyama. 2003. Differential expression of adrenomedullin and resistin in 3T3-L1 adipocytes treated with tumor necrosis factor-α. Eur. J. Endocr. 149: 231-238.

Lim, S.C. and N.G. Morgenthaler, T. Subramaniam, and Y.S. Wu. 2007. The relationship between adrenomedullin, metabolic factors, and vascular function in individuals with type 2 diabetes. Diabetes Care 30: 1513-1519.

Linscheid, P. and D. Seboek, H. Zulewski, and U. Keller. 2005. Autocrine/paracrine role of inflammation-mediated calcitonin gene-related pepetide and adrenomedullin expression in human adipose tissue. Endocrinology 146: 2699-2708.

Minami, J. and T. Nishikimi, T. Ishimitsu, and Y. Makino. 2000. Effect of a hypocaloric diet on adrenomedullin and natriuetic peptides in obese patients with essential hypertension. J. Cardiovasc. Pharmacol. 36: S83-S86.

Nishida, H. and T. Horio, Y. Suzuki, and Y. Iwashima. 2008. Plasma adrenomedullin as an independent predictor of future cardiovascular events in high-risk patients: comparison with C-reactive protein and adiponectin. Peptides 29: 599-605.

Nomura, I. and J. Kato, M. Tokashiki, and K. Kitamura. 2009. Increased plasma levels of the mature and intermediate forms of adrenomedullin in obesity. Regul. Pept. 158:127-31.

Paulmyer-Lacroix, O. and R. Desbriere, M. Poggi, and V. Achard. 2006. Expression of adrenomedullin in adipose tissue of lean and obese women. Eur. J. Endocr. 155: 177-185.

Shimosawa, T. and T. Ogihara, H. Matsui, and T. Asano. 2003. Deficiency of adrenomedullin induces insulin resistance by increasing oxidative stress. Hypertension 41: 1080-1085.

Takahashi, K. and K. Totsune, M. Sone, and K. Kikuchi. 2005. Effects of adipokines on expression of adrenomedullin and endothelin-1 in cultured vascular endothelial cells. Peptides 26: 845-851.G.

Temmesfeld-Wollbrück, B. and A.C. Hocke, N. Suttorp, and S. Hippenstiel. 2007. Adrenomedullin and endothelial barrier function. Thromb. Haemost. 98: 944-51.

Vila, G. and M. Riedl, C. Maier, and J. Struck. 2009. Plasma MR-proADM correlates to BMI and decreases in relation to leptin after gastric bypass surgery. Obesity 17: 1184-1188.

5

Angiotensinogen as an Adipokine

Miki Nagase[1] and Toshiro Fujita[1,*]

[1]Department of Nephrology and Endocrinology, University of Tokyo Graduate School of Medicine, 7-3-1 Hongo, Bunkyo-ku, Tokyo 113-8655, Japan

ABSTRACT

Although the liver is the primary site of angiotensinogen production, adipocytes have emerged as a major extrahepatic source of angiotensinogen. Angiotensinogen levels are higher in visceral adipose tissue than subcutaneous adipose tissue. Angiotensinogen gene expression is up-regulated by nutritional and hormonal stimuli, as well as during adipogenesis. Clinical and experimental studies have reported activation of circulating and adipose renin-angiotensin-aldosterone-system (RAAS) in obesity. Positive correlations were found between plasma or adipose angiotensinogen levels and BMI or waist-to-hip ratio in obese subjects. In addition, adipose angiotensinogen mRNA is increased in rodent models of diet-induced obesity. The impact of adipose angiotensinogen was supported by gain/loss of function animals, as well as in weight reduction studies. Angiotensinogen knockout mice were hypotensive, lean, and resistant to diet-induced obesity. Notably, adipocyte-specific over-expression of angiotensinogen in both wild-type and angiotensinogen knockout mice increased plasma angiotensinogen levels, BP, and adipose mass, suggesting that adipose angiotensinogen may be a potential link between obesity and hypertension. Weight loss in obese subjects lowered plasma and adipose angiotensinogen levels and decreased BP. Angiotensin II (Ang II), the product of angiotensinogen, modulates adipocyte functions, such as hypertrophy, differentiation, adipokine secretion, and induction of oxidative stress and inflammation. We demonstrated that chronic Ang II infusion in rats induced whole body insulin resistance. Ang II impaired insulin-induced glucose

*Corresponding author

uptake into skeletal muscle and adipocytes by inhibiting GLUT4 translocation to the membrane, possibly via the induction of oxidative stress. Compatible with these findings, several clinical trials have demonstrated that ACE inhibitors and ARBs improve insulin sensitivity and reduce the risk of the development of type 2 diabetes.

INTRODUCTION

Angiotensinogen, the precursor of Ang II, was discovered as a secretory product of adipocytes in the 1980s. Subsequent studies revealed that adipocytes synthesize and secrete all components of the renin-angiotensin-aldosterone-system (RAAS). Obesity is a condition of volume expansion and sodium retention, and there is accumulating evidence for the activation of circulating and adipose RAAS. The significance of adipose angiotensinogen was illustrated by gain/loss of function in animals as well as in weight reduction studies, shedding light on the possible role of adipose angiotensinogen in obesity-associated hypertension.

ANGIOTENSINOGEN EXPRESSION AND ITS REGULATION IN ADIPOCYTES

Although the liver is the primary site of angiotensinogen production, angiotensinogen mRNA expression in adipocytes is comparatively high (68% of that in the liver). Considering that the percentage of WAT accounts for up to 60% of body weight in obese subjects, adipocytes emerge as an important extrahepatic source of angiotensinogen, especially in obese individuals. The presence of angiotensinogen mRNA in adipocytes was first discovered in the 1980s (Campbell *et al.* 1987, Cassis *et al.* 1988). Angiotensinogen mRNA was found in perivascular adipose tissue and in WAT and brown adipose tissue depots, and angiotensinogen levels are higher in visceral than in subcutaneous adipose tissue.

Angiotensinogen mRNA levels are increased during adipogenesis. Angiotensinogen gene expression in adipocytes is positively regulated by fatty acids, dexamethasone, androgen, lipopolysaccharides, tumor necrosis factor α, and Ang II. With regard to the effect of insulin on angiotensinogen levels, both down-regulation and up-regulation have been reported. Adipocyte angiotensinogen formation appears to be enhanced in situations reflecting obesity-associated insulin resistance.

Angiotensinogen expression in rat adipose tissue is nutritionally regulated: fasting reduced angiotensinogen levels and refeeding increased its expression. These local changes were accompanied by parallel changes in BP, although plasma angiotensinogen levels and hepatic angiotensinogen expression were not affected.

ANGIOTENSINOGEN IN OBESITY

Several human studies have reported a positive correlation between plasma angiotensinogen concentrations and BMI (Bloem *et al.* 1995). A significant association was also found between adipose angiotensinogen expression and waist-to-hip ratio or BMI (Giacchetti *et al.* 2000, van Harmelen *et al.* 2000). The impact of obesity on RAAS was further evaluated in a weight reduction study (Engeli *et al.* 2005). In obese menopausal women, weight loss (-5%) lowered plasma angiotensinogen by 27% and adipose angiotensinogen expression by 20%, and was accompanied by a 7 mmHg reduction in systolic BP.

In rodent models of obesity, angiotensinogen mRNA expression was enhanced in visceral adipose tissue in rats with diet-induced obesity and hypertension (Boustany *et al.* 2004). In obese Zucker rats with a leptin receptor mutation, a significant increase in adipose angiotensinogen expression was found only in the early stages of obesity, suggesting that the duration of obesity may influence the degree of adipose angiotensinogen expression. Alternatively, expanded adipose mass as a whole, rather than enhanced expression in individual adipocytes, may be sufficient to activate the systemic RAAS. Treatment with ARB ameliorated adipokine dysregulation and reduced adipose oxidative stress in mice with diet-induced obesity.

EVIDENCE FROM GENETIC ENGINEERING STUDIES

Angiotensinogen-deficient mice were hypotensive and exhibited impaired sodium handling in the kidney. The mice were lean and partly protected from diet-induced obesity, despite normal food intake (Massiera *et al.* 2001b). Mice with deficiencies of other components of the RAAS, such as renin, ACE, and Ang II type 1a and type 2 receptors, also exhibited leanness and/or resistance to the development of obesity similar to angiotensinogen knockout mice, supporting a role for RAAS in the regulation of adipose mass.

The pathophysiological significance of adipose angiotensinogen in obesity-associated hypertension emerged from studies with adipocyte-specific angiotensinogen transgenic mice (Massiera *et al.* 2001a). Adipocyte-specific over-expression of angiotensinogen in wild-type mice resulted in elevated plasma angiotensinogen levels, increased BP, and adipose mass enlargement. Re-expression of angiotensinogen in the adipose tissue of angiotensinogen knockout mice led to detectable circulating angiotensinogen levels, and restoration of fat mass, BP and renal function.

In addition, adipocyte-specific over-expression of 11β-hydroxysteroid dehydrogenase-1, an enzyme that converts inactive glucocorticoids to their active forms, resulted in the development of metabolic syndrome, including visceral adiposity and hypertension, and elevation of circulating and adipose angiotensinogen (Masuzaki *et al.* 2003). Importantly, hypertension was ameliorated by treatment with ARB, indicating that RAAS activation played an important role in this model of obesity-associated hypertension. Together, these experimental models support the idea that adipocytes are an important source of elevated circulating angiotensinogen levels in obesity, thereby affecting BP.

EFFECTS OF ANG II ON ADIPOCYTES

Ang II, which is generated from angiotensinogen, regulates adipose functions via both Ang II type 1 and type 2 receptors. For example, Ang II inhibits differentiation of human preadipocytes and reduces the number of mature adipocytes available for fat storage. This can result in the deposition of excess fat in skeletal muscle, liver, and the pancreas, leading to increased insulin resistance. By contrast, ARB enhances preadipocyte differentiation and increases the number of small insulin-sensitive adipocytes.

Adipocytes store lipids either by the uptake of free fatty acids or by *de novo* synthesis of triacylglycerols via fatty acid synthase and glycerol-3-phosphate dehydrogenase. Ang II increases the activity and transcription of these enzymes and enlarges the size of adipocytes. In addition, Ang II modulates the secretion of several adipokines, such as adiponectin, tumor necrosis factor α, leptin, prostacyclin, nitric oxide, plasminogen activator inhibitor 1, and macrophage chemoattractant protein 1. Ang II can also induce oxidative stress and inflammation in adipose tissue. It should be noted, however, that there are species-specific differences concerning the effects of Ang II in adipose tissue.

ANG II AND INSULIN RESISTANCE

Ang II is believed to be an important mediator of insulin resistance. Crosstalk between the Ang II- and insulin- signaling cascades is implicated in the molecular mechanism of Ang II-induced insulin resistance.

We examined whether chronic Ang II excess induces insulin resistance *in vivo*, and determined the underlying mechanisms (Fig. 1) (Ogihara *et al.* 2002, 2003, Fujita 2007). Chronic Ang II infusion in rats (100 ng/kg/min for 12 d) resulted in whole body insulin resistance, as assessed by glucose infusion and utilization rates during hyperinsulinemic euglycemic clamp, and hepatic glucose production. Ang II impaired insulin-induced glucose uptake into skeletal muscle and adipocytes by inhibiting GLUT4 translocation to the membrane. Despite the development of insulin resistance, early insulin-signaling steps leading from activation of the insulin receptor and IRSs to activation of PI3-kinase and Akt phosphorylation were enhanced in muscle, liver, and adipose tissue in Ang II-infused rats. Oxidative stress is likely involved, because tempol, a membrane-permeable superoxide dismutase mimetic, normalized the above-mentioned Ang II-evoked insulin resistance. On the other hand, oxidative stress induced by glutathione depletion caused insulin resistance by impairing IRS-1 phosphorylation, PI3-kinase activation, and GLUT4 translocation in cultured adipocytes.

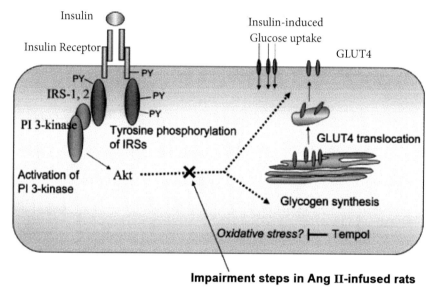

Impairment steps in Ang II-infused rats

Fig. 1 Angiotensin II induces insulin resistance through crosstalk with insulin signaling. Angiotensin II infusion into rats impaired insulin-induced glucose uptake into skeletal muscle and adipocytes by inhibiting GLUT4 translocation to the membrane. Early insulin-signaling steps, such as activation of the insulin receptor, IRSs, PI3-kinase, and Akt, were enhanced in angiotensin II-infused rats, indicating that angiotensin II disturbs later steps in the insulin signaling cascade (with permission from Ogihara *et al.* 2003).

CLINICAL TRIALS

Several large-scale clinical trials have shown that blockade of the RAAS with ACE inhibitors or ARBs, beyond lowering BP, reduces the incidence of type 2 diabetes in high-risk patients. A meta-analysis of these trials (~33,100 patients) indicated that RAAS blockade reduces the incidence of type 2 diabetes by an average of 22% ($P < 0.0001$), compared with other antihypertensive therapies (Scheen 2004). ACE inhibitors and ARBs appear to be equally effective. It is estimated that one new case of type 2 diabetes can be prevented by treating 45 patients for 5 yr.

Table I Key features of adipokine angiotensinogen as mediator of obesity-associated hypertension

- Adipocytes are a major extrahepatic source of angiotensinogen, especially in obesity.
- Plasma or adipose angiotensinogen was positively correlated with BMI or waist-to-hip ratio in obese subjects.
- Angiotensinogen mRNA expression was enhanced in visceral adipose tissue of rodents with diet-induced obesity.
- Adipocyte-specific over-expression of angiotensinogen increased plasma angiotensinogen levels, BP, and adipose mass, suggesting that adipose angiotensinogen is a potential link between obesity and hypertension.
- Weight reduction lowered plasma and adipose angiotensinogen, and decreased BP.
- ACE inhibitors or ARBs are promising drugs for the treatment of obesity-related disorders such as hypertension, type 2 diabetes, and metabolic syndrome.

CONCLUSION

Clinical and experimental evidence has shown that the adipose-tissue RAAS is up-regulated in obesity. Angiotensinogen is over-expressed, especially in visceral fat, and adipose-derived angiotensinogen contributes to elevated circulating angiotensinogen levels, providing a potential link between obesity and hypertension. Although some species-specific differences exist, Ang II was demonstrated to inhibit the differentiation of preadipocytes, regulate the production of several adipokines, promote oxidative stress and inflammation, and directly interfere with the insulin signaling pathway, culminating in insulin resistance. Finally, several randomized controlled trials support the efficacy of RAAS blockade by ACE inhibitors or ARBs to improve insulin sensitivity and reduce the risk of the development of type 2 diabetes.

Summary Points

- Visceral adipocytes are a major extrahepatic source of angiotensinogen.
- Circulating and local adipose-tissue RAAS are activated in obesity, whereas weight loss lowered plasma and adipose angiotensinogen levels and decreased BP. Adipocyte-specific over-expression of angiotensinogen increased plasma angiotensinogen levels, BP, and adipose mass, suggesting a link between adipose angiotensinogen production and obesity-associated hypertension.
- Angiotensinogen gene expression is up-regulated by nutritional and hormonal stimuli, and during adipogenesis.
- Ang II, the product of angiotensinogen, modulates adipocyte functions, such as hypertrophy and differentiation, adipokine secretion, and induction of oxidative stress and inflammation.
- Ang II infusion in rats caused insulin resistance by interfering with the insulin signaling cascade, possibly via induction of oxidative stress.

- Several large-scale clinical trials have shown that blockade of the RAAS with ACE inhibitors or ARBs, beyond lowering BP, reduces the incidence of type 2 diabetes in high-risk patients.

Abbreviations

ACE	:	angiotensin-converting enzyme
Ang II	:	angiotensin II
ARB	:	angiotensin II type 1 receptor blocker
BMI	:	body mass index
BP	:	blood pressure
GLUT4	:	glucose transporter 4
IRS	:	insulin receptor substrate
PI3-kinase	:	phosphatidylinositol-3 kinase
RAAS	:	renin-angiotensin-aldosterone system
WAT	:	white adipose tissue

Definition of Terms

Adipose-tissue RAAS: All components of RAAS, including renin, AGT, ACE, ACE2, prorenin receptor, AT1, AT2 receptors, and the mineralocorticoid receptor, are expressed in rodent and human adipocytes and act independently of systemic RAAS.

Insulin signaling: The binding of insulin to the insulin receptor stimulates the receptor tyrosine kinase, resulting in phosphorylation of the receptor and IRSs. Tyrosine-phosphorylated IRSs in turn activate PI3-kinase, which further activates the serine-threonine kinase, Akt.

Metabolic syndrome: A condition in which multiple risk factors for cardiovascular diseases such as hypertension, visceral obesity, insulin resistance, impaired glucose tolerance and dyslipidemia co-exist.

PI3-kinase: Activation of this kinase is implicated in insulin-induced glucose uptake into adipocytes and skeletal muscle, glycogen synthesis and inhibition of gluconeogenesis in liver.

References

Bloem, L.J. and A.K. Manatunga, D.A. Tewksbury, and J.H. Pratt. 1995. The serum angiotensinogen concentration and variants of the angiotensinogen gene in white and black children. J. Clin. Invest. 95: 948-953.

Boustany, C.M. and K. Bharadwaj, A. Daugherty, D.R. Brown, D.C. Randall, and L.A. Cassis. 2004. Activation of the systemic and adipose renin-angiotensin system in rats with diet-induced obesity and hypertension. Am. J. Physiol. Regul. Integr. Comp. Physiol. 287: R943-949.

Campbell, D.J. and J.F. Habener. 1987. Cellular localization of angiotensinogen gene expression in brown adipose tissue and mesentery: quantification of messenger ribonucleic acid abundance using hybridization *in situ*. Endocrinology 121: 1616-1626.

Cassis, L.A. and K.R. Lynch, and M.J. Peach. 1988. Localization of angiotensinogen messenger RNA in rat aorta. Circ. Res. 62: 1259-1262.

Engeli, S. and J. Bohnke, K. Gorzelniak, J. Janke, P. Schling, M. Bader, F.C. Luft, and A.M. Sharma. 2005. Weight loss and the renin-angiotensin-aldosterone system. Hypertension 45: 356-362.

Fujita, T. 2007. Insulin resistance and salt-sensitive hypertension in metabolic syndrome. Nephrol. Dial. Transplant. 22: 3102-3107.

Giacchetti, G. and E. Faloia, C. Sardu, M.A. Camilloni, B. Mariniello, C. Gatti, G.G. Garrapa, M. Guerrieri, and F. Mantero. 2000. Gene expression of angiotensinogen in adipose tissue of obese patients. Int. J. Obes. Relat. Metab. Disord. 24 Suppl. 2: S142-143.

Massiera, F. and M. Bloch-Faure, D. Ceiler, K. Murakami, A. Fukamizu, J.M. Gasc, A. Quignard-Boulange, R. Negrel, G. Ailhaud, J. Seydoux, *et al*. 2001a. Adipose angiotensinogen is involved in adipose tissue growth and blood pressure regulation. FASEB J. 15: 2727-2729.

Massiera, F. and J. Seydoux, A. Geloen, A. Quignard-Boulange, S. Turban, P. Saint-Marc, A. Fukamizu, R. Negrel, G. Ailhaud, and M. Teboul. 2001b. Angiotensinogen-deficient mice exhibit impairment of diet-induced weight gain with alteration in adipose tissue development and increased locomotor activity. Endocrinology 142: 5220-5225.

Masuzaki, H. and H. Yamamoto, C.J. Kenyon, J.K. Elmquist, N.M. Morton, J.M. Paterson, H. Shinyama, M.G. Sharp, S. Fleming, J.J. Mullins, *et al*. 2003. Transgenic amplification of glucocorticoid action in adipose tissue causes high blood pressure in mice. J. Clin. Invest. 112: 83-90.

Ogihara, T. and T. Asano, K. Ando, Y. Chiba, H. Sakoda, M. Anai, N. Shojima, H. Ono, Y. Onishi, M. Fujishiro *et al*. 2002. Angiotensin II-induced insulin resistance is associated with enhanced insulin signaling. Hypertension 40: 872-879.

Ogihara, T. and T. Asano, and T. Fujita. 2003. Contribution of salt intake to insulin resistance associated with hypertension. Life Sci. 73: 509-523.

Scheen, A.J. 2004. VALUE: analysis of results. Lancet 364: 932-933.

van Harmelen, V. and M. Elizalde, P. Ariapart, S. Bergstedt-Lindqvist, S. Reynisdottir, J. Hoffstedt, I. Lundkvist, S. Bringman, and P. Arner. 2000. The association of human adipose angiotensinogen gene expression with abdominal fat distribution in obesity. Int. J. Obes. Relat. Metab. Disord. 24: 673-678.

CHAPTER 6

Apelin as an Adipokine

Despina D. Briana[1] and Ariadne Malamitsi-Puchner[1,*]
[1]Neonatal Division, 2nd Department of Obstetrics and Gynecology, Athens University Medical School, Athens, Greece

ABSTRACT

The identification of the product of the gene *obese* threw light on the role of adipose tissue (AT) in the pathophysiology of obesity-related diseases. It has become increasingly evident that AT-derived cytokines mediate between obesity-related exogenous factors and the molecular events that lead to metabolic syndrome and inflammatory and/or autoimmune conditions. Furthermore, adipokines affect various biological processes, including metabolism, satiety, inflammation and cardiovascular function. Apelin is a 36 amino acid peptide identified as the endogenous ligand of the orphan G protein-coupled receptor APJ. Apelin and APJ mRNA are widely expressed in several rat and human tissues and have functional effects in both the central nervous system and periphery. The cardiovascular system appears to be a privileged source of apelin, since both apelin and its receptor are present in the heart, large and small conduit vessels, and endothelial cells. Apelin and its receptors are involved in the regulation of cardiovascular function, central fluid homeostasis, vessel formation and endothelial cell proliferation. Currently, roles have been established for the apelin system in lowering blood pressure, as a potent cardiac inotrope, in modulating pituitary hormone release and food intake, in stress activation, and as a novel adipokine that is excreted from fat cells and regulates insulin. Given its broad array of physiological roles, apelin has attracted much interest as a target for novel therapeutic approach and drug design.

*Corresponding author

INTRODUCTION

During the last decade, a growing number of adipocyte-derived hormones (adipokines) have been identified and documented to be differentially regulated during the onset of obesity and metabolic syndrome. By acting as circulating hormones systemically or locally on numerous cells types, adipokines are involved in physiological regulations (fat development, energy storage, metabolism, food intake) or in the promotion of obesity-associated disorders [type 2 diabetes mellitus (T2DM), cardiovascular disease] (Fasshauer and Paschke 2003).

Apelin is a bioactive peptide identified as the endogenous ligand of APJ, a G protein-coupled receptor (Tatemoto *et al.* 1998). Apelin peptides are derived from a 77 amino acid precursor, which is processed to several active molecular forms such as apelin-36 and apelin-13 in different tissues and in the bloodstream. Apelin and APJ are expressed in the hypothalamus, gastrointestinal tract, endothelial cells, vascular smooth muscle cells, cardiomyocytes, adipocytes and osteoblasts (Kleinz and Davenport 2005). Similarities between the structure and anatomical distribution of apelin and its receptor and that of angiotensin II and the angiotensin AY1 receptor provide clues about the physiological function of this novel signal-transduction system (Lee *et al.* 2000).

The most documented functions of apelin/APJ concern the regulation of fluid homeostasis and the modifications of cardiac contractility and blood pressure (Kleinz and Davenport 2005). In the heart, apelin potently stimulates heart rate and contraction (Szokodi *et al.* 2002, Berry *et al.* 2004). In the periphery, apelin causes vasodilatation via a nitric oxide–dependent mechanism (Tatemoto *et al.* 2001). In the hypothalamus, apelin inhibits the electrical activity of vasopressin-releasing neurons, suggesting an involvement in the regulation of vascular tone (De Mota *et al.* 2004). Moreover, apelin has recently been described as a novel adipokine, produced and secreted by human and mouse isolated mature adipocytes (Boucher *et al.* 2005).

So far, few data regarding the regulation of apelin or APJ are available, especially in humans. Apelin and APJ expression follow the same pattern of regulation in human heart failure (Berry *et al.* 2004). Regulation of apelin expression by insulin and tumor necrosis factor (TNF) α in human adipocytes has also been reported (Boucher *et al.* 2005, Daviaud *et al.* 2006). Moreover, the basal circulating apelin levels are higher in obese patients compared with control lean individuals, positively correlating with body mass index (BMI) (Boucher *et al.* 2005, Heinonen *et al.* 2005). These data suggest that apelin could play an important role in obesity-associated metabolic disorders and cardiovascular disease.

This chapter considers the main roles of apelin in pathophysiology with particular attention to its role in energy balance regulation and in obesity-associated disorders.

APELIN AND AT

In 2005 apelin was identified as a new adipokine expressed and secreted by human and mouse mature adipocytes (Boucher *et al.* 2005). Apelin mRNA is detectable in non-differentiated preadipocytes, but its production increases 4-fold upon differentiation of adipocytes, as previously described for the well-characterized adipokines leptin and adiponectin.

So far few data are available concerning apelin production and regulation in AT. A strong relationship exists between adipocyte-secreted apelin and insulin (Castan-Laurell *et al.* 2005). Adipocyte apelin mRNA expression and plasma apelin concentrations were increased in various mouse models of obesity associated with hyperinsulinemia. Accordingly, adipocytes of insulin-deficient mice (streptozocin-treated) had lower apelin mRNA levels than controls (Castan-Laurell *et al.* 2005). Direct action of insulin in the regulation of apelin expression in adipocytes was also shown *in vivo*. In humans, higher plasma apelin levels were found in obese, hyperinsulinemic subjects (Castan-Laurell *et al.* 2005). The exact origin of this overproduction of apelin still remains to be elucidated, but the striking up-regulation of apelin promoted in adipocytes by insulin suggests an adipocyte origin of the increased apelin in obesity.

On the other hand, since obesity and insulin resistance are associated with chronically elevated TNF-α levels, subsequent studies have shown a TNF-α-dependent up-regulation of apelin expression in AT from lean to severely obese subjects (Daviaud *et al.* 2006, Garcia-Diaz *et al.* 2007). Those reports suggested that in adipocytes there is a substantial regulation of apelin synthesis exerted by TNF-α, leading to sustained apelin secretion in obesity, probably representing an adaptive response that attempts to forestall the onset of obesity-related disorders, such as mild chronic inflammation, hypertension and cardiovascular dysfunctions. In line, an induction on adipocyte apelin gene expression by growth hormone (which antagonizes insulin signal transduction in insulin-sensitive tissues) was later described, supporting a link between apelin and the onset of insulin resistance (Kralisch *et al.* 2007). Accordingly, Garcia-Diaz *et al.* demonstrated significant associations among markers of adiposity and apelin mRNA expression from subcutaneous and retroperitoneal AT of rats (Garcia-Diaz *et al.* 2007).

Finally, more recent data provide evidence that apelin gene expression and release in AT is regulated by the transcriptional proliferator-activated receptor γ co-activator 1α, which is an important component of energy balance in human white AT (Mazzucotelli *et al.* 2008).

ROLES OF APELIN AS AN ADIPOKINE

The reported distribution of apelin and the apelin receptor in the hypothalamus, gastric mucosa and fat cells suggests that the apelin system has roles in modulating appetite, digestion and metabolism (Lee *et al.* 2000). In this respect, apelin may act as an "adiposity signal" generated in proportion to body fat stores and reducing food intake, which already includes leptin, insulin and amylin (Beltowski 2006). Interestingly, apelin and leptin also demonstrate several other similarities, including increases in core body temperature and locomotor activity which, apart from reducing food intake, may contribute to negative energy balance (Jaszberenyi *et al.* 2004). Moreover, since apelin has been recently described as a mitogenic agent in epithelial and endothelial cells, apelin could participate in the development of the fat pad, by acting on pre-adipocyte proliferation and/or angiogenesis (Boucher *et al.* 2005).

On the other hand, in an animal model of obesity, apelin treatment decreased the weight of white AT and serum levels of insulin and triglycerides, without influencing food intake (Higuchi *et al.* 2007). Additionally, apelin treatment increased serum adiponectin levels, mRNA expression of uncoupling protein 1 (a marker of peripheral energy expenditure in brown AT) and of uncoupling protein 3 (a regulator of fatty acid export in skeletal muscle) (Higuchi *et al.* 2007). These data suggest that apelin could be a novel adipokine involved in the regulation of adiposity and energy metabolism.

Finally, little is known about the role of the apelin/APJ system in human pregnancy, which is normally characterized by AT accretion and the development of insulin resistance (Barbour 2007). Cobellis *et al.* demonstrated that apelin and APJ are expressed in the human placenta and might play a role in the regulation of placental flow and vasculogenesis (Cobellis *et al.* 2007). However, a recent report documented that gestational diabetes mellitus had no impact on circulating apelin and apelin/APJ expression in both fat and placental tissue (Telejko *et al.* 2009).

Table I   Key endocrine roles of adipocyte-secreted apelin

- Inhibition of vasopressin release →reduction of Na and H_2O uptake
  →modulation of food intake
- Positive inotropic effect/improvement of heart function
- Lowering blood pressure
- Inhibition of insulin release/regulation of glucose homeostasis

APELIN IN OBESITY-RELATED DISORDERS

The potential role of apelin in whole-body glucose disposal in normal and insulin-resistant subjects remains largely unknown. Recent data show that in normal mice, acute apelin injection has a powerful glucose-lowering effect, associated with inhibition of glucose-stimulated insulin secretion in pancreatic islets and with enhanced glucose utilization in skeletal muscle and AT. In obese and insulin-resistant mice, apelin also improves glucose uptake in insulin-sensitive tissues (Dray *et al.* 2008).

Moreover, elevated apelin levels were shown in obese subjects (Boucher *et al.* 2005, Castan-Laurell *et al.* 2008, Heinonen *et al.* 2005), whereas in patients with T2DM both increased (Li *et al.* 206) and decreased (Erdem *et al.* 2008) plasma apelin concentrations were reported. Furthermore, in studies concerning obese individuals and patients with T2DM, circulating apelin was related to BMI, plasma glucose, insulin, HOMA-IR and lipid parameters (Heinonen et al 2005, Li *et al.* 2006); however, negative correlations with fasting glucose levels were also observed in patients with newly diagnosed T2DM (Erdem *et al.* 2008).

In a recent study, a diet-induced weight loss in obese women resulted in a reduction of both plasma apelin levels and AT apelin expression. This reduced apelin expression in AT could contribute to decreased circulating apelin levels (Castan-Laurell *et al.* 2008). By contrast, diet-induced weight loss in individuals with metabolic syndrome did not have an effect on plasma apelin levels, although apelin correlated with TNF-α and mean arterial pressure (Heinomen *et al.* 2009), probably suggesting that apelin may not be strongly related to fat mass as an adipokine, but it may be involved in the regulation of inflammation and cardiovascular tone (Hainomen *et al.* 2009).

Taken together, although apelin has been viewed as a beneficial adipokine up-regulated in obesity, it remains to be established whether the increased levels of apelin in obesity are an attempt to overcome either insulin resistance or obesity-associated cardiovascular diseases or another metabolic defect such as apelin resistance. Thus, understanding the contribution of such an adipokine in obesity-associated disorders appears to be of major importance.

Summary Points

- Apelin is the endogenous ligand of the orphan G protein-coupled receptor APJ.
- Apelin and APJ are widely expressed in several tissues and have functional effects in both the central nervous system and periphery.
- Apelin and its receptor are involved in the regulation of cardiovascular function, central fluid homeostasis, vessel formation and endothelial cell proliferation.
- Apelin is a novel adipokine that is excreted from fat cells and regulates insulin.
- Apelin may be involved in regulation of adiposity and energy metabolism.
- Apelin may be a target for novel therapeutic approaches and drug design in obesity-associated metabolic disorders and cardiovascular disease.

Abbreviations

APJ	:	orphan G protein-coupled receptor
AT	:	adipose tissue
BMI	:	body mass index
T2DM	:	type 2 diabetes mellitus
TNF-α	:	tumor necrosis factor α

Definition of Terms

Adipokines: Soluble mediators that are mainly, but not exclusively, produced by adipocytes and exert their biological function in an autocrine, paracrine or systemic manner.

APJ: The orphan G protein-coupled receptor.

Uncoupling protein 1: A protein of brown adipose tissue, which regulates energy expenditure and nonshivering thermogenesis.

Uncoupling protein 3: A protein of skeletal muscle and adipose tissue, which regulates fatty acid export.

References

Barbour, L.A. *et al.* 2007. Cellular mechanisms for insulin resistance in normal pregnancy and gestational diabetes. Diabetes Care 30 (Suppl 2): S112-S119.

Beltowski, J. 2006. Apelin and visfatin: unique "beneficial" adipokines upregulated in obesity? Med. Sci. Monit. 12: RA112-RA119.

Berry, M.F., *et al.* 2004. Apelin has *in vivo* inotropic effects on normal and failing hearts. Circulation. 110 (Suppl. 1):II187-II193.

Boucher, J. *et al.* 2005. Apelin, a newly identified adipokine up-regulated by insulin and obesity. Endocrinology 146: 1764-1771.

Castan-Laurell, I. *et al.* 2005. Apelin, a novel adipokine overproduced in obesity: friend or foe? Mol. Cell. Endocrinol. 245: 7-9.

Castan-Laurell, I. *et al.* 2008. Effect of hypocaloric diet-induced weight loss in obese women on plasma apelin and adipose tissue expression of apelin and APJ. Eur. J. Endocrinol. 158: 905-910.

Cobellis, L. *et al.* 2007. Modulation of apelin and APJ receptor in normal and preeclampsia-complicated placentas. Histol. Histopathol. 22: 1-8.

Daviaud, D. *et al.* 2006. TNFα up-regulates apelin expression in human and mouse adipose tissue. FASEB J. 20: E796-E802.

De Mota, N. *et al.* 2004. Apelin, a potent diuretic neuropeptide counteracting vasopressin actions through inhibition of vasopressin neuron activity and vasopressin release. Proc. Natl. Acad. Sci. USA 101: 10464-10469.

Dray, C. *et al.* 2008. Apelin stimulates glucose utilization in normal and obese insulin-resistant mice. Cell. Metab. 8: 437-445.

Erdem, G. *et al.* 2008. Low plasma apelin levels in newly diagnosed type 2 diabetes mellitus. Exp. Clin. Endocrinol. Diabetes 116: 289-292.

Fasshauer, M. and R. Paschke. 2003. Regulation of adipocytokines and insulin resistance. Diabetologia 46: 1594-1603.

Garcia-Diaz, D. *et al.* 2007. Adiposity dependent apelin gene expression: relationships with oxidative and inflammation markers. Mol. Cell. Biochem. 305: 87-94.

Heinonen, M.V. *et al.* 2005. Apelin, orexin-A and leptin plasma levels in morbid obesity and effect of gastric banding. Regul. Pept. 130: 7-13.

Heinonen, M.V. *et al.* 2009. Effect of diet-induced weight loss on plasma apelin and cytokine levels in individuals with the metabolic syndrome. Nutr. Metab. Cardiovasc. Dis. 19: 626-633.

Higuchi, K. *et al.* 2007. Apelin, an APJ receptor ligand, regulates body adiposity and favors the messenger ribonucleic acid expression of uncoupling proteins in mice. Endocrinology 148: 2690-2697.

Jaszberenyi, M. *et al.* 2004. Behavioral, neuroendocrine and thermoregulatory actions of apelin-13. Neuroscience 129: 811-816.

Kleinz, M.J. and A.P. Davenport. 2005. Emerging roles of apelin in biology and medicine. Pharmacol. Ther. 107: 198-211.

Kralisch, S. *et al.* 2007. Growth hormone induced apelin mRNA expression and secretion in mouse 3T3-L1 adipocytes. Regul. Pept. 139: 84-89.

Lee, D.K. *et al.* 2000. Characterization of apelin, the ligand for the APJ receptor. J. Neurochem. 74: 34-41.

Li, L. *et al.* 2006. Changes and relations of circulating visfatin, apelin, and resistin levels in normal, impaired glucose tolerance, and type 2 diabetic subjects. Exp. Clin. Endocrinol. Diabetes 114: 544-548.

Mazzucotelli, A. *et al.* 2008. The transcriptional co-activator PGC-1α up regulates apelin in human and mouse adipocytes. Regul. Pept. 150: 33-37.

Szokodi, I. *et al.* 2002. Apelin, the novel endogenous ligand of the orphan receptor APJ, regulates cardiac contractility. Circ. Res. 91: 434-440.

Tatemoto, K. *et al.* 1998. Isolation and characterization of a novel endogenous peptide ligand for the human APJ receptor. Biochem. Biophys. Res. Commun. 251: 471-476.

Tatemoto, K. *et al.* 2001. The novel peptide apelin lowers blood pressure via a nitric oxide-dependent mechanism. Regul. Pept. 99: 87-92.

Telejko, B. *et al.* 2009. Plasma apelin levels and apelin/APJ mRNA expression in patients with gestational diabetes mellitus. Diabetes Res. Clin. Pract. (in press).

CHAPTER 7

C1q/TNF-related Proteins (CTRPs) as Adipokines

Jonathan M. Peterson[1] and G. William Wong[1,*]

[1]Department of Physiology and Center for Metabolism and Obesity Research, Johns Hopkins University School of Medicine, 855 North Wolfe Street, Baltimore, MD 21205, USA

ABSTRACT

Adiponectin is well documented as an insulin sensitizer. However, variable and relatively mild metabolic dysfunctions in adiponectin knockout mice suggest the existence of compensatory mechanisms. Recently, a novel family of 10 secreted proteins (CTRP1 to CTRP10) sharing common structural and functional features with adiponectin was discovered. Most CTRPs are expressed by adipose tissue and circulate in plasma. Their circulating levels vary with the sex, genetic background and metabolic states of the animals. All CTRPs form trimers as their basic structural unit; some are further assembled into hexameric and HMW oligomeric complexes that may have distinct biological and signaling properties. Additionally, CTRPs form combinatorial associations, representing a potential mechanism to generate functionally distinct ligands with altered receptor specificity and/or function. To date, *in vitro* and/or *in vivo* metabolic functions have been demonstrated for CTRP1, CTRP2, CTRP3, CTRP5, and CTRP9. Additionally, CTRPs have been implicated in immune response, platelet aggregation, macular degeneration, bone formation and some types of cancers. Although the receptors for CTRPs have yet to be identified, liver, muscle and adipose tissues are likely targets of CTRPs. Future studies on these novel adipokines will provide new insights into physiological mechanisms that link adipose tissue to whole-body energy homeostasis.

*Corresponding author

INTRODUCTION

Adiponectin is one of the most highly expressed and intensely studied adipokines. Allelic polymorphisms and reduced plasma adiponectin levels are tightly linked to insulin resistance and type 2 diabetes (Trujillo and Scherer 2005). However, adiponectin-null mice have surprisingly mild metabolic phenotypes (Davis and Scherer 2008, Yano *et al.* 2008), suggesting the existence of compensatory mechanisms. Efforts to identify secreted proteins that share overlapping functions with adiponectin led to the identification and characterization of a highly conserved and novel family of 10 adiponectin structural and functional paralogs (CTRP1 to CTRP10; Fig. 1) (Wong *et al.* 2004, 2009) CTRPs and adiponectin are part of the expanding C1q/TNF superfamily of proteins (Kishore *et al.* 2004).

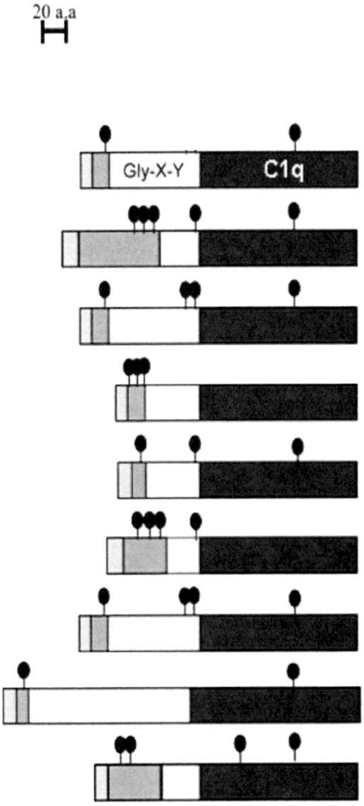

	Human	Mouse	% amino acid identity		
			Full-length	globular C1q domain	N-Terminus variable region
hAdipo	hAdipo	mAdipo	82	91	54
hCTRP1	hCTRP1	mCTRP1	77	86	65
hCTRP2	hCTRP2	mCTRP2	94	96	87
hCTRP3	hCTRP3	mCTRP3	95	99	95
hCTRP5	hCTRP5	mCTRP5	93	94	100
hCTRP6	hCTRP6	mCTRP6	67	82	53
hCTRP7	hCTRP7	mCTRP7	96	98	85
hCTRP9	hCTRP9	mCTRP9	84	89	100
hCTRP10	hCTRP10	mCTRP10	93	100	81

Fig. 1 Adiponectin and CTRPs share similar modular organization. Signal peptide for secretion, a unique N-terminus (light gray) with varying numbers of cysteine residue (ball-and-stick), a collagen domain with varying numbers of Gly-Xy-Y repeat (white) and a C-terminal globular domain homologous to immune complement C1q (dark gray). Numbers on the right refer to percentage amino acid identity between human and the corresponding mouse ortholog.

EXPRESSION AND REGULATION OF CTRP mRNAS IN ADIPOSE TISSUE AND IN MICE

Most CTRPs are expressed by adipose tissue (Table 1) but, in mice, the mRNA levels of multiple CTRPs are influenced by gender and metabolic states (Wong *et al.* 2009). A two-fold increase in plasma adiponectin levels is sufficient to normalize many aspects of metabolic defects in genetically obese (*ob/ob*) mice (Combs *et al.* 2001). Therefore, the increased expression of CTRP transcripts in relatively young *ob/ob* mice may represent a compensatory mechanism to counteract metabolic dysfunction in hyperphagic mice. Interestingly, the anti-diabetic drug rosiglitazone

Table I Overview of the function of adiponectin and CTRP family

	Expression (Major tissue)	Change w/ obesity	Change w/ caloric restriction	Immune system	Primary known function	Serum Concentration (Male vs. Female)	Role in Diseases
Adiponectin	Adipo	↑ then ↓	?	↓ pro-inflammatory mediator expression	Insulin Sensitization	↑ F	Diabetes
CTRP1	Adipo	↑	?	↑ by pro-inflammatory mediators	↓Thrombosis ↓ *	↔	?
CTRP2	Adipo	↑	↑ in old mice	?	↑ Glucose uptake ↑Fatty Acid oxidation	↔	?
CTRP3/ Cors26	Adipo	↑	?	↓ pro-inflammatory mediator expression	Promote Chondrocyte Proliferation	↑ F	Osteo-sarcoma
CTRP4	Brain	?	?	?	?	↔	?
CTRP5	Adipo	↔	?	?	?	↑ F	Macular Degeneration
CTRP6	placenta	↑	?	?	?	↑ F	ASFV replication
CTRP7	Adipo	↑	↑ in old mice	?	?	↔	?
CTRP8	Testis	?	?	?	?	?	?
CTRP9	Adipo	↑ then ↓	?	?	↓ plasma glucose	↑ F	?
CTRP10	Eye	↑	?	?	?	↔	?

↑, increased; ↓, decreased; ↔, no change; ?, unknown; Adipo, adipose tissue; F, female; *Plasma levels. (Unpublished.)

has been shown to up-regulate CTRP1 and down-regulate CTRP6 transcript levels in adipose tissue of *ob/ob* mice (Wong *et al.* 2008); these changes may mediate part of the beneficial metabolic effects of rosiglitazone *in vivo*.

GENDER AND GENETIC BACKGROUND INFLUENCE CIRCULATING LEVELS OF CTRPs

All CTRPs are secreted proteins and most circulate in plasma (Wong *et al.* 2008, 2009); therefore, they can act in an autocrine, paracrine and/or endocrine manner (Table 1). Age, sex and genetic background affect the levels of metabolic hormones and associated signaling pathways in animal models, differences that influence susceptibility to the development of obesity, insulin resistance and type 2 diabetes (West *et al.* 1992). As with adiponectin, female mice have higher circulating levels of CTRP5 and CTRP9 (Wong *et al.* 2008). Circulating levels of adiponectin and CTRPs also vary significantly in mice derived from six different genetic backgrounds: BALB/c, C57BL/6, 129SvJ, FVB, DBA and AKR mice are known to exhibit varying degrees of susceptibility or resistance to diet-induced obesity and diabetes (West *et al.* 1992). Intriguingly, in skeletal muscle, CTRP2 and CTRP7 mRNA and protein levels are increased in aging mice (Rohrbach *et al.* 2007), suggesting possible roles of CTRPs in aging physiology.

Table 2   Key points of C1q/TNF-related proteins (CTRPs) as adipokines

- Adipokines play major roles in regulating whole-body insulin sensitivity, as well as glucose and lipid metabolism.
- The expression of many adipokines is dysregulated in conditions of obesity and diabetes.
- Some adipokines promote insulin sensitivity; others promote insulin resistance.
- Some adipokines act locally; others act systemically.
- Some adipokines exert their effects through central mechanisms in the brain, while others act on peripheral tissues such as the muscle and liver.

CTRPs FORM HOMO- AND HETERO-OLIGOMERIC COMPLEXES

Adiponectin exists in three distinct oligomeric forms (trimers, hexamers, and HMW oligomers) with distinct biological properties (Pajvani *et al.* 2003, Tsao *et al.* 2002). The distribution of these complexes in serum varies with metabolic and disease states (Pajvani *et al.* 2004). Similar to adiponectin, all CTRPs form trimers as their basic structural unit (Wong *et al.* 2008, 2009). Several CTRPs are further assembled into hexamers and HMW oligomers via disulfide bonds involving their N-terminal cysteine residues. Different CTRP oligomeric forms may induce distinct signaling pathways in different cell types. In addition, several CTRPs and adiponectin can also form hetero-oligomeric complexes (Peterson *et al.* 2009, Wong *et al.* 2008, 2009). These combinatorial associations may represent a mechanism to generate functionally distinct ligands with altered signaling properties and/or receptor specificity.

METABOLIC FUNCTION OF CTRPs

Functional analysis has uncovered roles for some CTRPs. Adiponectin exerts an insulin-sensitizing effect on liver and skeletal muscle through activation of the AMPK signaling pathway (Tomas *et al.* 2002, Yamauchi *et al.* 2002). The same pathway is used by CTRP2 to enhance fatty acid oxidation and glycogen deposition in myotubes (Wong *et al.* 2004). Regarding CTRP1, transcript levels are increased by rosiglitazone treatment in mice (Wong *et al.* 2008). Further, circulating levels are elevated in adiponectin-null mice, suggesting this as a possible compensatory mechanism for adiponectin. Indeed, injection of recombinant CTRP1 into mice significantly lowered blood glucose levels (Wong *et al.* 2008). Because CTRP1 activates the Akt signaling pathway in myotubes, muscle may be an *in vivo* target tissue of CTRP1. CTRP9 forms hetero-oligomeric complexes with adiponectin. Adenovirus-mediated over-expression of CTRP9 *in vivo* modestly decreases serum glucose and insulin levels (Wong *et al.* 2009). Although CTRP9 activates multiple signaling pathways in myotubes, the *in vivo* target tissue(s) of CTRP9 is not known.

CTRPs, INFLAMMATION AND DISEASE

Links between inflammation, obesity and insulin resistance, mediated in part by TNF-α and other pro-inflammatory cytokines, have been well documented. TNF-α and IL-1β up-regulate CTRP1 expression in rat adipose tissue (Kim *et al.* 2006), though it is unclear whether CTRP1 promotes or suppresses inflammation. A recent study suggests a possible protective role for CTRP1 in the vasculature. In a monkey model, CTRP1 inhibits collagen-induced platelet thrombosis (Lasser *et al.* 2006), highlighting its potential therapeutic value for treating vascular disorder. In contrast, CTRP3 down-regulates the expression of pro-inflammatory mediators from primary human monocytes (Kopp *et al.* 2009, Weigert *et al.* 2005), suggesting an anti-inflammatory function, similar to adiponectin. Recently, CTRP3 (also known as CORS26/cartducin) has been shown to stimulate chondrogenic precursor cell proliferation (Akiyama *et al.* 2006) and promote angiogenesis (Akiyama *et al.* 2007) and is over-expressed in osteosarcoma (Akiyama *et al.* 2009). Whether CTRP3 promotes or suppresses tumorigenesis is unclear. Mutations in the CTRP5 gene that disrupt its proper assembly and secretion cause late-onset retinal macular degeneration in humans (Hayward *et al.* 2003). Recently, CTRP6 was identified as one of the genes required for the replication of African Swine Fever virus (Chang *et al.* 2006), though the mechanism behind this is unknown.

CONCLUSION

Many functional, physiological and mechanistic questions regarding the role of CTRPs in metabolism remain to be addressed. The establishment of transgenic and knockout mouse models for these novel adipokines will provide valuable tools to dissect their functions and mechanisms of action.

Summary Points

- CTRPs represent a novel family of adipokines homologous to adiponectin in both structure and function.
- CTRPs are secreted proteins and most are found circulating in plasma, with levels varying with gender, genetic background and metabolic states.
- CTRPs form combinatorial associations, representing a possible mechanism to generate functionally distinct ligands.
- Some CTRPs have been shown to have metabolic functions.
- Some CTRPs are associated with inflammation and/or disease.

Abbreviations

AMPK	:	AMP-activated protein kinase
CTRP	:	C1q/TNF-related protein
HMW	:	high molecular weight
TNF	:	tumor necrosis factor

Definition of Terms

Autocrine: Secreted factors that act on the cells that produce them.

Angiogenesis: The formation of new blood vessels.

Oligomer: A protein molecule made of two or more subunits. A hetero-oligomer is a complex made of different protein subunits, versus a homo-oligomer, made of only one type of protein.

ob/ob mice: Mice with a recessive mutation in the leptin gene. This results in a mouse that eats excessively and becomes profoundly obese.

References

Akiyama, H. and S. Furukawa, S. Wakisaka, and T. Maeda. 2006. Cartducin stimulates mesenchymal chondroprogenitor cell proliferation through both extracellular signal-regulated kinase and phosphatidylinositol 3-kinase/Akt pathways. FEBS J. 273: 2257-2263.

Akiyama, H. and S. Furukawa, S. Wakisaka, and T. Maeda. 2007. CTRP3/cartducin promotes proliferation and migration of endothelial cells. Mol. Cell. Biochem. 304: 243-248.

Akiyama, H. and S. Furukawa, S. Wakisaka, and T. Maeda. 2009. Elevated expression of CTRP3/cartducin contributes to promotion of osteosarcoma cell proliferation. Oncol. Rep. 21: 1477-1481.

Chang, A.C. and L. Zsak, Y. Feng, R. Mosseri, Q. Lu, P. Kowalski, A. Zsak, T.G. Burrage, J.G. Neilan, G.F. Kutish, Z. Lu, W. Laegreid, D.L. Rock, and S.N. Cohen. 2006. Phenotype-based identification of host genes required for replication of African swine fever virus. J. Virol. 80: 8705-8717.

Combs, T.P. and A.H. Berg, S. Obici, P.E. Scherer, and L. Rossetti. 2001. Endogenous glucose production is inhibited by the adipose-derived protein Acrp30. J. Clin. Invest. 108: 1875-1881.

Davis, K.E. and P.E. Scherer. 2008. Adiponectin: no longer the lone soul in the fight against insulin resistance? Biochem. J. 416: e7-9.

Hayward, C. *et al.* and A.F. Wright. 2003. Mutation in a short-chain collagen gene, CTRP5, results in extracellular deposit formation in late-onset retinal degeneration: a genetic model for age-related macular degeneration. Hum. Mol. Genet. 12: 2657-2667.

Kim, K.Y. and H.Y. Kim, J.H. Kim, C.H. Lee, D.H. Kim, Y.H. Lee, S.H. Han, J.S. Lim, D.H. Cho, M.S. Lee, S. Yoon, K.I. Kim, D.Y. Yoon, and Y. Yang. 2006. Tumor necrosis factor-alpha and interleukin-1beta increases CTRP1 expression in adipose tissue. FEBS Lett. 580: 3953-3960.

Kishore, U. and C. Gaboriaud, P. Waters, A.K. Shrive, T.J. Greenhough, K.B. Reid, R.B. Sim, and G.J. Arlaud. 2004. C1q and tumor necrosis factor superfamily: modularity and versatility. Trends Immunol. 25: 551-561.

Kopp, A. and M. Bala, J. Weigert, C. Buchler, M. Neumeier, C. Aslanidis, J. Scholmerich, and A. Schaffler. 2009. Effects of the new adiponectin paralogous protein CTRP-3 and of LPS on cytokine release from monocytes of patients with type 2 diabetes mellitus. Cytokine 49(1): 51-57.

Lasser, G. and P. Guchhait, J.L. Ellsworth, P. Sheppard, K. Lewis, P. Bishop, M.A. Cruz, J.A. Lopez, and J. Fruebis. 2006. C1qTNF-related protein-1 (CTRP-1): a vascular wall protein that inhibits collagen-induced platelet aggregation by blocking VWF binding to collagen. Blood 107: 423-430.

Pajvani, U.B. and X. Du, T.P. Combs, A.H. Berg, M.W. Rajala, T. Schulthess, J. Engel, M. Brownlee, and P.E. Scherer. 2003. Structure-function studies of the adipocyte-secreted hormone Acrp30/adiponectin. Implications for metabolic regulation and bioactivity. J. Biol. Chem. 278: 9073-9085.

Pajvani, U.B. *et al.* and P.E. Scherer. 2004. Complex distribution, not absolute amount of adiponectin, correlates with thiazolidinedione-mediated improvement in insulin sensitivity. J. Biol. Chem. 279: 12152-12162.

Rohrbach, S. and A.C. Aurich, L. Li, and B. Niemann. 2007. Age-associated loss in adiponectin-activation by caloric restriction: lack of compensation by enhanced inducibility of adiponectin paralogs CTRP2 and CTRP7. Mol. Cell. Endocrinol. 277: 26-34.

Tomas, E. and T.S. Tsao, A.K. Saha, H.E. Murrey, C.C. Zhang, S.I. Itani, H.F. Lodish, and N.B. Ruderman. 2002. Enhanced muscle fat oxidation and glucose transport by ACRP30 globular domain: acetyl-CoA carboxylase inhibition and AMP-activated protein kinase activation. Proc. Natl. Acad. Sci. USA 99: 16309-16313.

Trujillo, M.E. and P.E. Scherer. 2005. Adiponectin—journey from an adipocyte secretory protein to biomarker of the metabolic syndrome. J. Intern. Med. 257: 167-175.

Tsao, T.S. and H.E. Murrey, C. Hug, D.H. Lee, and H.F. Lodish. 2002. Oligomerization state-dependent activation of NF-kappa B signaling pathway by adipocyte complement-related protein of 30 kDa (Acrp30). J. Biol. Chem. 277: 29359-29362.

Weigert, J. and M. Neumeier, A. Schaffler, M. Fleck, J. Scholmerich, C. Schutz, and C. Buechler. 2005. The adiponectin paralog CORS-26 has anti-inflammatory properties and is produced by human monocytic cells. FEBS Lett. 579: 5565-5570.

West, D.B. and C.N. Boozer, D.L. Moody, and R.L. Atkinson. 1992. Dietary obesity in nine inbred mouse strains. Am. J. Physiol. 262: R1025-1032.

Wong, G.W. and S.A. Krawczyk, C. Kitidis-Mitrokostas, G. Ge, E. Spooner, C. Hug, R. Gimeno, and H.F. Lodish. 2009. Identification and characterization of CTRP9, a novel secreted glycoprotein, from adipose tissue that reduces serum glucose in mice and forms heterotrimers with adiponectin. FASEB J. 23: 241-258.

Wong, G.W. and S.A. Krawczyk, C. Kitidis-Mitrokostas, T. Revett, R. Gimeno, and H.F. Lodish. 2008. Molecular, biochemical and functional characterizations of C1q/TNF family members: adipose-tissue-selective expression patterns, regulation by PPAR-gamma agonist, cysteine-mediated oligomerizations, combinatorial associations and metabolic functions. Biochem. J. 416: 161-177.

Wong, G.W. and J. Wang, C. Hug, T.S. Tsao, and H.F. Lodish. 2004. A family of Acrp30/adiponectin structural and functional paralogs. Proc. Natl. Acad. Sci. USA 101: 10302-10307.

Yamauchi, T. and J. Kamon, Y. Minokoshi, Y. Ito, H. Waki, S. Uchida, S. Yamashita, M. Noda, S. Kita, K. Ueki, K. Eto, Y. Akanuma, P. Froguel, F. Foufelle, P. Ferre, D. Carling, S. Kimura, R. Nagai, B.B. Kahn, and T. Kadowaki. 2002. Adiponectin stimulates glucose utilization and fatty-acid oxidation by activating AMP-activated protein kinase. Nat. Med. 8: 1288-1295.

Yano, W. and N. Kubota, S. Itoh, T. Kubota, M. Awazawa, M. Moroi, K. Sugi, I. Takamoto, H. Ogata, K. Tokuyama, K. Noda, Y. Terauchi, K. Ueki, and T. Kadowaki. 2008. Molecular mechanism of moderate insulin resistance in adiponectin-knockout mice. Endocr. J. 55: 515-522.

CHAPTER 8

Fasting-induced Adipose Factor/ Angiopoietin-like Protein 4 as an Adipokine

Sander Kersten

Nutrition, Metabolism and Genomics Group, Division of Human Nutrition, Wageningen University, Wageningen, The Netherlands

ABSTRACT

The adipose tissue serves as a major endocrine organ by secreting a large number of metabolically and/or immunologically active factors. One of the proteins secreted by adipose and other tissues is fasting-induced adipose factor (FIAF)/ angiopoietin-like protein 4 (Angptl4), a member of the family of angiopoietins and angiopoietin-like protein. Angptl4 forms higher order oligomers and upon secretion undergoes proteolytic processing to yield N- and C-terminal fragments. The best-characterized function of Angptl4 is as inhibitor of the enzyme lipoprotein lipase, leading to decreased clearance of plasma triglycerides. Additionally, Angptl4 promotes adipose tissue lipolysis. Data suggest that Angptl4 is retained in the extracellular matrix and governs angiogenesis by influencing the function of endothelial cells. Expression of Angptl4 is stimulated by fatty acids via peroxisome proliferator-activated receptors and by hypoxia via hypoxia-inducible factor 1 alpha. In conclusion, Angptl4 is an adipokine with several biological activities with major impact on lipid metabolism.

INTRODUCTION

Until recently, adipose tissue was mainly regarded as an energy storage depot and was not considered to have a major endocrine function. It was the publication of leptin as the first adipokine in 1994 that unleashed a massive research effort aimed at the discovery of other adipokines, as well as at understanding their functional role and mechanism of action in the body. Most studies since then have been devoted to a limited number of adipokines for which a strong connection with obesity and diabetes could be established. Other adipokines transiently bloomed during a short-lived hype fed by the aspiration of uncovering a second leptin or adiponectin. The adipokine Angptl4 has followed a different trajectory where an almost hidden but concerted research effort among a limited number of dedicated laboratories contributed to an incremental understanding of the structural and functional properties of the protein and its importance for nutrient metabolism.

STRUCTURE AND FUNCTION OF ANGPTL4

Befitting its classification as adipokine, Angptl4 is produced by various adipose depots and is specifically well expressed in mouse adipocytes (Kersten *et al.* 2000, Yoon *et al.* 2000). Additionally, Angptl4 is expressed in numerous other organs and cell types including liver, muscle, intestine, and the endothelium. Angptl4 is part of a larger family of angiopoietins and angiopoietin-like proteins that share structural and functional similarities (Fig. 1). Many of these proteins, including Angptl4, undergo proteolytic processing to yield N- and C-terminal fragments via proprotein convertases (Chomel *et al.* 2009, Ge *et al.* 2004a, 2004b, Mandard *et al.* 2004). Moreover, Angptl4 forms higher order oligomeric structures via intermolecular disulfide bonds and hydrophobic interactions.

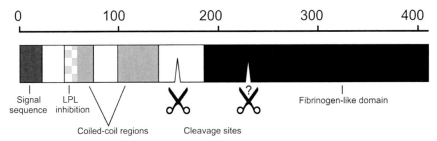

Fig. 1 Structural organization of the Angptl4 protein.

Since the protein was discovered in 2000 (Kersten *et al.* 2000, Kim *et al.* 2000, Yoon *et al.* 2000), it has become evident that Angptl4 carries both metabolic and non-metabolic functions. At the metabolic level, it has been clearly demonstrated that Angptl4 inhibits clearance of plasma triglycerides (Fig. 2) (Koster *et al.* 2005, Yu *et al.* 2005). Triglyceride clearance is mediated by lipoprotein lipase (LPL) (Table 1), which is tethered to the capillary endothelium via heparin sulfate proteoglycans (HSPG) and represents the rate-limiting enzyme for plasma TG hydrolysis. Inhibition of plasma TG clearance is achieved by permanently inactivating LPL via stimulation of the conversion of active LPL dimers into inactive LPL monomers (Lichtenstein *et al.* 2007, Sukonina *et al.* 2006, Yoshida *et al.* 2002). Consequently, uptake of fatty acids into LPL-dependent tissues is decreased and plasma TG levels are increased (Fig. 3). Inhibition of LPL is mediated by a small region of the protein near the N-terminus (Fig. 1). A portion of Angptl4 is bound to endothelial cells in tissues via HSPG and can be released into the bloodstream by heparin injection (Cazes *et al.* 2006, Chomel *et al.* 2009). Recent data suggest that in addition to binding to HSPG, LPL may also bind to the newly identified LPL anchor GPIHBP-1, which may protect it from inhibition by Angptl4 (Sonnenburg *et al.* 2009).

Apart from its ability to inhibit LPL, Angptl4 also induces adipose tissue lipolysis, thereby stimulating the release of fatty acids from adipose tissue (Fig. 2) (Yoshida *et al.* 2002). Indeed, the increase in plasma free fatty acids (FFA) during fasting is completely abolished in Angptl4-/- mice, whereas plasma FFA are elevated in mice over-expressing Angptl4 (Sanderson *et al.* 2009). By inhibiting LPL-mediated lipolysis and stimulating intracellular lipolysis, Angptl4 may cause a shift in fuel use from plasma triglycerides towards FFA. Although Angptl4 has been proposed to suppress hepatic glucose production and lower plasma glucose levels, presently its role in glucose metabolism remains controversial (Xu *et al.* 2005). Importantly, no changes in plasma glucose are detected in Angptl4-/- mice in either fed or fasting state (Koster *et al.* 2005).

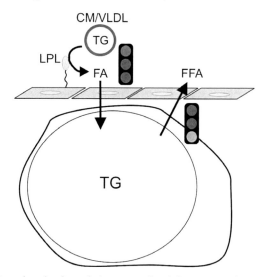

Fig. 2 Role of Angptl4 in lipid metabolism. Angptl4 inhibits LPL-mediated lipolytic processing of TG-rich lipoprotein particles. CM, chylomicron; VLDL, very low density lipoprotein; LPL, lipoprotein lipase; FA, fatty acid. Traffic lights indicate whether Angptl4 has a stimulatory (green) or inhibitory (red) effect.

Color image of this figure appears in the color plate section at the end of the book.

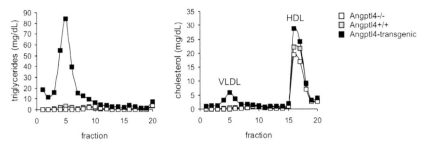

Fig. 3 Angptl4 influences levels of plasma lipoproteins. Lipoprotein profiles of fasting plasma from wild-type, Angptl4-/- and Angptl4-transgenic mice, as determined by HPLC.

Table I Key features of lipoprotein lipase

- Hydrolyses circulating triglycerides packages within lipoprotein particles.
- Is attached to the capillary endothelium of muscle and adipose tissue via heparin sulfate proteoglycans and GPIHBP-1.
- Is subject to extensive regulation at transcriptional and post-translational level.
- Is catalytically active as dimer.

NON-METABOLIC FUNCTIONS OF ANGPTL4

In addition to its metabolic functions in muscle, adipose tissue and possibly other tissues, Angptl4 influences non-metabolic processes via its expression in endothelial and tumor cells and its interaction with the ECM by associating with HSPG (Cazes *et al.* 2006). The interaction with the ECM protects Angptl4 from cleavage by proprotein convertases (Chomel *et al.* 2009). The majority of studies indicate that Angptl4 inhibits angiogenesis by reducing endothelial cell adhesion, migration, and tubule formation, which is mediated by the C-terminal portion of the protein (Yang *et al.* 2008). The exact molecular mechanism responsible for those effects remains to be elucidated. Data on the role of Angptl4 in metastasis and tumor cells are conflicting. Angptl4 was suggested to promote metastasis of breast cancer to lung by increasing capillary permeability in a transforming growth factor β–dependent fashion. However, other studies indicate that Angptl4 prevents metastasis and tumor growth and inhibits vascular permeability and angiogenesis. The reason behind this discrepancy is not known.

REGULATION OF ANGPTL4 EXPRESSION

As suggested by its alternative name, fasting-induced adipose factor, fasting induces expression of Angptl4 in murine adipose tissue as well as in several other tissues, including liver, skeletal muscle, heart, and brown adipose tissue (Kersten *et al.* 2000). Angptl4 is under transcriptional control of peroxisome proliferator activated receptor (PPAR) (Fig. 4). Depending on the tissue involved, Angptl4 expression is mainly regulated by PPARα (liver), PPARδ (muscle), or PPARγ (adipose tissue) (Kersten *et al.* 2000, Mandard *et al.* 2004, Staiger *et al.* 2009, Yoon

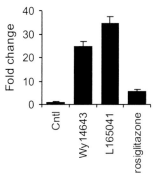

Fig. 4 Angptl4 is transcriptionally regulated by PPARs. Changes in Angptl4 expression upon treatment of C2C12 myotubes with synthetic agonists for PPARα (Wy14643), PPARδ (L165041), or PPARγ (rosiglitazone).

et al. 2000). What separates Angptl4 from other PPAR target genes is its extreme sensitivity to induction by fatty acids *in vitro* and *in vivo*. In human subjects the level of Angptl4 in plasma positively correlates with plasma FFA and is induced by caloric restriction and exercise via elevated FFA (Kersten *et al.* 2009). Furthermore, plasma Angptl4 levels are raised by treatment with fibrates and thiazolidinediones, which serve as synthetic agonists for PPARα and PPARγ, respectively (Kersten *et al.* 2009, Xu *et al.* 2005). Expression of Angptl4 is also governed by glucocorticoids (Koliwad *et al.* 2009).

Another major physiological stimulus of Angptl4 gene expression is hypoxia. In numerous cell types including adipocytes, endothelial cells, and cardiomyocytes, hypoxia dramatically induces Angptl4 mRNA via the HIF-1α transcription factor (Cazes *et al.* 2006). Currently, it is difficult to reconcile induction of Angptl4 under hypoxic conditions with a presumed anti-angiogenic action of Angptl4. From a metabolic point of view, induction of Angptl4 by hypoxia may be aimed at shifting fuel use away from fatty acids towards glucose, which can be metabolized anaerobically.

Finally, genetic studies support a role for Angptl4 in regulation of plasma lipid levels in humans. A rare loss of function sequence variant of Angptl4 (E40K) is associated with higher plasma HDL levels and decreased plasma TG levels, which may lead to decreased risk for coronary heart disease (Romeo *et al.* 2007). Effects of more common sequence variants within the Angptl4 gene are less well established and require further investigation.

In conclusion, Angptl4 represents a factor produced by adipose and other tissue that limits tissue uptake of plasma TG-derived fatty acids and promotes adipose tissue lipolysis. In addition, Angptl4 governs angiogenesis and endothelial cell function. Angptl4 can be considered as a pro-hormone whose pleiotropic actions are at least partly mediated by its N- and C-terminal fragments.

Summary Points

- Angptl4 is secreted from a variety of tissues including adipose tissue, liver and muscle.
- Angptl4 forms higher order oligomers and is proteolytically processed.
- Expression of Angptl4 is governed by PPARs.
- Angptl4 raises plasma TG and FFA levels by inhibiting lipoprotein lipase and adipose tissue lipolysis, respectively.
- Genetic variation within the human ANGPTL4 gene impacts plasma lipoprotein levels.
- Levels of Angptl4 in plasma are increased by fasting and exercise via elevated FFA.
- Angptl4 is retained in the extracellular matrix and governs angiogenesis by influencing the function of endothelial cells.

Abbreviations

Angptl4	:	angiopoietin-like protein 4
TG	:	triglycerides
LPL	:	lipoprotein lipase
HSPG	:	heparin sulfate proteoglycans
FFA	:	free fatty acids
GPIHBP-1	:	glycosylphosphatidylinositol-anchored high density lipoprotein-binding protein-1
ECM	:	extracellular matrix
PPAR	:	peroxisome proliferator-activated receptor
HIF-1α	:	hypoxia-inducibe factor 1-alpha

Definition of Terms

Adipose tissue lipolysis: Process describing the hydrolysis of stored triglycerides in adipose tissue to be released into the circulation as fatty acids.

Angiogenesis: Growth of new blood vessels from existing blood vessels.

Fasting: Condition of complete food deprivation that leads to major adaptive changes in energy metabolism.

Metastasis: Spread of disease, usually cancer, from one organ to another organ.

Plasma triglyceride clearance: Process describing the hydrolysis of circulating triglycerides to be taken up into underlying tissues as fatty acids.

PPARs: Group of ligand-activated transcription factors involved in regulation of metabolism and inflammation.

References

Cazes, A. and A. Galaup, C. Chomel, M. Bignon, N. Brechot, S. Le Jan, H. Weber, P. Corvol, L. Muller, S. Germain, and C. Monnot. 2006. Extracellular matrix-bound angiopoietin-like 4 inhibits endothelial cell adhesion, migration, and sprouting and alters actin cytoskeleton. Circ. Res. 99: 1207-1215.

Chomel, C. and A. Cazes, C. Faye, M. Bignon, E. Gomez, C. Ardidie-Robouant, A. Barret, S. Ricard-Blum, L. Muller, S. Germain, and C. Monnot. 2009. Interaction of the coiled-coil domain with glycosaminoglycans protects angiopoietin-like 4 from proteolysis and regulates its antiangiogenic activity. FASEB J. 23: 940-949.

Ge, H. and G. Yang, L. Huang, D.L. Motola, T. Pourbahrami, and C. Li. 2004a. Oligomerization and regulated proteolytic processing of angiopoietin-like protein 4. J. Biol. Chem. 279: 2038-2045.

Ge, H. and G. Yang, X. Yu, T. Pourbahrami, and C. Li. 2004b. Oligomerization state-dependent hyperlipidemic effect of angiopoietin-like protein 4. J. Lipid Res. 45: 2071-2079.

Kersten, S. and L. Lichtenstein, E. Steenbergen, K. Mudde, H.F. Hendriks, M.K. Hesselink, P. Schrauwen, and M. Muller. 2009. Caloric restriction and exercise increase plasma ANGPTL4 levels in humans via elevated free fatty acids. Arterioscler. Thromb. Vasc. Biol. 29: 969-974.

Kersten, S. and S. Mandard, N.S. Tan, P. Escher, D. Metzger, P. Chambon, F.J. Gonzalez, B. Desvergne, and W. Wahli. 2000. Characterization of the fasting-induced adipose factor FIAF, a novel peroxisome proliferator-activated receptor target gene. J. Biol. Chem. 275: 28488-28493.

Kim, I. and H.G. Kim, H. Kim, H.H. Kim, S.K. Park, C.S. Uhm, Z.H. Lee, and G.Y. Koh. 2000. Hepatic expression, synthesis and secretion of a novel fibrinogen/angiopoietin-related protein that prevents endothelial-cell apoptosis. Biochem. J. 346: 603-610.

Koliwad, S.K. and T. Kuo, L.E. Shipp, N.E. Gray, F. Backhed, A.Y. So, R.V. Farese, Jr., and J.C. Wang. 2009. Angiopoietin-like 4 (ANGPTL4, fasting-induced adipose factor) is a direct glucocorticoid receptor target and participates in glucocorticoid-regulated triglyceride metabolism. J. Biol. Chem. 284: 25593-25601.

Koster, A. and Y.B. Chao, M. Mosior, A. Ford, P.A. Gonzalez-DeWhitt, J.E. Hale, D. Li, Y. Qiu, C.C. Fraser, D.D. Yang, J.G. Heuer, S.R. Jaskunas, and P. Eacho. 2005. Transgenic angiopoietin-like (angptl)4 overexpression and targeted disruption of angptl4 and angptl3: regulation of triglyceride metabolism. Endocrinology 146: 4943-4950.

Lichtenstein, L. and J.F. Berbee, S.J. van Dijk, K.W. van Dijk, A. Bensadoun, I.P. Kema, P.J. Voshol, M. Muller, P.C. Rensen, and S. Kersten. 2007. Angptl4 upregulates cholesterol synthesis in liver via inhibition of LPL- and HL-dependent hepatic cholesterol uptake. Arterioscler. Thromb. Vasc. Biol. 27: 2420-2427.

Mandard, S. and F. Zandbergen, N.S. Tan, P. Escher, D. Patsouris, W. Koenig, R. Kleemann, A. Bakker, F. Veenman, W. Wahli, M. Muller, and S. Kersten. 2004. The direct peroxisome proliferator-activated receptor target fasting-induced adipose factor (FIAF/PGAR/ANGPTL4) is present in blood plasma as a truncated protein that is increased by fenofibrate treatment. J. Biol. Chem. 279: 34411-34420.

Romeo, S. and L.A. Pennacchio, Y. Fu, E. Boerwinkle, A. Tybjaerg-Hansen, H.H. Hobbs, and J.C. Cohen. 2007. Population-based resequencing of ANGPTL4 uncovers variations that reduce triglycerides and increase HDL. Nat. Genet. 39: 513-516.

Sanderson, L.M. and T. Degenhardt, A. Koppen, E. Kalkhoven, B. Desvergne, M. Muller, and S. Kersten. 2009. Peroxisome proliferator-activated receptor beta/delta (PPARbeta/delta) but not PPARalpha serves as a plasma free fatty acid sensor in liver. Mol. Cell. Biol. 29: 6257-6267.

Sonnenburg, W.K. and D. Yu, E.C. Lee, W. Xiong, G. Gololobov, B. Key, J. Gay, N. Wilganowski, Y. Hu, S. Zhao, M. Schneider, Z.M. Ding, B.P. Zambrowicz, G. Landes, D.R. Powell, and U. Desai. 2009. Glycosylphosphatidylinositol-anchored HDL-binding protein stabilizes lipoprotein lipase and prevents its inhibition by angiopoietin-like 3 and angiopoietin-like 4. J. Lipid Res. 50: 2421-2429.

Staiger, H. and C. Haas, J. Machann, R. Werner, M. Weisser, F. Schick, F. Machicao, N. Stefan, A. Fritsche, and H.U. Haring. 2009. Muscle-derived angiopoietin-like protein 4 is induced by fatty acids via peroxisome proliferator-activated receptor (PPAR)-delta and is of metabolic relevance in humans. Diabetes 58: 579-589.

Sukonina, V. and A. Lookene, T. Olivecrona, and G. Olivecrona. 2006. Angiopoietin-like protein 4 converts lipoprotein lipase to inactive monomers and modulates lipase activity in adipose tissue. Proc. Natl. Acad. Sci. USA 103: 17450-17455.

Xu, A. and M.C. Lam, K.W. Chan, Y. Wang, J. Zhang, R.L. Hoo, J.Y. Xu, B. Chen, W.S. Chow, A.W. Tso, and K.S. Lam. 2005. Angiopoietin-like protein 4 decreases blood glucose and improves glucose tolerance but induces hyperlipidemia and hepatic steatosis in mice. Proc. Natl. Acad. Sci. USA 102: 6086-6091.

Yang, Y.H. and Y. Wang, K.S. Lam, M.H. Yau, K.K. Cheng, J. Zhang, W. Zhu, D. Wu, and A. Xu. 2008. Suppression of the Raf/MEK/ERK signaling cascade and inhibition of angiogenesis by the carboxyl terminus of angiopoietin-like protein 4. Arterioscler. Thromb. Vasc. Biol. 28: 835-840.

Yoon, J.C. and T.W. Chickering, E.D. Rosen, B. Dussault, Y. Qin, A. Soukas, J.M. Friedman, W.E. Holmes, and B.M. Spiegelman. 2000. Peroxisome proliferator-activated receptor gamma target gene encoding a novel angiopoietin-related protein associated with adipose differentiation. Mol. Cell. Biol. 20: 5343-5349.

Yoshida, K. and T. Shimizugawa, M. Ono, and H. Furukawa. 2002. Angiopoietin-like protein 4 is a potent hyperlipidemia-inducing factor in mice and inhibitor of lipoprotein lipase. J. Lipid Res. 43: 1770-1772.

Yu, X. and S.C. Burgess, H. Ge, K.K. Wong, R.H. Nassem, D.J. Garry, A.D. Sherry, C.R. Malloy, J.P. Berger, and C. Li. 2005. Inhibition of cardiac lipoprotein utilization by transgenic overexpression of Angptl4 in the heart. Proc. Natl. Acad. Sci. USA 102: 1767-1772.

CHAPTER 9

Interleukins as Adipokines

Kirsten Prüfer Stone
[1]Pennington Biomedical Research Center, 6400 Perkins Rd., Baton Rouge,
LA 70808, USA

ABSTRACT

The major functions of interleukins (IL) are to regulate via cell-cell signaling the responses of an organism to an immune challenge. Whereas the major site of production and secretion of interleukins as well as their major targets are leukocytes, some interleukins are also produced and secreted by adipocytes, defining them as another class of adipokines. Obesity puts an organism into a chronic inflammatory state. The correlation of obesity with the presence of pro-inflammatory interleukins such as IL-1 in adipose tissue is linked to insulin resistance, likely via direct crosstalk between IL-1 and insulin signaling pathways. Production and secretion by adipocytes of IL-6, another pro-inflammatory interleukin, also correlate with obesity, but its functions in adipose tissue are more complex. Whereas short-term increase of IL-6 appears to have positive effects on fat oxidation and glucose transport, chronic exposure to elevated IL-6 induces insulin resistance similar to effects seen with high exposure to IL-1. Production and secretion by adipocytes of other pro-inflammatory interleukins, IL-8 and IL-18, also correlate with obesity. Studies on mice genetically modified to lack IL-18, however, show that IL-18 is important for maintaining normal weight. The production and secretion by adipocytes of the anti-inflammatory interleukin, IL-10, are inversely correlated to obesity. Whereas few interleukins are adipokines, many interleukins have direct effects on adipose tissue.

INTRODUCTION

Cytokines are molecules that are secreted by cells and play crucial roles in cell-cell-signaling. Interleukins (IL) are cytokines of the immune system. The function of the immune system depends in large part on interleukins, and deficiencies of a number of them can cause autoimmune diseases or immune deficiency. The majority of interleukins are synthesized by leukocytes and endothelial cells. Their main functions are to promote the development and differentiation of leukocytes and blood cell precursors. Recent findings, however, showed that interleukins are produced by a wide variety of body cells.

The accumulation of adipose tissue inside the abdominal cavity is associated with chronic, systemic low-grade inflammation including altered cytokine production. Studies have shown that several interleukins are produced by adipocytes and affect the pathology and physiology of type 2 diabetes mellitus (T2DM) and obesity.

TYPES OF INTERLEUKINS

Interleukins can be classified according to their function in inflammation. The main pro-inflammatory interleukins are IL-1 and IL-6; anti-inflammatory interleukins include IL-10 and the interleukin 1 receptor antagonist (IL-1ra). Interleukins have partly overlapping functions in activating either cells of the immediate response to infection (T cells; IL-2, IL-4, IL-9, IL-12, IL-15) or of the adaptive response (B cells carrying antibodies to specific antigens; IL-4, IL-5, IL-6, IL-21).

FUNCTIONS AND SIGNALING OF INTERLEUKINS

Interleukins are secreted as a response to an immune challenge. Interleukins then can activate signaling pathways in the same cell (autocrine) or a cell in close proximity (paracrine). During severe infection, interleukins can also be transported in the blood circulation and act on many target cells. Binding of interleukins to their cell surface receptors induces signaling pathways described in Fig. 1, resulting in the activation of either gene expression or regulatory kinases. Such pathways are partly overlapping with and can therefore crosstalk with those of other cytokines, including the insulin signaling pathway.

INTERLEUKINS AS ADIPOKINES (TABLE I)

IL-6

Adipose tissue is a major source (up to 35%) of IL-6 in the absence of exercise or acute inflammation (Mohamed-Ali *et al.* 1997). Serum IL-6 is increased in obesity and decreases with weight loss (Table 2). The current model of IL-6 actions, albeit

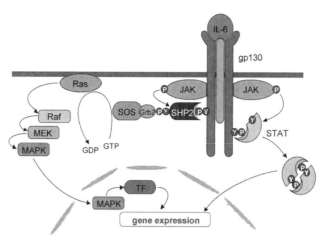

Fig. 1 Example for signaling pathways of interleukins. (Modified from and reproduced with permission from Heinrich et al. 2003, © the Biochemical Society.) Binding of IL-6 causes dimerization of the IL-6 receptor gp130. Janus kinases (Jak) bind to specific residues in the cytoplasmic tail of the IL-6 receptor and are phosphorylated (P). Signal transducing activators of transcription (STATs) are recruited to phosphorylated residues in the cytoplasmic tail of the IL-6 receptor and tyrosine (Y) phosphorylated (P) by Jak. Phosphorylated STATs dimerize and translocate into the nucleus to regulate gene expression. Alternatively, phosphorylation of a specific tyrosine in the cytoplasmic tail of the IL-6 receptor causes binding of protein-tyrosine phosphatase SHP2. Tyrosine phosphorylation of SHP2 by Jak allows binding of adapter "growth factor receptor-bound protein" (Grb)2 and guanine nucleotide exchange factor "son of sevenless" (SOS). Activation of Ras by phosphorylation induces the MAPK pathway.

Table I   Interleukins as adipokines

	Function in immune system	Function as adipokine
IL-6	Pro- and anti-inflammatory, pyrogen, stimulates proliferation of activated B cells and differentiation of hematopoetic cells	Short-term exposure beneficial by stimulating glucose uptake, lipolysis, and fat oxidation; long-term exposure contributing to insulin resistance
IL-1	Pro-inflammatory; pyrogen, stimulates proliferation of T and B cells	Causes insulin resistance, elevated in patients with T2DM
IL-1ra	Competitive inhibitor of IL-1 for the IL-1 receptor, antagonizes activities of IL-1	Increased levels correlate with higher risk for obesity; ratio of IL1/IL1ra is important for normal weight
IL-8	Pro-inflammatory; induction of chemotaxis of T cells and neutrophils	Production in muscle induced by exercise (adipogenic induction unknown)
IL-18	Pro-inflammatory; stimulates natural killer cells and certain T cells	Role in maintaining normal weight
IL-10	Anti-inflammatory; acts on T cells as cytokine inhibitory factor, inhibits macrophage functions	Negatively correlates with obesity, increases with weight loss and after exercise

controversial, links obesity-derived chronic inflammation with insulin resistance. Whereas acute elevation of IL-6 is beneficial, chronic elevation of IL-6 is not. Short-term IL-6 administration increases lipolysis and fat oxidation (Petersen et al. 2005) and stimulates glucose transport in adipocytes (Stouthard *et al.* 1996). The need of IL-6 for normal insulin signaling was confirmed in IL-6-deficient mice that develop mature onset obesity and insulin resistance (Fig. 2). On the

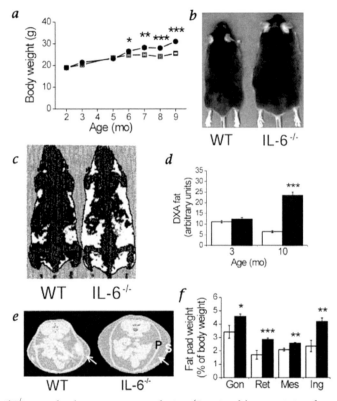

Fig. 2 *Il6*[-/-] mice develop mature-onset obesity. (Reprinted by permission from Macmillan Publishers Ltd: Nature Medicine, Wallenius et al. 2002, copyright 2002.) Mice lacking the gene for IL-6 were generated and characterized. (a) Body weight of wild-type ($\square$) and *Il6*[-/-] ($\bullet$) females at various ages. *Il6*[-/-] mice have significantly higher body weight from the age of 6 mo. (b) 11-mo-old male mice. (c) Adiposity in 10-mo-old male mice measured by dual energy X-ray absorptiometry (DXA); white represents areas with more than 50% fat. (d) Quantification of several DXA measurements for 3- and 10-mo-old wild-type ($\square$) and *Il6*[-/-] ($\blacksquare$, $n = 6$) males. (e) Representative computerized tomography (CT) images of 11-mo-old wild-type and *Il6*[-/-] male mice. White arrows indicate s.c. adipose tissue between the skin (S) and the peritoneum (P). (f) Relative weights of different fat depots (% of body weight) in wild-type ($\square$) and *Il6*[-/-] ($\blacksquare$) mice. 3 intra-abdominal fat pads (gonadal (Gon), retroperitoneal (Ret) and mesenteric (Mes)) and the inguinal fat pad (Ing: a s.c. fat pad in the groin) were dissected and weighed in 18-mo-old female mice. $P < 0.05$; $P < 0.01$; $P < 0.001$, versus corresponding wild-type mice.

other hand, long-term IL-6 treatment induces insulin resistance in adipocytes by decreasing transcription and activation of molecules crucial for insulin signaling, glucose transport, and lipogenesis (Fig. 3).

Table 2   Anthropometric and biologic parameters of a study population exposed to reduction of their calorie intake by 500 kcal/day, the appetite suppressant sibutramine, and a program of exercises (assembled from Jung *et al.* 2008, with permission from Elsevier).

	Baseline	After 12 weeks	*P*
n	78	78	
Sex (male:female)	37:41		
Age (years)	38.5±11.8		
Weight (kg)	87.9±15.4	81.7±16.2	<.001
BMI (kg/m²)	32.2±3.5	29.9±4.0	<.001
IL-6 (pg/ml)	2.50±0.81	2.20±0.86	.001
IL-10 (pg/ml)	12.02±4.44	13.50±5.49	.041

Serum IL-6 decreased, whereas IL-10 increased concomitantly with weight loss in obese study participants. Data are presented as mean±S.D.

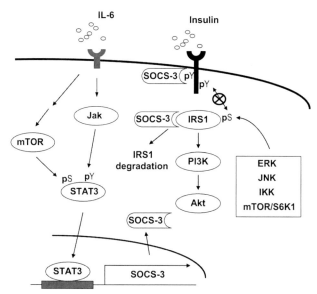

Fig. 3   Crosstalk of interleukin with insulin signaling pathways. (Reprinted from Kim *et al.* 2009, with permission from Elsevier.) Normal insulin signaling includes tyrosine phosphorylation of the insulin-bound insulin receptor followed by activation of the insulin receptor substrate (IRS1), phosphoinositide 3-kinase (PI3K), and protein kinase B (Akt). Activation of STATs by IL-6 induces expression of suppressor of cytokine signaling (SOCS)-3. SOCS-3 directly binds to and inhibits both the insulin receptor and IRS1. IL-1α treatment causes interference with insulin signaling through serine phosphorylation of IRS-1 by activated kinases such as extracellular signal-regulated kinase (ERK), c-jun N-terminal kinase (JNK), inhibitor kB kinase (IKK), and ribosomal S6 kinase (S6K1). Such alterations in IRS1 signaling (including sorting serine phosphorylated IRS1 to degradation) ultimately inhibit protein kinase B (Akt), leading to insulin resistance.

IL-1 and IL-1ra

Both IL-1 and its anti-inflammatory IL-1 receptor antagonist (IL-1ra) are adipokines. Moreover, IL-1 can induce IL-1ra (Juge-Aubry et al. 2003). Functions of these adipokines in obesity were shown in genetic studies: A polymorphism in the gene encoding IL-1ra causes increased plasma levels of IL-1ra and is associated with an increased risk of obesity (Strandberg *et al.* 2006).

Mechanistic studies exploring the effect of IL-1 on adipocytes showed that both IL-1α and IL-1β treatment causes transient insulin resistance through interference with and inhibition of a crucial signaling step downstream of insulin binding to the insulin receptor (He *et al.* 2006, Jager *et al.* 2007) as described in the caption to Fig. 3.

IL-8, IL-18 and IL-10

In obesity and T2DM, production and secretion by adipocytes of IL-8 (Bruun *et al.* 2001) and IL-18 (Bruun *et al.* 2007) is increased. The role of IL-18 in maintaining normal weight was shown in mice genetically modified to lack IL-18 (Fig. 4). Another adipokine is the anti-inflammatory cytokine IL-10 (Juge-Aubry *et al.*

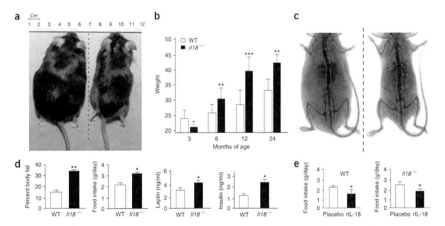

Fig. 4 *Il18*$^{-/-}$ mice develop spontaneous obesity and diabetes mellitus. (Reprinted by permission from Macmillan Publishers Ltd: Nature Medicine, Netea et al. 2006, copyright 2006, http://www.nature.com/nm/index.html.) Mice lacking the gene for IL-18 were generated and characterized. (a) Spontaneous obesity in an *Il18*$^{-/-}$ mouse photographed at age 1 yr (left), compared with a wild-type (*Il18*$^{+/+}$; WT) control mouse (right). (b) Evolution of body weight in *Il18*$^{+/+}$ and *Il18*$^{-/-}$ mice. (c) DXA measurements of *Il18*$^{-/-}$ (left) and *Il18*$^{+/+}$ (right) mice at age 6 mo. (d) Metabolic and hormonal parameters in *Il18*$^{+/+}$ and *Il18*$^{-/-}$ mice at 6 mo. (e) Food intake of *Il18*$^{+/+}$ and *Il18*$^{-/-}$ mice given an intracerebral injection of 10µg/ml rIL-18 or vehicle twice weekly. Data show that lack of IL-18 leads to obesity correlating with increased food intake. Intracerebral injection of IL-18 inhibits food intake in both wild-type and *Il18*$^{-/-}$ mice. Data are presented as mean ±s.d. (*n* = 10 mice per group). $^{*}P < 0.05$, $^{**}P < 0.01$, $^{***}P < 0.005$.

2005). IL-10 production increases with both exercise and weight loss (Table 2), linking its anti-inflammatory functions to the known effects of exercise to reduce chronic inflammation.

Taken together, adipose tissue is a source of interleukins. The correlation between body weight and production/secretion by adipocytes of some interleukins indicates their role in the inflammatory processes in obesity.

Summary Points

- Obesity is associated with chronic low-grade inflammation including changed cytokine production.
- Pro-inflammatory interleukins IL-1, IL-1ra, IL-6, IL-8, and IL-18 and anti-inflammatory IL-10 are produced and secreted by adipocytes and can therefore be called adipokines.
- Interleukin production is modulated in obesity and T2DM.
- IL-1 and IL-6 can directly cause insulin resistance via crosstalk of signaling pathways.
- IL-6 has both positive and negative effects on insulin signaling dependent on acute or chronic changes in IL-6 production and secretion.
- Other interleukins albeit not produced by adipocytes exert effects on adipose tissue metabolism.

Abbreviations

ERK	:	extracellular signal-regulated kinase
IL	:	interleukin
IRS	:	insulin receptor substrate
Jak	:	Janus kinase
JNK	:	c-jun N-terminal kinase
MAPK	:	mitogen-activated protein kinase
SOCS-3	:	suppressor of cytokine signaling 3
STAT	:	signal transducing activators of transcription
T2DM	:	type 2 diabetes mellitus

Definition of Terms

Chronic inflammation: A condition that leads to a progressive shift in the type of cells present at the site of inflammation and characterized by simultaneous destruction and healing of the tissue from the inflammatory process.

Inflammation: The complex biological response of vascular tissues to harmful stimuli.

Interleukins: The means by which leukocytes (-leukin) "communicate" with each other (inter).

References

Bruun, J.M. and S.B. Pedersen, and B. Richelsen. 2001. Regulation of interleukin 8 production and gene expression in human adipose tissue *in vitro*. J. Clin. Endocrinol. Metab. 86: 1267-1273.

Bruun, J.M. and B. Stallknecht, J.W. Helge, and B. Richelsen. 2007. Interleukin-18 in plasma and adipose tissue: effects of obesity, insulin resistance, and weight loss. Eur. J. Endocrinol. 157: 465-471.

He, J. and I. Usui, K. Ishizuka, Y. Kanatani, K. Hiratani, M. Iwata, A. Bukhari, T. Haruta, T. Sasaoka, and M. Kobayashi. 2006. Interleukin-1alpha inhibits insulin signaling with phosphorylating insulin receptor substrate-1 on serine residues in 3T3-L1 adipocytes. Mol. Endocrinol. 20: 114-124.

Heinrich, P.C. and I. Behrmann, S. Haan, H.M. Hermanns, G. Muller-Newen, and F. Schaper. 2003. Principles of interleukin (IL)-6-type cytokine signalling and its regulation. Biochem. J. 374: 1-20.

Jager, J. and T. Gremeaux, M. Cormont, Y. Le Marchand-Brustel, and J.F. Tanti. 2007. Interleukin-1beta-induced insulin resistance in adipocytes through down-regulation of insulin receptor substrate-1 expression. Endocrinology 148: 241-251.

Juge-Aubry, C.E. and E. Somm, V. Giusti, A. Pernin, R. Chicheportiche, C. Verdumo, F. Rohner-Jeanrenaud, D. Burger, J.M. Dayer, and C.A. Meier. 2003. Adipose tissue is a major source of interleukin-1 receptor antagonist: upregulation in obesity and inflammation. Diabetes 52: 1104-1110.

Juge-Aubry, C.E. and E. Somm, A. Pernin, N. Alizadeh, V. Giusti, J.M. Dayer, and C.A. Meier. 2005. Adipose tissue is a regulated source of interleukin-10. Cytokine 29: 270-274.

Jung, S.H. and H.S. Park, K.S. Kim, W.H. Choi, C.W. Ahn, B.T. Kim, S.M. Kim, S.Y. Lee, S.M. Ahn, Y.K. Kim, H.J. Kim, D.J. Kim, and K.W. Lee. 2008. Effect of weight loss on some serum cytokines in human obesity: increase in IL-10 after weight loss. J. Nutr. Biochem. 19: 371-375.

Kim, J.H. and R.A. Bachmann, and Chen. 2009. Interleukin-6 and insulin resistance. Vitam. Horm. 80: 613-633.

Mohamed-Ali, V. and S. Goodrick, A. Rawesh, D.R. Katz, J.M. Miles, J.S. Yudkin, S. Klein, and S.W. Coppack. 1997. Subcutaneous adipose tissue releases interleukin-6, but not tumor necrosis factor-alpha, *in vivo*. J. Clin. Endocrinol. Metab. 82: 4196-4200.

Netea, M.G. and L.A. Joosten, E. Lewis, D.R. Jensen, P.J. Voshol, B.J. Kullberg, C.J. Tack, K.H. van, S.H. Kim, A.F. Stalenhoef, F.A. van de Loo, I. Verschueren, L. Pulawa, S. Akira, R.H. Eckel, C.A. Dinarello, W. van den Berg, and J.W. van der Meer. 2006. Deficiency of interleukin-18 in mice leads to hyperphagia, obesity and insulin resistance. Nat. Med. 12: 650-656.

Petersen, E.W. and A.L. Carey, M. Sacchetti, G.R. Steinberg, S.L. Macaulay, M.A. Febbraio, and B.K. Pedersen. 2005. Acute IL-6 treatment increases fatty acid turnover in elderly humans *in vivo* and in tissue culture *in vitro*. Am. J. Physiol. Endocrinol. Metab. 288: E155-E162.

Stouthard, J.M. and R.P. Oude Elferink, and H.P. Sauerwein. 1996. Interleukin-6 enhances glucose transport in 3T3-L1 adipocytes. Biochem. Biophys. Res. Commun. 220: 241-245.

Strandberg, L. and M. Lorentzon, A. Hellqvist, S. Nilsson, V. Wallenius, C. Ohlsson, and J.O. Jansson. 2006. Interleukin-1 system gene polymorphisms are associated with fat mass in young men. J. Clin. Endocrinol. Metab. 91: 2749-2754.

Wallenius, V. and K. Wallenius, B. Ahren, M. Rudling, H. Carlsten, S.L. Dickson, C. Ohlsson, and J.O. Jansson. 2002. Interleukin-6-deficient mice develop mature-onset obesity. Nat. Med. 8: 75-79.

CHAPTER 10

Leptin as an Adipokine: Important Definitions and Applications for Cancer Research

Barbara Stadterman,[1,*] Anna Lokshin,[2] Robert P. Edwards,[3] and
Faina Linkov[4]

[1]Graduate School of Public Health, University of Pittsburgh,
127 Parran Hall, Pittsburgh, PA, USA
[2]Magee-Womens Hospital of UPMC, 300 Halket Street, Pittsburgh, PA, USA
[3]University of Pittsburgh Cancer Institute, CNPAV 466, Pittsburgh, PA, USA
[4]University of Pittsburgh Cancer Institute, HCCLB 1.19d, Pittsburgh, PA, USA

ABSTRACT

White adipose tissue (WAT) was once regarded as simply a fat depository in animals that released fatty acids and glycerol in times of starvation. However, WAT is now recognized as an active endocrine organ that secretes various hormones and protein factors. The name *adipokines* is given to a group of cytokine proteins, which are produced mainly by the adipocytes of WAT. Adipokines mediate and regulate immunity, inflammation, and hematopoiesis; thus, they potentially influence a variety of physiological processes. Leptin is a small, 166 amino acid peptide hormone primarily secreted by subcutaneous adipose tissue (and somewhat by visceral adipose tissue) crucial to proper energy homeostasis and has a putative signal sequence. Leptin augments the action of insulin on both the inhibition of hepatic glucose production and in stimulating glucose uptake. However, in the obese state, leptin acts to sensitize the effect of insulin in target tissues. Studies examining the effects of the administration of recombinant leptin to both animals

*Corresponding author

and humans have shown a marginal ability to reverse the phenotype associated with leptin deficiency. Leptin administration has proven to be advantageous in the treatment of hyperphagia (disorder marked by an abnormal appetite and excessive ingestion of food), reducing appetite and normalizing neuroendocrine activity in some leptin-deficient patients. However, because human obesity is most often characterized by elevated serum leptin levels (hyperleptinemia), further study of leptin resistance could be potentially beneficial in developing a therapeutic treatment. Additionally, further research studies on leptin's role in cancer development would be needed, as preliminary evidence suggests that leptin may play an important role in the development of various malignancies.

INTRODUCTION

Leptin was discovered in 1994 as a regulator of body weight and energy balance, manufactured primarily in the adipocytes of white adipose tissue (WAT). Leptin plays an important role in obesity, as there is a direct relationship between body mass index and the levels of leptin protein in circulating serum (Fig. 1). Obesity is becoming a significant health problem, with over 60% of the US population suffering from overweight or obesity (Fig. 2).

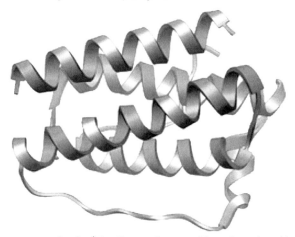

Fig. 1 Leptin protein molecule. (This figure is from a public website, *http://en.wikipedia.org/wiki/Leptin.*)

Obesity is characterized by chronic, low-grade systemic inflammation (Trayhurn and Wood 2005), potentially leading to the development of chronic disease. Increased levels of the inflammatory markers C-reactive protein, interleukin-6 and tumor necrosis factor alpha are observed in obese humans. The association between obesity and insulin resistance and metabolic syndrome is believed to result from increased serum concentrations of both pro-inflammatory cytokines

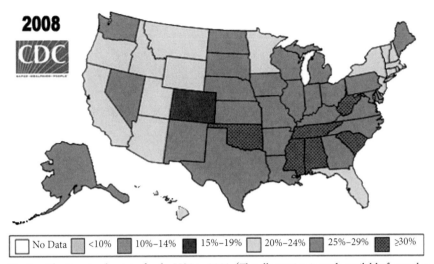

Fig. 2   Obesity rates, by state, for the USA in 2008. (This illustration is made available for use by the CDC website.)

Color image of this figure appears in the color plate section at the end of the book.

and acute phase proteins produced by WAT. Recent studies indicate a relationship between increased inflammation and increased circulating levels of leptin along central leptin resistance during obesity (Trayhurn and Wood 2005). In most cases, weight loss reverses this trend, improving insulin sensitivity (Manco *et al.* 2007).

There is substantial evidence that leptin works in an intermediary fashion between the peripheral and central nervous systems, thus allowing metabolic and neuroendocrine function to adapt to variations in an animal's nutritional status. Research with animal models confirmed the hypothalamus to be the primary center for regulation of food intake and body weight (Hill *et al.* 2008). Upon release by the WAT into the bloodstream, leptin crosses the blood-brain barrier and binds to the hypothalamic leptin receptors, thus relaying information regarding the body's energy reserves (Banks 2001). Leptin inhibits appetite and prevents excessive fat storage by acting on receptors in the hypothalamus, where it counteracts the effects of *neuropeptide Y*, a protein that stimulates hunger and is produced by cells in the gut (Banks 2001). Leptin also neutralizes the effects of the appetite stimulant *anandamide* and promotes the synthesis of the appetite suppressant *α-MSH* (melanocyte stimulating hormone), a product of (increased) POMC (proopiomelanocortin) production resulting from the activation of leptin receptors (Nisoli and Carruba 2004).

LEPTIN AND OBESITY

Leptin resistance, a state in which the effect of leptin is blocked, is common in the extremely obese but rare in subjects of normal weight. This concept appears contradictory to nature, as the physiological and evolutionary role of leptin is to signal the need for nutrition with low levels, versus preventing over-eating when levels are elevated. However, during times of preparation for hibernation and gestation, *temporary* leptin resistance appears to occur in animals, as feeding increases regardless of the amount of WAT present or the levels of circulating leptin (Munzberg *et al.* 2005).

Studies done in the early part of the decade suggest that central leptin resistance develops with age and/or obesity and that the subsequent increase in WAT mass leads to hyperleptinemia (Scarpace and Zhang 2009). With this, the lipopenic effect of leptin would be obstructed and thus probable inhibition of insulin signaling in fat tissue would result, possibly leading to overall insulin resistance. Obese animals in this state overeat, despite marked hyperleptinemia. While this trend can potentially be reversed with weight loss, previous research demonstrated that leptin signaling is crucial for the preservation of lean body mass during caloric restriction (Buettner *et al.* 2008). When leptin or leptin receptors are absent, food intake is unregulated and obesity results. Circulating leptin levels are also influenced by gender, exercise and glucose uptake (Klok *et al.* 2007).

Early studies in laboratory mice conducted in 1950 indicated a genetic link to obesity (Ingalls *et al.* 1996). It was found that animals lacking the *ob* gene had a body weight approximately three times that of their normal counterparts. In these animals, single-gene, homozygous mutations were found in both the gene coding for the leptin protein itself (*ob/ob*) and in the gene coding for the leptin receptor (*db/db*), conditions associated with a phenotype of extreme hunger and morbid obesity (Zhang *et al.* 1994). Later studies similarly found that humans lacking the *ob* gene were morbidly obese and consumed large quantities of food. Congenital leptin deficiency resulting from defects in the *OB* gene (a homozygous frameshift mutation) show a phenotype of normal birth weight, succeeded by rapid early-onset obesity associated with hyperphagia due to impaired satiety, decreased energy expenditure, and endocrine abnormalities (Farooqi *et al.* 1998). Low leptin levels are also seen in patients with congenital or acquired lipodystrophy associated with insulin resistance and dyslipidemia (Blüher and Mantzoros 2009). Subsequent studies in humans found leptin to be encoded by a gene located on chromosome 7q31.3, now known as the "Ob(Lep)" gene (Ob for obese, Lep for leptin), which spans 20 kilobases. Only one *OB* mRNA species has been indentified in abundance in human WAT (Isse *et al.* 1995). The gene coding for the leptin receptor, human homolog LEPR, is located on chromosome 1p31, the mutation of which is due to transgene insertion (Rankinen *et al.* 2006).

APPLICATIONS TO CANCER

The effect of leptin on cancer cells has been explored in several *in vitro* studies of breast, colorectal, prostate, pancreatic, ovarian, and lung cancers. Leptin and

ObR appear to be significantly over-expressed in cancer tissue relative to non-cancer epithelial tissues. In the area of endometrial cancer, researchers in Greece found evidence that leptin is strongly positively associated with endometrial cancer. However, it could not be conclusively inferred whether leptin elevation, as a consequence of obesity, plays a role in endometrial carcinogenesis or whether it is a simple correlate of obesity (Petridou *et al.* 2002). Additionally, leptin may affect the risk of clinically relevant prostate cancer through testosterone and factors related to stature and obesity (Chang *et al.* 2001). High leptin levels have been associated with an increased risk of intestinal metaplasia. Moreover, serum leptin levels are a significant independent marker for the presence of this malignancy (Capelle *et al.* 2009).

Obesity is a major risk factor for the development and progression of breast cancer. Increased circulating levels of the obesity-associated hormones leptin and insulin-like growth factor-I (IGF-I) and over-expression of the leptin receptor (Ob-R) and IGF-I receptor (IGF-IR) have been detected in a majority of breast cancer cases (Ozbay and Nahta 2006). Leptin has been found to play an important role in the proliferation of breast cancer cells (Ren *et al.* 2010). In breast tumors, high levels of leptin have been associated with increased incidence of breast cancer metastasis.

Leptin signaling could have a key role in renal cell carcinoma invasion (Horiguchi *et al.* 2006). Pro-inflammatory, angiogenic and mitogenic properties of leptin do not seem to be important for esophageal cancer development but hypoleptinemia, independently from co-occurring reduction of adiposity, appears to be strongly associated with esophageal cancer-related cachexia-anorexia syndrome and non-malignant CAS of the alimentary tract. While leptin has been explored in the context of cancer, existing data about the role of leptin in cancer development is somewhat inconclusive. Additional studies, especially large-scale population-based investigations, are needed to shed more light on the nature of the relationship between leptin and cancer.

Summary Points

- Leptin is a cytokine that was discovered in 1994 as a regulator of body weight and energy balance.
- Leptin is manufactured primarily in the adipocytes of white adipose tissue.
- Leptin is also produced in much smaller amounts by gastric, mammary epithelium, placental, cardiac and other human tissues.
- Leptin resistance, a state in which the effect of leptin is blocked, appears to be common in the extremely obese but rare in subjects of normal weight.
- Leptin has been implicated in the development of malignancies associated with obesity.

- Leptin appears to play a role in breast, endometrial, renal, and several other cancers.
- Additional studies, especially large-scale population-based investigations, are needed to shed more light on the relationship between leptin and cancer.

Abbreviations

α-MSH   :   alpha melanocyte-stimulating hormone
POMC   :   pro-opiomelanocortin
WAT   :   white adipose tissue

Definition of Terms

Adipose tissue: Tissue made up of mainly adipose (fat) cells such as the yellow layer of fat beneath the skin.

Dyslipidemia: A disorder of lipoprotein metabolism, including lipoprotein overproduction or deficiency.

Insulin resistance: The diminished ability of cells to respond to the action of insulin in transporting glucose (sugar) from the bloodstream into muscle and other tissues.

Leptin resistance: Impaired leptin transport across the blood-brain barrier involving defects in leptin receptor signaling.

Lipodystrophy: A disorder of adipose (fatty) tissue characterized by a selective loss of body fat.

"Ob(Lep)" gene: The obesity gene.

References

Banks, W.A. 2001. Leptin transport across the blood-brain barrier: implications for the cause and treatment of obesity. Curr. Pharm. Des. 7: 125-133.

Blüher, S. and C. Mantzoros. 2009. Leptin in humans: lessons from translational research. Am. J. Clin. Nutr. 89: 991S-997S.

Buettner, C. and E.D. Muse, A. Cheng, L. Chen, T. Scherer, A. Pocai, K. Su, B. Cheng, X. Li, J. Harvey-White, G.J. Schwartz, G. Kunos, L. Rossetti, and C. Buettner. 2008. Leptin controls adipose tissue lipogenesis via central, STAT3–independent mechanisms. *Nat. Med.* 14: 667-675.

Capelle, L.G. and A.C. de Vries, J. Haringsma, E.W. Steyerberg, C.W. Looman, N.M. Nagtzaam, H. van Dekken, F. ter Borg, R.A. de Vries, and E.J. Kuipers. 2009. Serum levels of leptin as marker for patients at high risk of gastric cancer. Helicobacter 14(6): 596-604.

Chang, S. and S.D. Hursting, J.H. Contois, S.S. Strom, Y. Yamamura, and R.J. Babaian. 2001. Leptin and prostate cancer. Prostate 46(1): 62-67.

Farooqi, S. and H. Rau, J. Whitehead, and S. O'Rahilly. 1998. Ob gene mutations and human obesity. (1998). Proc. Nutr. Soc. 57: 471-475.

Hill, J.W. and J.K. Elmquist and C.F. Elias. 2008. Hypothalamic pathways linking energy balance and reproduction. Am. J. Physiol. Endocrinol. Metab. 294: 827-E832.

Horiguchi, A. and M. Sumitomo, J. Asakuma, T. Asano, R. Zheng, T. Asano, D.M. Nanus, and M. Hayakawa. 2006. Leptin promotes invasiveness of murine renal cancer cells via extracellular signal-regulated kinases and rho dependent pathway. J. Urol. 176(4 Pt 1): 1636-1641.

Ingalls, A.M. and M.M. Dickie, and G.D. Snell. 1996. Obese, a new mutation in the house mouse. Obes. Res. 4(1): 101.

Isse, N. and Y. Ogawa, N. Tamura, H. Masuzaki, K. Mori, T. Okazaki, N. Satoh, M. Shigemoto, Y. Yoshimasa, S. Nishi *et al.* 1995. Structural organization and chromosomal assignment of the human obese gene. J. Biol. Chem. 270: 27728-27733.

Klok, M.D. and S. Jakobsdottir, and M.L. Drent. 2007. The role of leptin and ghrelin in the regulation of food intake and body weight in humans: a review. Obes. Rev. 8(1): 21-34.

Manco, J. and J.M. Fernandez-Real, F. Equitani, J. Vendrell, M.E. Valera Mora, G. Nanni, V. Tondolo, M. Calvani, W. Ricart, M. Castagneto, and G. Mingrone. 2007. Effect of massive weight loss on inflammatory adipocytokines and the innate immune system in morbidly obese women. J. Clin. Endocrinol. Metab. 92: 483-490.

Münzberg, H. and M. Björnholm, S.H. Bates, and M.G. Myers, Jr. 2005. Leptin receptor action and mechanisms of leptin resistance. CMLS, Cell. Mol. Life Sci. 62: 642-652.

Nisoli, E. and M.O. Carruba. 2004. Emerging aspects of pharmacotherapy for obesity and metabolic syndrome. Pharmacol. Res. 50(5): 453-469.

Ozbay, T. and R. Nahta. 2008. A novel unidirectional cross-talk from the insulin-like growth factor-I receptor to leptin receptor in human breast cancer cells. Mol. Cancer Res. (6):1052-8. Epub 2008 May 30.

Petridou, E. and S. Kedikoglou, P. Koukoulomatis, N. Dessypris, and D. Trichopoulos. 2002. Diet in relation to endometrial cancer risk: a case-control study in Greece. Nutr. Cancer 44(1): 16-22.

Rankinen, T. and A. Zuberi, Y.C. Chagnon, S.J. Weisnagel, G. Argyropoulos, B. Walts, L. Pérusse, and C. Bouchard. 2006. The human obesity gene map: the 2005 update. Obesity (Silver Spring) (4): 529-644.

Ren, H. and T. Zhao, X. Wang, C. Gao, J. Wang, M. Yu, and J. Hao. 2010. Leptin upregulates telomerase activity and transcription of human telomerase reverse transcriptase in MCF-7 breast cancer cells. Biochem. Biophys. Res. Commun. 26: 394(1): 59-63. Epub 2010 Feb 18.

Scarpace, P.J. and Yi Zhang. 2009. Leptin resistance: a prediposing factor for diet-induced obesity. Am. J. Physiol. Regul. Integr. Comp. Physiol. 296: R493-R500.

Trayhurn, P. and I.S. Wood. 2005. Signalling role of adipose tissue: adipokines and inflammation in obesity. Biochem. Soc. Trans. 33: 1078-1081.

Zhang, Y. and R. Proenca, M. Maffei, M. Barone, L. Leopold, and J.M.Friedman. 1994. Positional cloning of the mouse obese gene and its human homologue. Nature 372(6505): 425-432.

CHAPTER 11

Omentin as an Adipokine: Role in Health and Disease

Anne M. Lenz[1,*] and Frank Diamond[1]
[1]University of South Florida College of Medicine, Tampa, FL, All Children's Hospital, 501 Sixth Avenue South, Box 6900, St. Petersburg, FL 33701

ABSTRACT

First described as intelectin, an intestinal protein involved in the recognition of bacterial cell wall carbohydrates, human omentin is an adipokine preferentially secreted by visceral adipose tissue. Omentin expression is primarily from stromal vascular cells of adipose tissue rather than adipocytes themselves. The major circulating isoform in humans is omentin-1, which localizes to chromosome 1q22-q23, a region linked to increased susceptibility to type 2 diabetes. Human omentin is measurable in plasma and has also been detected in the peritoneal fluid of some patients with ascites. Omentin is an insulin-sensitizing agent, increasing insulin-mediated glucose uptake via the phosphorylation of Akt. Omentin-l correlates negatively with fasting plasma levels of insulin and determinations of insulin resistance (HOMA), body mass index, waist circumference and leptin, a pattern similar to that seen with the insulin-sensitizing, anti-inflammatory adipokine adiponectin. Omentin-1 is positively correlated with adiponectin and HDL cholesterol, while levels are decreased in obesity and polycystic ovarian syndrome. Omentin also may play a role in inflammation, particularly in the gastrointestinal tract, where it has been described in creeping fat associated with inflammatory bowel disease. Omentin is emerging as an important adipokine in the regulation of glucose homeostasis, body composition, and inflammation.

*Corresponding author

INTRODUCTION

Omentin is an adipokine expressed preferentially by the visceral adipose tissue, which acts as an insulin-sensitizing agent (Yang *et al.* 2006). The omentin gene localizes to chromosome 1q22-q23, a region linked to increased susceptibility to type 2 diabetes (Elbein *et al.* 1999). Omentin is emerging as an important adipokine in the regulation of glucose homeostasis, body composition, and inflammation.

GENETICS, PROTEOMICS AND EXPRESSION

Omentin was first discovered by expressed sequence tag analysis of human omental fat and was later found to be 100% homologous to an intestinal protein, intelectin (Yang *et al.* 2006, Tsuji *et al.* 2001). Intelectin is a receptor for galactofuranose, a carbohydrate moiety present in bacterial cell walls, and may have some role in gastrointestinal immune surveillance (Tsuji *et al.* 2001). Omentin cDNA is also homologous to XL35, a cortical granule lectin expressed by Xenopus oocytes, which prevents polyspermy during fertilization (Lee *et al.* 2001). In human, omentin is expressed as two homologous proteins, omentin-1 and -2, encoded by two separate genes located adjacent to one another on 1q22-q23. Omentin-1 is the major circulating isoform in human plasma. Omentin-2 shows 83% amino acid (aa) identity with omentin-1 but demonstrates greater intestinal expression (de Souza Batista *et al.* 2007). Omentin cDNA (1,269 base pairs, eight exons, seven introns) encodes for a 313 aa protein. The 5' adjacent region of the omentin gene contains putative binding sites for transcription factors involved in regulation of adipocytokine expression (Schäffler *et al.* 2005a). Though single nucleotide polymorphisms have been identified, no disease-causing mutation in omentin has been described (Schäffler *et al.* 2007).

Human omental fat expresses omentin mRNA 350 times that of subcutaneous fat. Low levels of omentin expression have also been detected in intestine, lung, heart, kidney, endothelial cells, and muscle (Yang *et al.* 2006). Omentin mRNA expression in epicardial fat suggests that, like other periadventitial epicardial adipokines, omentin could participate in coronary atherogenesis. (Fain *et al.* 2008, Sacks and Fain 2007). Fractionating the cellular components of adipose tissue reveals that mRNA expression of omentin is from the stromal vascular cells of adipose tissue, whose cell fraction consists of pre-adipocytes, fibroblasts, endothelial cells and macrophages. The omentin mRNA content of isolated adipocytes is only 9% of that in nonfat cells (Fain *et al.* 2008).

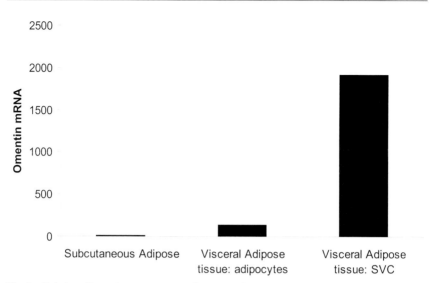

Fig. 1   Relative adipose tissue expression of omentin. Omentin mRNA expression via quantitative real-time PCR in sub-compartments of adipose tissue. (Adapted from Yang *et al.* 2006.)

The initial translated protein includes a signal sequence characteristic of secreted peptides. Once the signal sequence is cleaved, the mature secreted peptide measures 296 amino acids, with a calculated molecular weight of 33 kilo Daltons (kDa) (Yang *et al.* 2006). SDS-PAGE analysis localizes at a band measuring 38-40 kDa, because of glycosylation of the mature protein (Tsuji *et al.* 2001). The amino terminus region is homologous to a fibrinogen-related domain (Schäffler *et al.* 2005a). Omentin protein has been detected in media of *in vitro* models and found circulating in humans at a range of 100 ng/mL to 1 ug/mL (Yang *et al.* 2006). Omentin has also been identified in peritoneal fluid of patients with ascites (Wiest *et al.* 2009). The estimated half-life of human omentin is 30 h (Schäffler *et al.* 2005a). In a small study (N=20) of slim subjects, baseline plasma omentin levels were highly variable, but variation was not indicative of a gender dimorphism, whereas other investigators have identified higher levels in women (Wurm *et al.* 2007, de Souza Batista *et al.* 2007). Omentin's receptor, target tissues, and signaling mechanism remain undescribed.

OMENTIN IN GLUCOSE HOMEOSTASIS AND OBESITY

The presence of visceral adiposity correlates strongly with insulin resistance and the metabolic syndrome. Yang *et al.* hypothesized that omentin may play a role in insulin responsiveness. At concentrations of 150-300 ng/mL, omentin did not increase basal glucose uptake, but did increase insulin-stimulated glucose uptake in explants of subcutaneous adipocytes by approximately 50%. This insulin-stimulated

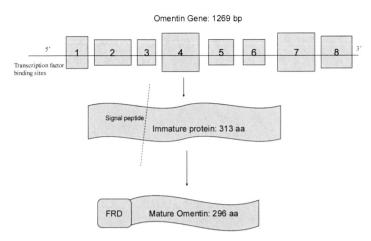

Fig. 2 Schematic of omentin gene and gene products. Initial translated protein undergoes cleavage of signal peptide to form mature omentin. aa, amino acid; bp, base pair; FRD, fibrinogen-related domain.

Table 1 Key features of visceral adipocytokines: a comparison of omentin and visfatin

	Omentin	Visfatin
Alternate names	Intelectin XL35	Nicotinamide phosphoribosyltransferase; Pre-B-Cell Colony-Enhancing Factor 1
Gene location	1q22-q23	7q22.2
Mature aa length	296 aa	473 aa
Molecular weight	33 kD	52 kD
Expressed by	Visceral and epicardial adipose tissue stromal vascular cells	Visceral and subcutaneous adipose tissue, Lymphocytes
Function in Glucose homeostasis	Insulin sensitizing	Insulin sensitizing, possible insulinomimetic
Correlation with markers of metabolic syndrome	Negative	Positive

glucose uptake occurs via the phosphorylation of Akt. Therefore, omentin appears to act as an insulin-sensitizing agent at physiological concentrations (Yang *et al.* 2006, Schäffler *et al.* 2005a). Omentin acts by binding and activating the insulin receptor via a different pathway from insulin itself. Both glucose and insulin cause dose-dependent declines in omentin-1 net protein production and secretion into conditioned media from human omental adipose tissue explants *in vitro* (Tan *et al.* 2008a). *In vivo*, hyperinsulinemic clamp significantly lowers concentrations of omentin-1 with maximal suppression 4 h after initiation of insulin infusion

(Tan *et al.* 2008a). A study of slim, insulin-sensitive subjects before and 120 min after 75 g glucose load demonstrated no significant change in plasma omentin levels (Wurm *et al.* 2007).

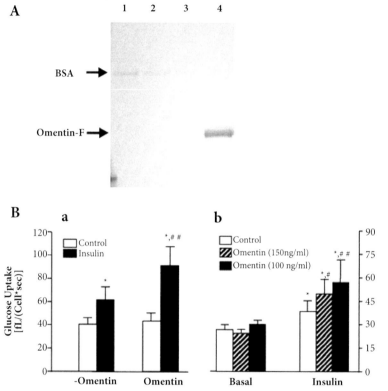

Fig. 3 Omentin enhances insulin-mediated glucose uptake. (A) Purity and quantification of omentin produced from HEK-293T cells. *Lanes 1, 2,* and *3* contain 200, 100, and 50 ng of bovine serum albumin (BSA), respectively, and *lane 4* contains 500 ng of omentin-F. (B) Omentin increases insulin-mediated glucose uptake by human adipocytes: *a,* human abdominal subcutaneous adipocytes were treated with omentin, with or without insulin (3 nM, black bar), and glucose uptake was measured; *b,* dose responsiveness of omentin's effect. Human adipocytes (4 abdominal subcutaneous and 1 omental) were treated with omentin [150 ng/ml (hatched bars), 300 ng/ml (black bars)] or vehicle (white bars) with or without insulin (3 nM), and glucose uptake was measured. Values represent means ± SE. *$P < 0.01$, vs. the control without insulin; #$P < 0.05$; ##$P < 0.01$ vs. insulin-only treatment, paired *t*-test; $n = 9$ in *a, n = 5* in *b.* (Reprinted with permission, Yang et al. 2006.)

Obesity lowers omentin levels. Higher levels of omentin-l are present in lean versus obese and overweight subjects, independent of age and gender. Omentin-l correlates negatively with fasting plasma levels of insulin and determinations of insulin resistance (HOMA), BMI, waist circumference and leptin, a pattern similar to that seen with the insulin-sensitizing, anti-inflammatory adipokine adiponectin. Omentin-1 is positively correlated with adiponectin and high density lipoprotein (HDL) cholesterol. When plasma omentin levels are adjusted for auxologic differences, there is loss of association of most of these correlates; however, the correlation with plasma leptin remains significant, suggesting a regulatory relationship between omentin and leptin, independent of the effects of obesity. Omentin-1 is regulated by body fat mass, perhaps via other adipokine actions or inflammatory cytokines (de Souza Batista *et al.* 2007).

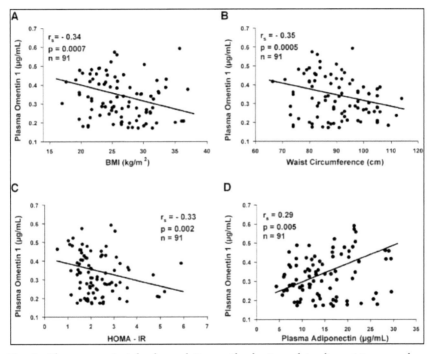

Fig. 4 Plasma omentin-1 level correlations with obesity and insulin resistance markers. Significant correlation coefficients based on variance components analysis adjusted for sex, age, and family structure were found between plasma omentin-1 levels with BMI (n = 91, r = -0.34, P = 0.0007) (A); waist circumference (n = 91, r = -0.35, P = 0.0005) (B); Ln HOMA index (n = 91, r = -0.33, P = 0.002) (C); and plasma adiponectin levels (n = 91, r = 0.29, P = 0.005) (D). (Reprinted with permission from de Souza Batista *et al.* 2007.)

Adipose tissue omentin mRNA expression and circulating concentrations of omentin-1 are reduced in overweight insulin-resistant women with polycystic ovarian syndrome, although this finding may reflect the effects of obesity (Tan *et al.* 2008a). In a small study of omentin mRNA expression in surgically obtained omental fat, subjects with Type 2 diabetes had lower mRNA expression than obese subjects with normal glucose regulation, whose levels were less than those of normal weight and normal glucose regulation controls (Cai *et al.* 2009).

Omentin-1 concentrations have also been shown to be lower in adults with type 1 diabetes compared to matched controls (Tan *et al.* 2008b). In this population, omentin-1 levels inversely related to BMI and waist circumference and did not differ between fasting and postprandial specimens (Tan *et al.* 2008b).

OMENTIN AND INFLAMMATION

In inflammatory bowel disease, omental, "creeping fat" may be involved in transmural and intra-abdominal inflammation (Schäffler *et al.* 2005b). Omentin could participate in intestinal immune defense with mesenteric fat serving as a barrier against bacterial invasion (Schäffler *et al.* 2005b). Omental adipokines released into the portal system reach the liver in higher concentrations than those circulating peripherally. Portal hypertension could contribute to the appearance of adipokines in peritoneal fluid. Omental adipocytes from patients with Crohn's disease express varying concentrations of omentin mRNA (Schäffler *et al.* 2005a). In patients with hepatic cirrhosis, omentin demonstrates a broad range of expression in ascitic fluid, and ascitic leukocyte count and albumin are significantly higher in patients expressing omentin than in those who do not. There is a strong inverse correlation between the quantity of omentin protein and the albumin serum to ascites ratio (Wiest *et al.* 2009). These findings suggest a possible role of omentin as a marker of peritoneal inflammation.

Summary Points

- Omentin is an adipokine preferentially secreted by visceral adipose stromal vascular cells and shares 100% homology with intelectin, an intestinal protein that recognizes bacterial cell wall carbohydrates.
- The human omentin gene is located on 1q22-q23, a region associated with increased susceptibility to type 2 diabetes.
- Human omentin circulates in plasma and has also been detected in the peritoneal fluid of some patients with ascites.
- Omentin increases insulin-mediated glucose uptake via the phosphorylation of Akt.
- Omentin-1 correlates negatively with fasting plasma levels of insulin and determinations of insulin resistance (HOMA), BMI, waist circumference and leptin.
- Omentin-1 is positively correlated with adiponectin and HDL cholesterol.
- Omentin also may play a role in inflammation, particularly in the gastrointestinal tract, where it has been described in creeping fat associated with inflammatory bowel disease.

Abbreviations

aa	:	amino acid
Akt	:	Akt or protein kinase B
BMI	:	body mass index
cDNA	:	coding deoxyribonucelic acid
HDL	:	high density lipoprotein
HOMA	:	homeostasis model assessment
kDa	:	kilo Daltons
mL	:	microliter
mRNA	:	messenger ribonucleic acid
ng	:	nanogram
ug	:	microgram

Definition of Terms

Ascites: An accumulation of fluid within the peritoneum, most commonly associated with hepatic or gastrointestinal disease.

Chromosome 1q22-q23: A genetic susceptibility locus for type 2 diabetes as well as the location of the omentin gene.

Intelectin: A protein homologous to omentin, which was first described in small intestine.

Peritoneal fluid: Liquid contained within the abdominal cavity that surrounds the abdominal organs.

Polycystic ovarian syndrome: A condition in women characterized by irregular menstrual periods, anovulation, elevated androgen levels, multiple small cysts within the ovaries, and insulin resistance.

Stromal vascular cells: Connective tissue cells of adipose tissue containing pre-adipocytes, fibroblasts, endothelial cells and macrophages.

References

Cai, R.C. and L. Wei, J.Z. Dl, H.Y. Yu, Y.Q. Bao, and W.P. Jai. 2009. Expression of omentin in adipose tissues in obese and type 2 diabetic patients. Zhonghua Yi Xue Za Zhi 89: 381-384.

de Souza Batista, C.M. and R.Z. Yang, M.J. Lee, N.M. Glynn, D.Z. Yu, J. Pray, K. Ndubuizu, S. Patil, A. Schwartz, M. Kligman, S.K. Fried, D.W. Gong, A.R. Shuldiner, T.I. Pollin, and J.C. McLenithan. 2007. Omentin plasma levels and gene expression are decreased in obesity. Diabetes 56: 1655-1661.

Elbein, S.C. and M.D. Hoffman, K. Teng, M.F. Leppert, and S.J. Hasstedt. 1999. A genome-wide search for type 2 diabetes susceptibility genes in Utah Caucasians. Diabetes 48: 1175-1182.

Fain, J.N. and H.S. Sacks, B. Buehrer, S.W. Bahouth, E. Garrett, R.Y. Wolf, R.A. Carter, D.S. Tichansky, and A.K. Madan. 2008. Identification of omentin mRNA in human epicardial adipose tissue: comparison to omentin in subcutaneous, internal mammary artery periadventitial and visceral abdominal depots. Int. J. Obes. 32: 810-815.

Lee, J.K. and J. Schnee, M. Pang, M. Wolfert, L.G. Baum, K.W. Moremen, and M. Pierce. 2001. Human homologues of the Xenopus oocyte cortical granule lectin XL35. Glycobiology 11: 65-73.

Sacks, H.S. and J.N. Fain. 2007. Human epicardial adipose tissue: A review. Am. Heart J. 153: 907-917.

Schäffler, A. and M. Neumeier, H. Herfarth, A. Fürst, J. Schölmerich, and C. Büchler. 2005a. Genomic structure of human omentin, a new adipocytokine expressed in omental adipose tissue. Biochim. Biophys. Acta. 1732: 96-102.

Schäffler, A. and J. Scholmerich, and C. Büchler. 2005b. Mechanisms of disease: adipocytokines and visceral adipose tissue-emerging role in intestinal and mesenteric diseases. Nat. Clin. Pract. Gastroenterol. Hepatol. 2: 103-111.

Schäffler, A. and M. Zeitoun, H. Wobser, C. Buechler, C. Aslanidis, and H. Herfarth. 2007. Frequency and significance of the novel single nucleotide missense polymorphism Val109Asp in the human gene encoding omentin in Caucasian patients with type 2 diabetes or chronic inflammatory bowel diseases. Cardiovasc. Diabetol. 6: 3.

Tan, B.K. and R. Adya, S. Farhatullah, K.C. Lewandowski, P. O'Hare, H. Lehnert, and H.S. Randeva. 2008a. Omentin-1, a novel adipokine, is decreased in overweight insulin resistant women with polycystic ovary syndrome: *ex vivo* and *in vivo* regulation of omentin-1 by insulin and glucose. Diabetes 57: 801-808.

Tan, B.K. and S. Pua, F. Syed, K.C. Lewandowski, J.P. O'Hare, and H.S. Randeva. 2008b. Decreased plasma omentin-1 levels in type 1 diabetes mellitus. Diabet. Med. 25: 1254-1255.

Tsuji, S. and J. Uehori, M. Matsumoto, Y. Suzuki, A. Matsuhisa, K. Toyoshima, and T. Seya. 2001. Human intelectin is a novel soluble lectin that recognizes galactofuranose in carbohydrate chains of bacterial cell walls. J. Biol. Chem. 276: 23456-23463.

Wiest, R. and F. Leidl, A. Koop, M. Weigert, M. Neumeier, C. Buechler, J. Schoelmerich, and A. Schäffler. 2009. Peritoneal fluid adipocytokines: ready for prime time? Eur. J. Clin. Invest. 39: 219-229.

Wurm, S. and M. Neumeier, J. Weigert, A. Schäffler, and C. Buechler. 2007. Plasma levels of leptin, omentin, collagenous repeat-containing sequence of 26-kDa protein (CORS-26) and adiponectin before and after oral glucose uptake in slim adults. Cardiovasc. Diabetol. 6: 7.

Yang, R.Z. and M.J. Lee, H. Hu, J. Pray, H.B. Wu, B.C. Hansen, A.R. Shuldiner, S.K. Fried, J.C. McLenithan, and D.W. Gong. 2006. Identification of omentin as a novel depot-specific adipokine in human adipose tissue: possible role in modulation insulin action. Am. J. Physiol. Endcrinol. Metab. 290: E1253-E1261.

CHAPTER 12

Plasminogen Activator Inhibitor 1 (Pai-1) as an Adipokine

Francis Agyemang-Yeboah

Department of Molecular Medicine, School of Medical Science, Kwame Nkrumah University of Science and Technology, Kumasi, Ghana

ABSTRACT

Plasminogen activator inhibitor 1 (PAI-1) is a fast-acting inhibitor of tissue-type plasminogen activator (t-PA) and urokinase-type plasminogen activator (u-PA), the activators of plasminogen and hence fibrinolysis. It is mainly produced by the endothelium, but also secreted by other tissue types, such as adipose tissue. It is a serine protease inhibitor (serpin) protein. Like other serpins, PAI-1 reacts with its cognate proteases to form 1:1 enzyme-inhibitor complexes that are enzymatically inactive. PAI-1 also inhibits the activity of matrix metalloproteinases, which play a crucial role in invasion of malignant cells across the basal lamina.

PAI-1 is coded by the *PLANH1* gene, which is located on the seventh chromosome (7q21.3-q22). There is a common polymorphism known as 4G/5G in the promoter region. The 5G allele is slightly less transcriptionally active than the 4G. This polymorphism in *PLANH1* gene is the subject of debate in various studies as to whether or not it is associated with an increased risk of venous thromboembolism. Congenital deficiency of PAI-1 has been reported to lead to a haemorrhagic diathesis. PAI-1 is present in increased levels in various disease states as well as in obesity and metabolic syndrome.

INTRODUCTION

A pro-thrombotic state involving disturbances in the haemostatic and fibrinolytic pathways is increasingly being recognized as contributing to the excess cardiovascular risk in obesity. The disturbances in this coagulation pathway include platelet hyper-aggregability, hypercoagulability, and hypofibrinolysis. Fibrinolysis is normally mediated by plasmin, which circulates in blood as its pro-enzyme, plasminogen. Conversion and activation of plasminogen into plasmin is effected by two plasminogen activators: t-PA and u-PA. The process of fibrinolysis can thus be inhibited by inhibition plasmin itself by α-2 anti-plasmin, or by inhibition of the plasminogen activators by PAI-1 (Fay *et al.* 1997) (Fig. 3).

PAI-1 is present in increased levels in various disease states as well as in obesity and metabolic syndrome (Skurk and Hauner 2004). It has been linked to the increased occurrence of thrombosis (Minowa *et al.* 1999).

NATURE, DISTRIBUTION AND CELLULAR FUNCTION OF PAI-I

PAI-1 is a serine protease inhibitor that inhibits fibrinolysis by inactivating urokinase-type and tissue-type plasminogen activator. It is a serine protease inhibitor (serpin) protein (Fig. 1). It is mainly produced by the endothelium as well as by other tissue types, such as adipose tissue, platelets, vascular endothelial cells, and vascular smooth muscle cells. Several non-vascular cell types express PAI-1, and it is abundant in the extracellular matrix (Podor and Loskuff 1992). Therefore, PAI-1 may regulate proteolysis within the extravascular space by inhibiting plasminogen activators and secondary target proteases, such as thrombin (Ehrlich *et al.* 1991). The majority of the circulating PAI-1 is contributed by adipose tissue (Mertens and Van Gaal 2002). Thus, obese subjects have higher levels of PAI-1 (De Pergolada *et al.* 1997).

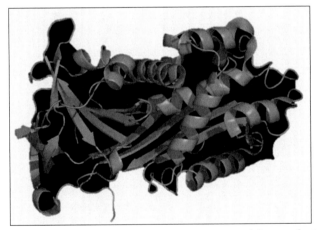

Fig. 1 Three-dimensional structure of plasminogen activator inhibitor 1 (PAI-1) (*Source:* Wikipedia, the Free Encyclopedia). PAI-1 in complex with somatomed in B domain of Vitronectin.

ROLE OF PAI-I IN DISEASE STATES

PAI-I and Cardiovascular Disease

PAI-1 levels can predict future risk for diabetes and cardiovascular disease. A lowered fibrinolytic state (Meade *et al.* 1993) as well as an increased plasma PAI-1 activity (Wimam *et al.* 2000) have both been shown to be independently associated with increased risk of future coronary heart disease events. High levels of plasma PAI-1 activity were also shown to be associated with a higher risk for first myocardial infarction in prospective studies on healthy individuals (ThÃ¶gersen *et al.* 1998). Physical exercise protects against the development of cardiovascular disease, partly by lowering plasmatic total cholesterol and LDL cholesterol (Fig. 2) and increasing HDL cholesterol levels. In addition, it is now established that reduction of plasmatic adiponectin and increased C-reactive

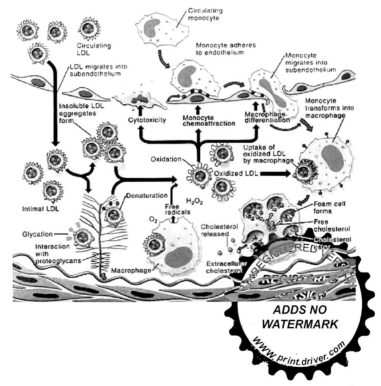

Fig. 2 Cellular pathway of atherogenesis. Cascade of events in which lipoprotein (oxidized LDL) is coupled to other cellular events, such as cytotoxity, macrophage differentiation, oxidative stress, which are associated with an increased endothelial expression of PAI-1.

Source: Wikipedia, the Free Encyclopedia.

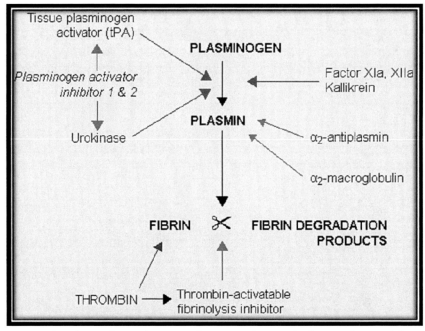

Fig. 3   Fibrinolytic pathway. Blue arrows denote stimulation and red arrows denote inhibition (*Source:* Wikipedia, the Free Encyclopedia). Fibrinolysis is normally mediated by plasmin, which circulates in blood as its pro-enzyme, plasminogen. Conversion and activation of plasminogen into plasmin is effected by two plasminogen activators: tissue-type plasminogen activator (t-PA) and urokinase-type plasminogen activator (u-PA). The process of fibrinolysis can thus be inhibited by inhibition plasmin itself, by α-2 anti-plasmin, or by inhibition of the plasminogen activators by PAI-1.

Color image of this figure appears in the color plate section at the end of the book.

protein and PAI-1 levels play a role in the maintenance of an inflammatory state and in the development of cardiovascular disease (Lira *et al.* 2010). Angiotensin II increases synthesis of PAI-1, so it accelerates the development of atherosclerosis.

PAI-I in Fibrinolysis, Obesity and Metabolic Syndrome

Macrophages are more abundant in adipose tissue from obese individuals than from those of normal weight and may contribute to the metabolic consequences of obesity by producing various circulating factors including PAI-1 (Kishore *et al.* 2010). There is now compelling evidence that obesity and, in particular, an abdominal type of body fat distribution are associated with elevated PAI-1 antigen and activity levels (Skurk and Hauner 2004). The greater the fat-cell size and the adipose tissue mass, the greater the contribution of adipose production to circulating PAI-1. Levels of PAI-1 have also been shown to be associated with metabolic syndrome (Huotari *et al.* 2010). Impaired fibrinolysis in obesity is probably also due to an increased expression of PAI-1 in adipose tissue. Studies have demonstrated that weight reduction induced by modest caloric restriction in elderly obese subjects leads to a decline in prevailing concentrations of PAI-1 in blood paralleled by augmentation of functional activity of the fibrinolytic system (Calles-Escandon *et al.* 1996).

Table I Key facts about PAI-I

- Plasminogen activator inhibitor 1 (PAI-1) is a fast-acting inhibitor of tissue-type plasminogen activator (t-PA) and urokinase-type plasminogen activator (u-PA) produced by the endothelium and the adipose tissue.
- It is a serine protease inhibitor (serpin) protein whose expression is associated with insulin resistance, characterized by visceral fat accumulation.
- Obesity and abdominal fat distribution are associated with elevated PAI-1 levels.
- A lowered fibrinolytic-induced expression of PAI-1 activity positively correlates with increased risk of future coronary heart disease.

PAI-I in Cancer Cell Invasion and Angiogenesis

Acquisition of invasive/metastatic potential through protease expression is an essential event in tumour progression. High levels of components of the plasminogen activation system, including urokinase, but paradoxically also its inhibitor, PAI-1, have been correlated with a poor prognosis for some cancers (Jänicke *et al.* 2001). The study concluded that PAI-1 levels measured in routinely prepared cytosols are an important parameter to predict a metastatic potential in both node-negative and node-positive human primary breast cancer.

PAI-I MODULATORS

Age-related changes in PAI-1 have not been consistent in studies. Younger men have higher PAI-1 than women (Margaglione *et al.* 1998). A study by Gebara *et al.* (1995) has suggested that post-menopausal women have higher plasma PAI-1 than pre-menopausal women, which protects the latter from cardiovascular disease. Meilahn *et al.* (1996) in their study also reported that older women receiving post-menopausal hormone therapy had more favourable plasma levels of the haemostatic factors than those not receiving therapy, strongly suggesting that hormone replacement therapy seems to be associated with lower PAI-1 levels in post-menopausal women. Plasma testosterone has been shown to correlate negatively with high levels of PAI-1 (Caron *et al.* 1989).

PAI-1 production is increased by glucocorticoids (Morange *et al.* 1999) and insulin, while PAI-1 expression and secretion by adipocytes is decreased by catecholamines (Halleux *et al.* 1999). Studies have suggested that chronic hyperinsulinism in obese diabetic patients may explain the raised expression of PAI-1 in adipose tissue (Koistinen *et al.* 2000). The potential role of very low density lipoprotein (VLDL) receptor in mediating VLDL-induced PAI-1 expression has been studied *in vitro* (Nilsson *et al.* 1999). Polymorphisms of the PAI-1 gene have been described, but despite some studies showing an association with cardiac events (Ossei-Gerning *et al.* 1997), their impact on cardiovascular risk remains to be defined.

PAI-1 GENE IS ASSOCIATED WITH MULTIPLE ORGAN DYSFUNCTION AND SEPTIC SHOCK

Activation of inflammation and coagulation are closely related and mutually interdependent in sepsis. Elevated levels of PAI-1 have been related to worse outcome in pneumonia. 4G/5G polymorphism of PAI-1 gene has been shown to be associated with multiple organ dysfunction and septic shock in pneumonia-induced severe sepsis (Madach *et al.* 2010).

FUNCTIONAL ACTIVITY OF PAI-1 AND BLEEDING DIATHESIS

Various studies have shown a strong link between declining levels of PAI-1 and bleeding diathesis (Minowa *et al.* 1999). Indeed, it is now certain that congenital PAI-1 deficiency, an extremely rare disorder, is characterized by a bleeding diathesis that begins in childhood because of hyperfibrinolysis.

CONCLUSION

PAI-1 expression in the adipose tissue is related to adipose mass and fat-cell size. The greater the fat-cell size and the adipose tissue mass, the greater the contribution of adipose production to circulating PAI-1. PAI-1 synthesis is inducible by TNF and TGF. Thus, it could be speculated that obesity with insulin resistance represents a favourable condition for expression of the inducers of PAI-1 synthesis. The relations described among TNF, obesity, and insulin resistance suggest a contribution of TNF in explaining the role of PAI-1 in obesity, metabolic syndrome and cardiovascular disease. Thus, obesity-associated fibrinolysis is due to an increased expression of PAI-1 in adipose tissue.

Summary Points

- Plasminogen activator inhibitor 1 expression is up-regulated in the adipose tissue.
- PAI-1 expression is associated with insulin resistance characterized by visceral fat accumulation.

- Elevated expression has been shown to be partly controlled by tumour necrosis factor.
- Obesity-associated fibrinolysis is due to an increased expression of PAI-1 in adipose tissue.
- Expression of PAI-1 is associated with obesity, type 2 diabetes mellitus and cardiovascular disease.

Abbreviations

BMI	:	body mass index
HRT	:	hormone replacement therapy
LDL	:	low density lipoprotein
PAI-1	:	plasminogen activator inhibitor 1
TGF	:	transforming growth factor
TNF	:	tumour necrosis factor
t-PA	:	tissue type plasminogen activator
u-PA	:	urokinase-type plasminogen activator
VLDL	:	very low density lipoprotein

Definition of Terms

Adiposity: Having the property of containing fat.

Angiogenesis: A physiological process involving the growth of new blood vessels from pre-existing vessels.

Atherogenesis: The formation of atheromas on the walls of the arteries as in atherosclerosis

Atherosclerosis: The condition in which an artery wall thickens as the result of a build-up of fatty materials such as cholesterol.

Diathesis: A bleeding tendency.

Fibrinolysis: The process wherein a fibrin clot, the product of coagulation, is broken down.

Hypofibrinolysis: A decreased capacity to dissolve a blood clot.

Hyperinsulinism: An above-normal level of insulin in the blood of a person or animal.

References

Andersen, P. and I. Seljeflot, M. Abdelnoor, H. Arnesen, P.O. Dale, A. LÃ¸vik, and K. Birkeland. 1995. Increased insulin sensitivity and fibrinolytic capacity after dietary intervention in obese women with polycystic ovary syndrome. Metabolism 44: 611-616.

Calles-Escandon, J. and D. Ballor, J. Harvey-Berino, P. Ades, R. Tracy and B. Sobel. 1996. Amelioration of the inhibition of fibrinolysis in elderly, obese subjects by moderate energy intake restriction. Am. J. Clin. Nutr. 64: 7-11.

Caron, P. and A. Bennet, R. Camera, J.P. Louvet, B. Boneu, and P. Siã©. 1989. Plasminogen activator inhibitor in plasma is related to testosterone in men. Metabolism 38: 1010-1015.

De Pergolada, G. and V. De Mitrio, F. Giorgino, M. Sciaraffia, A. Minenna, L. Di Bari, and R. Giorgino. 1997. Increase in both pro-thrombotic and anti-thrombotic factors in obese premenopausal women: relationship with body fat distribution. Int. J. Obes. Relat. Metab. Disord. 7: 527-535.

Ehrlich, H.J. and R.K. Gebbink, J. Keijer, M. Linders, K.T. Preissner, and H. Pannekoek. 1991. Thrombin neutralizes plasminogen activator inhibitor 1 (PAI-1) that is complexed with vitronectin in the endothelial cell matrix. J. Cell. Biol. 115: 1773.

Fay, W.P. and A.C. Parker, L.R. Condrey, and A.D. Shapiro. 1997. Human plasminogen activator inhibitor-1 (PAI-1) deficiency: Characterization of a large lindred with a null mutation in the PAI-1 gene. Blood 90: 204-208.

Halleux, C.M. and P.J. Declerck, S.L. Tran, R. Detry, and S.M. Brichard. 1999. Hormonal control of plasminogen activator inhibitor-1 gene expression and production in human adipose tissue: stimulation by glucocorticoids and inhibition by catecholamines. J. Clin. Endocrinol. Metab. 84: 4097-4105.

Huotari, A. and S.M. Lehto, L. Niskanen, K.H. Herzig, J. Hintikka, H. Koivumaa-Honkanen, T. Tolmunen, K. Honkalampi, *et al.* 2010. Increased serum pai-1 levels in subjects with metabolic syndrome and long-term adverse mental symptoms: a population-based study. Cardiovascular Psychiatry and Neurology, Vol. 2010.

Jänicke F. and A. Prechtl, C. Thomssen, N. Harbeck, C. Meisner, M. Untch, C.G.J.F. Sweep, H.K. Selbmann, *et al.* 2001. Randomized adjuvant chemotherapy trial in high-risk, lymph node-negative breast cancer patients identified by urokinase-type plasminogen activator and plasminogen activator inhibitor Type 1. J. Natl. Cancer Inst. 93: 913-920.

Kishore, P. and W. Li, J. Tonelli, D.E. Lee, S. Koppaka, K. Zhang, Y. Lin, and S. Kehlenbrink *et al.* 2010. Adipocyte-derived factors potentiate nutrient-induced production of plasminogen activator inhibitor-1 by macrophages. Sci. Transl. Med. Vol. 2: p. 20ra15.

Koistinen, H.A. and E. Dusserre, P. Ebeling, P. Vallier, V.A. Koivisto, and H. Vidal. 2000. Subcutaneous adipose tissue expression of plasminogen activator inhibitor-1 (PAI-1) in nondiabetic and Type 2 diabetic subjects. Daibetes Metab. Res. Rev. 16: 364-369.

Lira, F.S. and J.C. Rosa, A.E. Lima-Silva, H.A. Souza, E.C. Caperuto, M.C. Seelaender, A.R. Damaso, L.M. Oyama, *et al.* 2010. Sedentary subjects have higher PAI-1 and lipoproteins levels than highly trained athletes. Diabetol. Metabol. Syndr. 2:7doi:10.1186/1758-5996-2-7.

Lowe, G.D. and J.W. Yarnell, P.M. Sweetnam, A. Rumley, H.F. Thomas, and P.C. Elwood. 1998. Fibrin D-dimer, tissue plasminogen activator, plasminogen activator inhibitor, and the risk of major ischaemic heart disease in the Caerphilly Study. Thromb. Haemost. 79: 129-133.

Madách, K. and I. Aladzsity, Á. Szilágyi, G. Fust, J.Gál, I. Pénzes, and Z. Prohászka. 2010. 4G/5G polymorphism of *PAI-1* gene is associated with multiple organ dysfunction and septic shock in pneumonia induced severe sepsis: prospective, observational, genetic study. Crit. Care 14: R79.

Meade, T.W. and V. Ruddock, Y. Stirling, R. Chakrabarti, and G.J. Miller. 1993. Fibrinolytic activity, clotting factors, and long-term incidence of ischaemic heart disease in the Northwick Park Heart Study. Lancet 342: 1076-1079.

Mertens, I. and L.F. Van Gaal. 2002. Obesity, haemostasis and fibronolytic system. Obes. Rev. 2: 85-101.

Meilahn, E.N. and J.A. Cauley, R.P. Tracy, E.O. Macy, J.P. Gutai, and L.H. Kuller. 1996. Association of sex hormones and adiposity with plasma levels of fibrinogen and PAI-1 in postmenopausal women. Am. J. Epidemiol. 143: 159-166.

Minowa, H. and Y. Takahashi, T. Tanaka, K. Naganuma, S. Ida, I. Maki, and A. Yoshioka. 1999. Four cases of bleeding diathesis in children due to congenital plasminogen activator inhibitor-1 deficiency. Haemostasis 29: 286-291.

Morange, P.E and H.R. Lijnen, M. Verdier, R. Negrel, I.Juhan-Vague, and M.C. Alessi. 1999. Glucorticoids and insulin promote plasminogen activator inhibitor 1 production by human adipose tissue. Diabetes 48: 890-895.

Nilsson, L. and M. Gåfvels, L. Musakka, K. Ensler, D.K. Strickland, B. Angelin, A. Hamsten, and P. Eriksson. 1999. VLDL activation of plasminogen activator inhibitor-1 (PAI-1) expression: involvement of the VLDL receptor. J. Lipid Res. 40: 913-919.

Ossei-Gerning, N. and M.W. Mansfield, M.H. Stickland, I.J. Wilson, and P.J. Grant. 1997. Plasminogen activator inhibitor-1 promoter 4G/5G genotype and plasma levels in relation to a history of myocardial infarction in patients characterized by coronary angiography. Arterioscler. Thromb. Vasc. Biol. 17: 33-37.

Podor, T.J. and D.J. Loskutoff. 1992. Immunoelectron microscopic localization of type 1 plasminogen activator inhibitor in the extracellular matrix of transforming growth factor-beta-activated endothelial cells. Ann. NY Acad. Sci. 667: 46.

Skurk, T. and H. Haune. 2004. Obesity and impaired fibrinolysis: role of adipose production of plasminogen activator inhibitor-1. Intl. J. Obes. 28: 1357-1364.

ThÃ¶gersen, A.M. and J.H. Jansson, K. Boman, T.K. Nilsson, L. Weinehall, F. Huhtasaari, and G. Hallmans. 1998. High plasminogen activator inhibitor and tissue plasminogen activator levels in plasma precede a first acute myocardial infarction in both men and women: evidence for the fibrinolytic system as an independent primary risk factor. Circulation 98: 2241-2247.

Wiman, B. and T. Andersson, J. Hallqvist, C. Reuterwall, A. Ahlbom and UdeFaire. 2000. Plasma levels of tissue plasminogen activator/plasminogen activator inhibitor-1 complex and von Willebrand factor are significant risk markers for recurrent myocardial infarction in the Stockholm Heart Epidemiology Program (SHEEP) study. Arterioscler. Thromb. Vasc. Biol. 20: 2019-2023.

CHAPTER 13

Resistin as an Adipokine

Zhikui Wei[1] and G. William Wong[1,*]

[1]Department of Physiology and Center for Metabolism and Obesity Research, Johns Hopkins University School of Medicine, 855 North Wolfe Street, Baltimore, MD 21205, USA

ABSTRACT

Obesity is tightly linked to type 2 diabetes. However, the molecular mechanisms underpinning this observation remain incompletely understood. One proposed mechanism is that obesity alters the expression and secretion of polypeptide hormones by adipocytes, actively contributing to insulin resistance in multiple tissues. Resistin is one candidate molecule in this process, as a protein that has been shown to disturb glucose homeostasis and impair insulin action in rodent models of obesity and diabetes. Additionally, resistin inhibits adipogenesis and promotes inflammation, linking excess fat mass to inflammation, impaired systemic insulin sensitivity and glucose metabolism. Much work remains to be done, but future studies on resistin will likely provide novel therapeutic strategies for treating metabolic and inflammatory diseases associated with obesity.

INTRODUCTION

Resistin/ADSF/FIZZ3 is a small polypeptide hormone secreted by adipose tissue and macrophages (Patel *et al.* 2003, Steppan *et al.* 2001). This protein belongs to the resistin-like molecule (RELM) or FIZZ protein family, which includes resistin, RELMα/FIZZ1, RELMβ/FIZZ2 and RELMγ (Fig. 1). Since its discovery a decade ago (Holcomb *et al.* 2000, Kim *et al.* 2001, Steppan *et al.* 2001), many studies have provided evidence for the biological function of resistin in glucose homeostasis,

---

*Corresponding author

insulin resistance, adipogenesis and inflammation (Lazar 2007, Steppan and Lazar 2004). These studies establish resistin as a potential mediator linking obesity to inflammation and peripheral tissue insulin resistance.

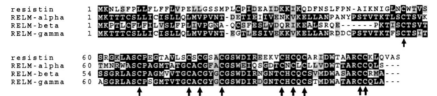

Fig. 1 Sequence alignment of mouse resistin, RELM-α, RELM-β, RELM-γ. The GenBank protein accession numbers for these proteins are NP_075360, NP_065255, NP_076370 and NP_853627, respectively. Identical amino acids are shaded, gaps are indicated by broken line, highly conserved cysteine residues are indicated by arrows.

SYNTHESIS AND STRUCTURE

In mice, resistin is produced almost exclusively by adipocytes within the white adipose tissue (Steppan *et al.* 2001). Humans, however, express resistin at very low levels in adipose tissue; rather, it is predominantly expressed and secreted by monocytes and macrophages (Patel *et al.* 2003). In rodents, resistin is preferentially expressed at higher levels by visceral fat associated with insulin resistance (Gabriely *et al.* 2002). Serum levels of resistin in humans are inversely correlated to waist-to-hip ratio and positively correlated with adipose tissue mass (Yannakoulia *et al.* 2003).These findings are consistent with the known biological function of resistin derived from animal studies.

Resistin is an unusual hormone, rich in cysteine residues. In fact, of the 114 amino acids comprising the resistin polypeptide, 11 are cysteines. The crystal structure of resistin reveals a "coiled-coil" trimer and a "tail-to-tail" hexamer, formed by disulfide bonds that link together protomer containing a β-sandwich "head" domain and α-helical "tail" region (Fig. 2) (Patel *et al.* 2004). In serum, resistin circulates predominantly as trimers and hexamers, with trimer being the more bioactive form (Patel *et al.* 2004).

PHYSIOLOGICAL FUNCTION

Multiple gain- and loss-of-function studies show that resistin causes peripheral tissue insulin resistance (Fig. 3). Administration of recombinant resistin to normal and healthy mice impairs glucose tolerance and insulin action (Steppan *et al.* 2001). Similarly, infusion of resistin protein in Sprague-Dawley rats impairs hepatic insulin sensitivity and glucose metabolism (Rajala *et al.* 2003). Over-expression of resistin using adenovirus in liver results in insulin resistance in male Wistar rats (Satoh *et al.* 2004). Transgenic rats over-expressing mouse resistin have impaired skeletal muscle metabolism and glucose tolerance (Pravenec *et al.* 2003).

Fig. 2 Crystal structure of mouse resistin. The homotrimeric crystal structure of resistin was downloaded from the PDB protein databank (structure 1RGX) (Patel *et al.* 2004).
Color image of this figure appears in the color plate section at the end of the book.

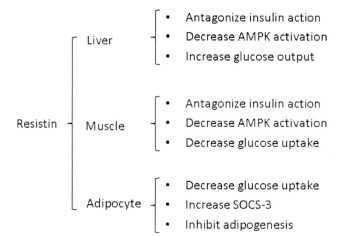

Fig. 3 Physiological function of mouse resistin and its mechanisms of action. Resistin exerts its biological effects in muscle, liver and adipocytes through the suppression of AMPK activity and the up-regulation of SOCS-3.

Likewise, a transgenic mouse model with chronic elevated circulating resistin has higher fasting glucose, impaired glucose tolerance and decreased hepatic insulin sensitivity (Rangwala *et al.* 2004). *In vitro*, recombinant resistin acts directly on adipocytes and myotubes to decrease insulin-stimulated glucose uptake (Moon *et al.* 2003, Steppan *et al.* 2001).

In contrast to gain-of-function studies, blocking resistin action with neutralizing antibody improves whole-body insulin sensitivity and decreases blood glucose in diet-induced obese (DIO) mice (Steppan *et al.* 2001). Indeed, resistin knockout mice have reduced fasting blood glucose, decreased hepatic glucose output and better glucose tolerance when challenged with a high-fat diet (Banerjee *et al.* 2004). Normalizing plasma resistin levels in DIO mice using antisense oligodeoxynucleotide against resistin mRNA completely reverses hepatic insulin resistance in these animals (Muse *et al.* 2004). Further, transgenic mice over-expressing a dominant negative resistin show enhanced glucose tolerance and insulin sensitivity on either a chow or high-fat diet (Kim *et al.* 2004). Additionally, resistin deficiency in *ob/ob* mice also improves glucose homeostasis and insulin sensitivity, an effect reversed by resistin administration (Qi *et al.* 2006).

Aside from regulating metabolism in liver and muscle, resistin also inhibits 3T3-L1 adipocyte differentiation (Kim *et al.* 2001). Consistent with *in vitro* data, mice over-expressing a dominant-negative resistin have increased fat mass resulting from increased adipocyte differentiation and hypertrophy (Kim *et al.* 2004). In contrast, resistin deficiency in obese (*ob/ob*) mice causes further gain in body weight and adiposity owing to decreased energy expenditure (Qi *et al.* 2006).

Several lines of evidence indicate possible roles for resistin in inflammation and vascular diseases. Pro-inflammatory cytokines, such as TNF-α and IL-6, induce resistin expression in human macrophages (Kaser *et al.* 2003, Lehrke *et al.* 2004). Resistin itself induces the expression of pro-inflammatory cytokines in human mononuclear cells (Silswal *et al.* 2005). Further, resistin induces the expression of monocyte chemoattractant protein (MCP) 1 and adhesion molecules in aortic endothelial cells (Burnett *et al.* 2005), likely increasing the recruitment of monocytes and macrophages to the inflamed vasculature. These cells eventually may become the cholesterol-laden foam cells found in atherosclerotic lesions. Indeed, in atherogenic-prone *apoE-/-* mice, the levels of resistin circulating and in the aorta are increased (Burnett *et al.* 2005).

MECHANISMS OF ACTION

The mechanisms by which resistin impairs insulin action in multiple tissues are poorly defined, in part because receptors for resistin have not been identified. Regardless, two signaling pathways appear to be involved in antagonizing insulin signaling and action. Resistin inhibits the AMPK signaling pathway in liver and muscle (Banerjee *et al.* 2004, Muse *et al.* 2004, Satoh *et al.* 2004). In contrast, this same pathway is activated by leptin and adiponectin to promote systemic insulin sensitivity. Additionally, resistin induces the expression of suppressor of cytokine signaling (SOCS) 3 in adipocytes, a signaling molecule known to antagonize insulin signaling (Steppan *et al.* 2005). Little else is known about how resistin exerts its biological effects on cells and tissues. Further, the mechanisms by which resistin inhibits adipogenesis are unknown.

CONCLUSION

Resistin is a cytokine produced by adipocytes and macrophages. Three functional effects have been indicated for resistin: inducing insulin resistance, inhibiting adipocyte differentiation and promoting inflammation. Several outstanding questions concerning resistin biology remain to be addressed. These include a better understanding of its mechanisms of action, identification of a specific receptor(s) that mediates resistin action, and the biological relevance and function of human resistin. Answers to these questions will likely shed light on the interplay between obesity, inflammation and diabetes and provide potential new therapeutic targets.

Table I   Key features of resistin

- Insulin signaling and actions regulate many metabolic pathways. These include stimulation of glucose uptake into the insulin-sensitive tissues, suppression of endogenous glucose production and promotion of anabolic processes. Insulin resistance occurs when insulin-sensitive tissues have a diminished response to insulin.
- Adipogenesis is the cellular differentiation process by which preadipocytes become mature adipocytes. This process includes a defined series of events that result in morphological changes, cessation of cell growth, expression of lipogenic enzymes and adipocyte-specific genes, extensive triglyceride accumulation in lipid droplets and an acquired ability to respond to hormones such as insulin.
- AMPK is a highly conserved energy gauge that is activated by high AMP/ATP ratio and other stimuli, such as cellular stress, hormonal stimulation and pharmacological agents. Activated AMPK turns off anabolic processes that consume ATP and turns on catabolic processes that generate ATP.
- SOCS proteins are induced by cytokine stimulation and function to attenuate or turn off cytokine-activated signal transductions. SOCS-3 modulates insulin signaling by targeting insulin receptor substrate 1 (IRS-1) and IRS-2, two key signaling proteins in insulin action, for degradation by the proteasome.

Summary Points

- Resistin is a secreted hormone produced by adipocytes and macrophages.
- Rodent and human resistin exhibit significant sequence divergence; the biological relevance and function of human resistin is poorly defined.
- Resistin causes insulin resistance by impairing insulin action.

- Resistin inhibits the AMPK signaling pathway in liver and muscle and activates SOCS-3 in adipose tissue, accounting for its antagonizing effects on insulin signaling and action.
- Resistin inhibits adipogenesis *in vitro* and appears to do so *in vivo*.
- Resistin regulates the expression of inflammatory cytokines, chemokines and adhesion molecules, together promoting inflammation and vascular diseases.

Abbreviations

ADSF   :   adipocyte-specific secretory factor
AMPK   :   AMP-activated protein kinase
ApoE   :   apolipoprotein E
ASO   :   antisense oligodeoxynucleotide
DIO   :   diet-induced obesity
FIZZ   :   found in inflammatory zone
IL-6   :   interleukin-6
MCP-1   :   monocyte chemoattractant protein
RELM   :   resistin-like molecule
TNF-α   :   tumor necrosis factor α

Definition of Terms

Adipogenesis: The process by which a preadipocyte becomes a mature adipocyte. This process is tightly regulated by hormones and cellular signaling pathways.

Antisense oligodeoxynucleotide: A short DNA sequence complementary to a specific mRNA sequence; it is used to block the translation of cognate mRNA, thereby inhibiting protein synthesis.

Glucose tolerance: A test used to assess how quickly glucose is cleared from the blood after acute administration of a bolus of glucose. Individuals with impaired glucose tolerance have slower rates of glucose clearance and are predisposed to the development of type 2 diabetes and cardiovascular diseases.

Macrophage: A type of white blood cell derived from monocytes. These cells express scavenger receptor and LDL receptor and can take up lipids from the vasculature. A cholesterol-laden macrophage is called a foam cell.

References

Banerjee, R.R. and S.M. Rangwala, J.S. Shapiro, A.S. Rich, B. Rhoades, Y. Qi, J. Wang, M.W. Rajala, A. Pocai, P.E. Scherer, *et al.* 2004. Regulation of fasted blood glucose by resistin. Science 303: 1195-1198.

Burnett, M.S. and C.W. Lee, T.D. Kinnaird, E. Stabile, S. Durrani, M.K. Dullum, J.M. Devaney, C. Fishman, S. Stamou, D. Canos, *et al.* 2005. The potential role of resistin in atherogenesis. Atherosclerosis 182: 241-248.

Gabriely, I. and X.H. Ma, X.M. Yang, G. Atzmon, M.W. Rajala, A.H. Berg, P. Scherer, L. Rossetti, and N. Barzilai. 2002. Removal of visceral fat prevents insulin resistance and glucose intolerance of aging: an adipokine-mediated process? Diabetes 51: 2951-2958.

Holcomb, I.N. and R.C. Kabakoff, B. Chan, T.W. Baker, A. Gurney, W. Henzel, C. Nelson, H.B. Lowman, B.D. Wright, N.J. Skelton, *et al.* 2000. FIZZ1, a novel cysteine-rich secreted protein associated with pulmonary inflammation, defines a new gene family. EMBO J. 19: 4046-4055.

Kaser, S. and A. Kaser, A. Sandhofer, C.F. Ebenbichler, H. Tilg, and J.R. Patsch. 2003. Resistin messenger-RNA expression is increased by proinflammatory cytokines *in vitro*. Biochem. Biophys. Res. Commun. 309: 286-290.

Kim, K.H. and K. Lee, Y.S. Moon, and H.S. Sul. 2001. A cysteine-rich adipose tissue-specific secretory factor inhibits adipocyte differentiation. J. Biol. Chem. 276: 11252-11256.

Kim, K.H. and L. Zhao, Y. Moon, C. Kang, and H.S. Sul. 2004. Dominant inhibitory adipocyte-specific secretory factor (ADSF)/resistin enhances adipogenesis and improves insulin sensitivity. Proc. Natl. Acad. Sci. USA 101: 6780-6785.

Lazar, M.A. 2007. Resistin- and obesity-associated metabolic diseases. Horm. Metab. Res. 39: 710-716.

Lehrke, M. and M.P. Reilly, S.C. Millington, N. Iqbal, D.J. Rader, and M.A. Lazar. 2004. An inflammatory cascade leading to hyperresistinemia in humans. PLoS Med. 1: e45.

Moon, B. and J.J. Kwan, N. Duddy, G. Sweeney, and N. Begum. 2003. Resistin inhibits glucose uptake in L6 cells independently of changes in insulin signaling and GLUT4 translocation. Am. J. Physiol. Endocrinol. Metab. 285: E106-15.

Muse, E.D. and S. Obici, S. Bhanot, B.P. Monia, R.A. McKay, M.W. Rajala, P.E. Scherer, and L. Rossetti. 2004. Role of resistin in diet-induced hepatic insulin resistance. J. Clin. Invest. 114: 232-239.

Patel, L. and A.C. Buckels, I.J. Kinghorn, P.R. Murdock, J.D. Holbrook, C. Plumpton, C.H. Macphee, and S.A. Smith. 2003. Resistin is expressed in human macrophages and directly regulated by PPAR gamma activators. Biochem. Biophys. Res. Commun. 300: 472-476.

Patel, S.D. and M.W. Rajala, L. Rossetti, P.E. Scherer, and L. Shapiro. 2004. Disulfide-dependent multimeric assembly of resistin family hormones. Science 304: 1154-1158.

Pravenec, M. and L. Kazdova, V. Landa, V. Zidek, P. Mlejnek, P. Jansa, J. Wang, N. Qi, and T.W. Kurtz. 2003. Transgenic and recombinant resistin impair skeletal muscle glucose metabolism in the spontaneously hypertensive rat. J. Biol. Chem. 278: 45209-45215.

Qi, Y. and Z. Nie, Y.S. Lee, N.S. Singhal, P.E. Scherer, M.A. Lazar, and R.S. Ahima. 2006. Loss of resistin improves glucose homeostasis in leptin deficiency. Diabetes 55: 3083-3090.

Rajala, M.W. and S. Obici, P.E. Scherer, and L. Rossetti. 2003. Adipose-derived resistin and gut-derived resistin-like molecule-beta selectively impair insulin action on glucose production. J. Clin. Invest. 111: 225-230.

Rangwala, S.M. and A.S. Rich, B. Rhoades, J.S. Shapiro, S. Obici, L. Rossetti, and M.A. Lazar. 2004. Abnormal glucose homeostasis due to chronic hyperresistinemia. Diabetes 53: 1937-1941.

Satoh, H. and M.T. Nguyen, P.D. Miles, T. Imamura, I. Usui, and J.M. Olefsky. 2004. Adenovirus-mediated chronic "hyper-resistinemia" leads to *in vivo* insulin resistance in normal rats. J. Clin. Invest. 114: 224-231.

Silswal, N. and A.K. Singh, B. Aruna, S. Mukhopadhyay, S. Ghosh, and N.Z. Ehtesham. 2005. Human resistin stimulates the pro-inflammatory cytokines TNF-alpha and IL-12 in macrophages by NF-kappaB-dependent pathway. Biochem. Biophys. Res. Commun. 334: 1092-1101.

Steppan, C.M. and S.T. Bailey, S. Bhat, E.J. Brown, R.R. Banerjee, C.M. Wright, H.R. Patel, R.S. Ahima, and M.A. Lazar. 2001. The hormone resistin links obesity to diabetes. Nature 409: 307-312.

Steppan, C.M. and M.A. Lazar. 2004. The current biology of resistin. J. Intern. Med. 255: 439-447.

Steppan, C.M. and J. Wang, E.L. Whiteman, M.J. Birnbaum, and M.A. Lazar (2005. Activation of SOCS-3 by resistin. Mol. Cell. Biol. 25: 1569-1575.

Yannakoulia, M. and N. Yiannakouris, S. Bluher, A.L. Matalas, D. Klimis-Zacas, and C.S. Mantzoros. 2003. Body fat mass and macronutrient intake in relation to circulating soluble leptin receptor, free leptin index, adiponectin, and resistin concentrations in healthy humans. J. Clin. Endocrinol. Metab. 88: 1730-1736.

CHAPTER 14

Retinol-Binding Protein 4 as an Adipokine

Elmar Aigner[1,*] **and Christian Datz**[1]

[1]Department of Internal Medicine, Oberndorf Hospital, Paracelsusstrasse 37, A-5110 Oberndorf, Austria

ABSTRACT

Retinol-binding protein 4 (RBP4) is synthesized in the liver as the carrier for Vitamin A. RBP4 was identified as an adipokine when adipose tissue expression in adipose-tissue-specific GLUT4 knockout mice was increased and transgenic over-expression of RBP4 induced muscle and liver insulin resistance (IR). In humans, its expression was found increased in obese visceral adipose tissue along with IR. Clinical investigations have yielded conflicting results on the relationship between RBP4 and IR. Renal function as assessed by glomular filtration rate seems to be an important variable influencing RBP4 serum concentrations. Likewise, interactions between RBP4 and steroid hormone abnormalities need to be taken into account as potential confounders when studying RBP4 as an adipokine. Clinically, measurement of RBP4 serum concentrations has been used to assess nutritional status, and calorie excess as well as deficit may influence serum and adipose tissue concentrations independently of IR. RBP4 gene transcription is regulated in a complex manner where the involved transcription factors also influence glucose metabolism genes and steroid hormones. In summary, although there is evidence that RBP4 serves as an adipokine in mice, its role in human health or disease is not clear.

*Corresponding author

INTRODUCTION: PHYSIOLOGY OF RBP4

Retinol-binding protein 4 (RBP4) has long been known as the carrier for vitamin A (retinol; Fig. 1). Retinol is bound to apo-RBP4, which is attached to transthyretin in a 1:1 ratio aggregating to a large homotetramer to prevent glomerular filtration (Blaner 1989). Delivery of retinol to target tissues such as the retinal pigment epithelium is facilitated by the cell surface receptor STRA6.

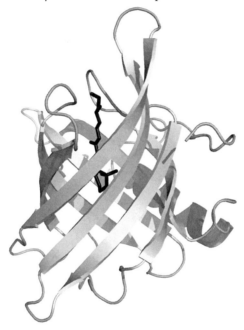

Fig. 1 The three-dimensional structure of RBP4. Human RBP4 in complex with centrally bound retinol (from wikipremed.com, free copyright under the terms of the GNU Free Documentation License, Version 1.2).

DISCOVERY OF RBP4 AS AN ADIPOKINE

To clarify the role of low adipose tissue GLUT4 in insulin resistance (IR), adipose tissue–specific GLUT4 knockout mice were generated. This led to liver and muscle IR. RBP4 was increased in these adipocytes and RBP4 over-expression induced IR (Yang *et al.* 2005). In humans, serum RBP4 was linked to IR and type 2 diabetes mellitus (T2DM) in non-obese and obese men and women. Particularly high RBP4 was found in visceral adipose tissue (VAT). (Graham *et al.* 2006). These observations identified RBP4 as an adipokine.

HOW RBP4 INDUCES IR

In the liver, RBP4 up-regulates phosphoenolpyruvate kinase (PEPCK) increasing fasting glucose production. In muscle, decreased expression and phosphorylation of insulin receptor substrates (IRS1 and IRS2) were found in response to RBP4. In adipocytes, RBP4 treatment led to attenuated IRS1 signalling via blocking serine(307) phosphorylation and inhibiting MAP kinase ERK1/2 phosphorylation. These changes resemble insulin signaling abnormalities observed in T2DM.

CLINICAL EVIDENCE FROM ADULT AND PEDIATRIC POPULATIONS

Clinical research on RBP4 and IR was conducted in males and females from all age groups, ranging from lean to morbidly obese study subjects with various degrees of insulin sensitivity. IR-related conditions such as gestational hypertension or diabetes, preeclampsia and the polycystic ovary syndrome (PCOS) were studied. The results of these studies in humans with regard to the relationship of RBP4 with IR or metabolic syndrome are conflicting. The most consistent link may exist between RBP4 and VAT mass or triglycerides. Investigations in children found an association with general obesity, central obesity and components of the metabolic syndrome but not IR. Although no association with the change of IR during pubertal development was observed, baseline RBP4 was associated with worsening of pubertal IR, suggesting usefulness as a biomarker for the development of IR in the young. In summary, clinical evidence suggests that RBP4 is likely linked to IR in a secondary, non-causal manner.

RBP4 AND LIFESTYLE INTERVENTION

The effects of diet or exercise are of interest in diseases caused by lifestyle habits. RBP4 concentrations in healthy young men remained unchanged during a 7 d overfeeding protocol although baseline RBP4 was negatively linked to the change in IR in normal-weight but not overweight men (Shea *et al.* 2007). On a very low calorie diet, subcutaneous adipose tissue RBP4 and GLUT4 mRNA concentrations concomitantly decreased in obese women (an inverse relationship was expected, Fig. 2). The authors concluded that RBP4 was suitable for assessment of nutritional status but not of IR (Vitkova *et al.* 2007). Exercise training in obese women lowered adipocyte fatty acid–binding protein and BMI but RBP4 remained unchanged. The switch from standard to a Mediterranean pattern diet over 8 wk led to a decrease in RBP4, as did weight loss after bariatric surgery (Hermsdorff *et al.* 2009). In conclusion, lifestyle-intervention studies suggest that RBP4 is linked to nutritional status, especially energy deficit, but do not support a role as a mediator of IR.

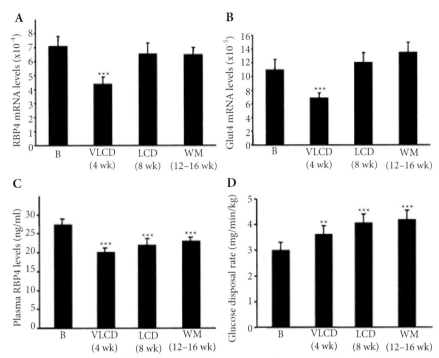

Fig. 2 Subcutaneous adipose tissue mRNA expression of RBP4 and Glut4, plasma level of RBP4, and glucose disposal rate during a weight reduction program in obese women (n=24). (A) Adipose tissue RBP4 mRNA levels. (B) Glut4 mRNA levels. (C) RBP4 plasma levels. (D) Glucose disposal rate. Values are means ± SEM. **P < 0.01. ***P < 0.001 compared with basal values. B, Basal conditions; LCD, low calorie diet, end of the LCD; VLCD, very low calorie diet, end of the VLCD; WM, weight maintenance, end of the WM phase (with permission from Vitkova *et al.* 2007, © The Endocrine Society).

RBP4 AND FATTY LIVER (NAFLD)

NAFLD is the hepatic manifestation of IR, and the liver is the site of retinol-RBP4 biosynthesis. In adult and pediatric NAFLD patients, serum RBP4 decreased from steatosis to NASH to cirrhosis and was not linked to BMI, IR or glucose (Fig. 3, Alkhouri *et al.* 2009). Hepatic RBP4 immunostaining is higher in progressed forms of NAFLD, suggesting a secretory defect of RBP4 in NAFLD (Schina *et al.* 2009).

RBP4 AND RENAL FUNCTION

After delivery of Vitamin A to target tissues, apo-RBP4 is glomerularly filtered and degraded in the proximal tubular system. In patients with or without diabetes, irrespective of obesity, glomerular filtration rate determined RBP4 concentrations

(Fig. 4; Hentze *et al.* 2008). Similar results were reported for microalbuminuria, macroalbuminuria, end-stage renal disease or patients on chronic hemodialysis. Thus, kidney function is an important variable influencing serum RBP4 levels.

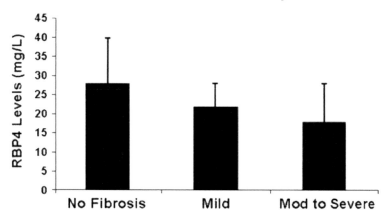

Fig. 3   Serum RBP4 in NAFLD according to histological severity. A stepwise decrease in RBP4 levels was noted from patients without fibrosis (28 mg/L) to patients with mild fibrosis (22 mg/L) to patients with moderate to severe fibrosis (18 mg/L) (P value <0.05) (with permission from Alkouri *et al.* 2009, © Lippincott Williams Wilkins, Inc.).

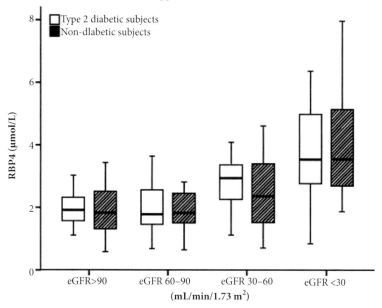

Fig. 4   RBP4 serum concentration in type 2 diabetic and non-diabetic subjects dependent on renal function. eGFR, estimated glomerular filtration rate. (Reprinted with permission from The American Diabetes Association 2008; from Diabetes, Vol. 57, 2008; 3328-3326.)

RBP4 AND STEROID HOROMONES

IR and obesity are linked to PCOS in young women and links between RBP4 and steroid hormone metabolism have been described. The use of hormonal contraception increased RBP4 concentrations and serum RBP4 concentrations have been associated with estradiol levels in PCOS patients. Molecular support for these findings was provided when 17β-estradiol increased RBP4 mRNA and protein in adipocytes from PCOS patients (Tan *et al.* 2007). The potential genetic basis for these findings is discussed below.

RBP4 GENE ASSOCIATIONS

The RBP4 gene is located on chromosome 10q23-q24, which was linked to T2DM prior to the identification of RBP4 as an adipokine. Associations of RBP4 single nucleotide polymorphisms (SNPs) with diabetes or prediabetic traits were described in Mongolian and Caucasian populations. A common six SNP haplotype was linked to RBP4 mRNA levels in VAT and diabetes (Kovacs *et al.* 2007). However, similar to clinical studies, these results were not confirmed in consecutive studies. Genetic variation in the APOA5 gene enhanced the RBP4 triglyceride association, suggesting that RBP4 may be part of a complex regulatory network linked to lipid metabolism or IR.

RBP4 GENE REGULATION

Discoveries in RBP4 gene regulation may explain incongruous murine and human results and shed light on the role of RBP4 in obesity-related IR. Activity of the RBP4 promoter is regulated in a bipartite manner where a proximal region is required for basal RBP4 transcription and a distal region for cAMP-mediated induction. cAMP-mediated induction is linked to HMGA1 (high mobility group A1). HMGA1 serves as a co-factor for proper insulin receptor gene (INSR) activation and also RBP4 transcription by recruiting transcription factors including steroidogenic factor 1 (SF1) to the respective promoter. This multi-protein aggregate structurally and functionally resembles transcription factor complexes modulating steroid hormone biosynthesis and glucose metabolism genes. Thus, RBP4 expression may be coordinately regulated in a transcription network relevant to glucose, lipid and steroid hormone metabolism (Bianconcini *et al.* 2009). The insulin-sensitizing thiazolidinediones that act through PPARγ activation decrease RBP4 in mice and humans in some studies.

Recently, insulin-sensitizing properties of fenretinide apart from RBP4 were reported and effective lowering of RBP4 by attaching the non-retinoid component A1120 to RBP4 did not affect insulin sensitivity. Such findings challenge the role of RBP4 in IR and highlight the need to decipher its complex genetic background and the multi-faceted activities of Vitamin A and its metabolites in insulin sensitivity and adipogenesis.

Table I Basic facts of insulin and insulin resistance

- Insulin is a polypeptide secreted from pancreatic beta-cells. Insulin is the key hormone facilitating the uptake of glucose (to a lesser degree lipids and amino acids) into adipose tissue, muscle or liver.
- Insulin receptors are present on most cells of the body. Tyrosine autophosphorylation of its intracellular domain activates a signaling cascade leading to expression of glucose transporters on the cell surface.
- Signal transduction may be impaired at multiple steps accounting for impaired insulin activity, i.e., insulin resistance. Beta-cells respond by increasing insulin production to achieve normal glucose. Failure to maintain normal glucose is called diabetes.
- The development of insulin resistance involves a genetic background; it is linked to obesity and a sedentary lifestyle.

Summary Points

- RBP4 is secreted from the liver as the carrier for Vitamin A.
- In obesity, apo-RBP4 is also synthesized in adipocytes.
- In adipoGLUT4$^{-/-}$ mice, RBP4 is increased and over-expression induced liver and muscle IR.
- The results from clinical and genetic studies on the relationship of RBP4 and IR are conflicting.
- The gene-regulatory network of RBP4 transcription may be a promising area to define the role of RBP4 in human disease.

Abbreviations

GLUT4	:	glucose transporter 4
HMGA1	:	high mobility group A1 gene
IR	:	insulin resistance
NAFLD	:	non-alcoholic fatty liver disease
PCOS	:	polycystic ovary syndrome
RBP4	:	retinol binding protein 4
SNP	:	single nucleotide polymorphism
T2DM	:	type 2 diabetes mellitus
VAT	:	visceral adipose tissue

Definition of Terms

GLUT4 and GLUT1: The membrane channels facilitating glucose uptake into liver (GLUT1) or other body cells (GLUT4). A characteristic of IR is low activity of GLUT4 in muscle or adipose tissue, the main glucose metabolizing organs, leading to hyperglycemia.

HMGA1 gene: A gene that encodes for a small basic protein involved in regulation of inducible gene transcription and the metastatic progression of cancer cells. The protein preferentially binds to the minor groove of A/T-rich regions in double-stranded DNA and is the second most abundant DNA-associated protein next to histone proteins. Loss of function of HMGA1 leads to severe IR in humans with the phenotype of T2DM at a young age. HMGA1 serves as a cofactor in gene activation for proper insulin receptor gene (INSR) and RBP4 transcription by recruiting transcription factors to the respective promoter.

Steroid hormones: Structurally related messengers derived from the cholesterol structure. Steroid hormones can be allocated to five groups by their receptors: glucorcorticoids, mineral corticoids, androgens, estrogens and progestagens. Vitamin D and A share structural homologies with steroids. The PCOS is a common cause of infertility in healthy young women. It is characterized by androgen excess, anovulatory menstrual cycles, polycystic ovaries and IR.

Vitamin A metabolites and analogues: Retinaldehyde, the first cellular metabolite of retinol, can inhibit adipogenesis and suppress PPAR-γ-mediated responses. *All-trans retinoic acid* (treatment of acne vulgaris, acute promyelocytic leukemia) decreases adipose tissue RBP4 production. *Fenretinide* is a synthetic retinoid that disrupts binding of RBP4 to transthyretin leading to renal loss; it has insulin-sensitizing properties independent of RBP4. The *non-retinoid RBP4-ligand A1120* leads to effective renal excretion of RBP4 without affecting insulin sensitivity.

References

Alkhouri, N. and R. Lopez, M. Berk, and A.E. Feldstein. 2009. Serum retinol-binding protein 4 levels in patients with nonalcoholic fatty liver disease. J. Clin. Gastro. 43: 985-989.

Bianconcini, A. and A. Lupo, S. Capone, L. Quadro, M. Monti, D. Zurlo, A. Fucci, L. Sabatino, A. Brunetti, E. Chiefari, M.E. Gottesman, W.S. Blaner, and V. Colantuoni. 2009. Transcriptional activity of the murine retinol-binding protein gene is regulated by a multiprotein complex containing HMGA1, p54 nrb/NonO, protein-associated splicing factor (PSF) and steroidogenic factor 1 (SF1)/liver receptor homologue 1 (LRH-1). Int. J. Biochem. Cell. Biol. 41: 2189-2203.

Blaner, W.S. 1989. Retinol-binding protein: the serum transport protein for vitamin A. Endocr. Rev. 10: 308-316.

Graham, T.E. and Q. Yang, M. Blüher, A. Hammarstedt, T.P. Ciaraldi, R.R. Henry, C.J. Wason, A. Oberbach, P.A. Jansson, U. Smith, and B.B. Kahn. 2006. Retinol-binding protein 4 and insulin resistance in lean, obese, and diabetic subjects. N. Engl. J. Med. 354: 2552-2563.

Henze, A. and S.K. Frey, J. Raila, M. Tepel, A. Scholze, A.F. Pfeiffer, M.O. Weickert, J. Spranger, and F.J. Schweigert. 2008. Evidence that kidney function but not type 2 diabetes determines retinol-binding protein 4 serum levels. Diabetes 57: 3323-3326.

Hermsdorff, H.H. and M.A. Zulet, I. Abete, and J.A. Martínez. 2010. Discriminated benefits of a Mediterranean dietary pattern within a hypocaloric diet program on plasma RBP4 concentrations and other inflammatory markers in obese subjects. Endocrine (in press).

Kovacs, P. *et al.* 2007. Effects of genetic variation in the human retinol binding protein-4 gene (RBP4) on insulin resistance and fat depot-specific mRNA expression. Diabetes 56: 3095-3100.

Schina, M. and J. Koskinas, D. Tiniakos, E. Hadziyannis, S. Savvas, B. Karamanos, E. Manesis, and A. Archimandritis. 2009. Circulating and liver tissue levels of retinol-binding protein-4 in non-alcoholic fatty liver disease. Hepatol. Res. 39: 972-978.

Shea, J. and E. Randell, S. Vasdev, P.P. Wang, B. Roebothan, and G. Sun. 2007. Serum retinol-binding protein 4 concentrations in response to short-term overfeeding in normal-weight, overweight, and obese men. Am. J. Clin. Nutr. 86: 1310-1315.

Tan, B.K. and J. Chen, H. Lehnert, R. Kennedy, and H.S. Randeva. 2007. Raised serum, adipocyte, and adipose tissue retinol-binding protein 4 in overweight women with polycystic ovary syndrome: effects of gonadal and adrenal steroids. J. Clin. Endocrinol. Metab. 92: 2764-2772.

Vitkova, M. and E. Klimcakova, M. Kovacikova, C. Valle, C. Moro, J. Polak, J. Hanacek, F. Capel, N. Viguerie, B. Richterova, M. Bajzova, J. Hejnova, V. Stich, and D. Langin. 2007. Plasma levels and adipose tissue messenger ribonucleic acid expression of retinol-binding protein 4 are reduced during calorie restriction in obese subjects but are not related to diet-induced changes in insulin sensitivity. J. Clin. Endocrinol. Metab. 92: 2330-2335.

Yang, Q. and T.E. Graham, N. Mody, F. Preitner, O.D. Peroni, J.M. Zabolotny, K. Kotani, L. Quadro, and B.B. Kahn. 2005. Serum retinol binding protein 4 contributes to insulin resistance in obesity and type 2 diabetes. Nature 436: 356-362.

CHAPTER 15

TNF-α as an Adipokine

Theron C. Gilliland, Jr.[1] and G. William Wong[1,2,*]

[1]Department of Physiology John Hopkins University School of Medicine, 202 Physiology, 725 North Wolfe Street Baltimore, MD 21205, USA

[2]Center for Metabolism and Obesity Research, Johns Hopkins University School of Medicine, 855 North Wolfe Street, Baltimore, MD 21205, USA

ABSTRACT

Tumor necrosis factor alpha (TNF-α), a multifunctional cytokine and adipokine, exerts pleiotropic effects on many cell types and tissues. As an adipokine, this protein negatively regulates many aspects of glucose and lipid metabolism. Downstream targets of TNF-α down-regulate genes involved in the uptake and storage of free fatty acids and glucose in adipocytes. Additionally, TNF-α promotes insulin resistance by inducing serine phosphorylation of insulin receptor substrate 1 (IRS-1) and reducing the expression of IRS-1 mRNA, attenuating downstream insulin signaling. Reduction in insulin signaling results in decreased insulin-stimulated GLUT4 translocation to the plasma membrane, reducing glucose uptake into muscle and adipocytes. TNF-α also decreases the uptake of fatty acids into adipocytes while promoting lipolysis, thus contributing to dyslipidemia and exacerbating tissue insulin resistance. The increased levels of TNF-α found in obesity promote inflammation within the adipose compartment, which dysregulates the expression of adipokines (e.g., adiponectin) that promote insulin sensitivity. Collectively, TNF-α exerts a negative impact on whole-body insulin sensitivity and metabolism. Despite its importance, the therapeutic efficacy of blocking TNF-α is currently disputed. Understanding the many mechanisms of action of TNF-α will likely provide new avenues in treating metabolic disorders associated with diabetes and obesity.

*Corresponding author

INTRODUCTION

Tumor necrosis factor alpha (TNF-α) functions as a pleiotropic cytokine and adipokine. This protein is an important cell signal involved in many processes, including cell proliferation, differentiation, inflammation, immunity and metabolism. This chapter discusses TNF-α adipokine functions in metabolism.

STRUCTURE AND SYNTHESIS

TNF-α is synthesized as a 26 kDa plasma membrane-bound monomer and assembled into a homotrimer (Tang *et al.* 1996). The disintegrin ADAM metallopeptidase domain 17 (ADAM17) proteolytically cleaves the membrane-bound TNF-α precursor to release the secreted trimer (Figs. 1 and 2) (Black *et al.* 1997). Despite little sequence homology between TNF-α and the C1q family of proteins, the trimeric structure of TNF-α strikingly resembles the trimeric structure of adiponectin globular C1q domain. Therefore, TNF-α is now considered part of the growing C1q/TNF superfamily of proteins (Shapiro and Scherer 1998, Kishore *et al.* 2004).

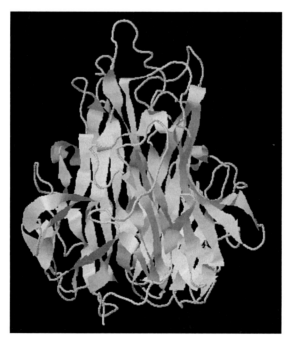

Fig. 1 Structure of TNF-α. Note the homotrimeric structure and jelly-roll β-sandwich fold characteristic of the C1q/TNF superfamily (Eck and Sprang 1989).

Color image of this figure appears in the color plate section at the end of the book.

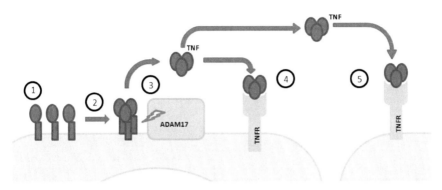

Fig. 2 Synthesis and secretion of TNF-α. (1) TNF-α is synthesized as a membrane-bound 26 kDa monomer. (2) TNF-α is assembled into a homotrimer. (3) The membrane-bound form is cleaved by ADAM17 to release the secreted trimer. (4) Within adipose tissue, TNF-α acts in an autocrine fashion. (5) TNF-α may also act in a paracrine fashion.

Macrophages, natural killer cells, T and B cells, and adipocytes produce TNF-α (Kishore *et al.* 2004). Within adipose tissue, both adipocyte and stromal vascular cells express TNF-α (Ruan and Lodish 2003). The adipokine function of TNF-α was identified because its levels were increased in obesity and linked to adipocyte insulin resistance (Hotamisligil *et al.* 1993, 1995).

MECHANISM OF ACTION

The effects of TNF-α are mediated through two receptors: TNFR1 and TNFR2. Divergent cytoplasmic tails indicate that these receptors transduce distinct signals (Tartaglia and Goeddel 1992). Most of the metabolic effects of TNF-α are mediated by TNFR1.

Some major signaling pathways activated by TNF-α in adipose tissue include NF-κB, JNK1, p38-MAPK, IKK2 and Erk1/2 (Fig. 3) (Cawthorn and Sethi 2008). Activation of these pathways adversely affects adipocyte function by altering the uptake, synthesis and storage of triglycerides and the synthesis and secretion of adipokines. For example, NF-κB activation by TNF-α inhibits many genes involved in uptake and storage of free fatty acids (FFA) and glucose in adipocytes (Ruan and Lodish 2003). Inhibition can be blocked by members of the thiazolidinedione class of drugs, which restore insulin sensitivity through agonist activity on the PPAR-γ nuclear receptor. Further, knocking out IκB kinase 2 (IKK2), a direct activator of NF-κB, in mice significantly improves insulin responses (Yuan *et al.* 2001).

TNF-α creates an insulin-resistant state in adipose and other tissues through modification of the insulin signaling pathway (White 2002). Binding of insulin to the IR, a receptor tyrosine kinase, stimulates both autophosphorylation and phosphorylation and activation of IRS-1, an important proximal component of

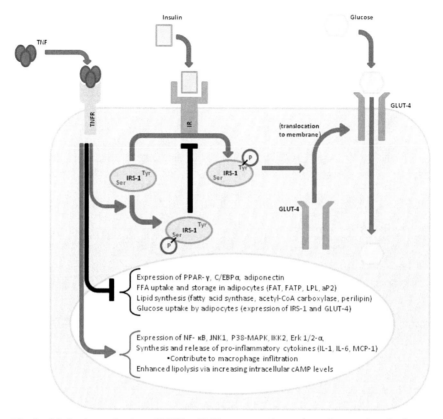

Fig. 3 Mechanisms of action of TNF-α. TNF-α can interfere with insulin signaling, leading to decreased translocation of GLUT4 to the plasma membrane. It also influences transcription of many metabolic genes.

Color image of this figure appears in the color plate section at the end of the book.

the insulin signaling pathway. Tyrosine-phosphorylated IRS-1 serves as a docking site to recruit many downstream signaling proteins. TNF-α also induces serine phosphorylation of IRS-1 by activating multiple serine/threonine kinases such as JNK1, S6K1, IKK and PKC-ζ (Gao *et al.* 2002, Zhang *et al.* 2008, Boura-Halfon and Zick 2009). Serine phosphorylation uncouples IRS-1 from the IR, resulting in reduced tyrosine phosphorylation and decreased insulin signaling (White 2002). Additionally, TNF-α reduces the expression of IRS-1 mRNA. These combined effects of TNF-α on IRS-1 lead to a reduction in insulin receptor signaling. Reduced insulin-stimulated GLUT4 translocation to the plasma membrane follows, resulting in decreased glucose uptake into muscle and adipose tissue. Indeed, cells lacking endogenous IRS-1 display resistance to TNF-α-mediated IR signaling inhibition; conversely, ectopic expression of IRS-1 restores TNF-α sensitivity, highlighting the importance of IRS-1 in this mechanism (Hotamisligil 1999).

TNF-α inhibits adipogenesis via down-regulation of two dominant transcription factors, PPAR-γ and C/EBPα, which drive preadipocyte differentiation into mature adipocytes (Farmer 2006). PPAR-γ and C/EBPα binding sites are found in promoters of many metabolic genes. Consequently, down-regulation of PPAR-γ and C/EBPα expression by TNF-α decreases synthesis of many adipocyte-specific genes, such as GLUT4 and aP2, a fatty acid carrier protein (Sethi and Hotamisligil 1999, Farmer 2006).

Finally, TNF-α potently regulates the expression of other cytokines and adipokines. For example, TNF-α inhibits production of adiponectin—which promotes systemic insulin sensitivity—while it stimulates release of pro-inflammatory cytokine IL-6 from adipocytes (Ruan and Lodish 2003, Arner 2005). By modulating these and other signaling molecules, TNF-α achieves many indirect effects on whole-body metabolism.

PHYSIOLOGICAL EFFECTS

Circulating levels of TNF-α do not rise appreciably in obesity and diabetes. Therefore, most excess TNF-α produced in obese adipose tissue is thought to act in autocrine and paracrine manners, directly impacting adipocytes and other stromal vascular cells within the adipose compartment (Cawthorn and Sethi 2008). By interfering with gene transcription and insulin signaling, TNF-α prevents adipose tissue from carrying out essential functions as a fat storage depot and an endocrine organ. Impairing adipose functions compromises homeostatic mechanisms that maintain whole-body energy balance, since adipose tissue is one of the central players regulating whole-body glucose and lipid metabolism (Rosen and Spiegelman 2006).

TNF-α profoundly alters adipocyte gene expression, leading to a down-regulation of genes (e.g., IRS-1, GLUT4) critical for insulin responsiveness (Ruan and Lodish 2004). Consequently, deletion of the TNF-α gene or both TNF-α receptors in animal models of obesity and diabetes restores insulin sensitivity (Moller 2000, Ruan and Lodish 2003). Because insulin promotes triglyceride synthesis and storage in adipocytes, TNF-α-induced insulin resistance promotes lipolysis in adipocytes and contributes to dyslipidemia. These effects on lipid metabolism occur through several mechanisms. First, TNF-α decreases FFA uptake in adipocytes by down-regulating expression and/or activity of genes involved in this process, such as FAT, FATP and LPL (Ruan and Lodish 2003). Second, TNF-α down-regulates genes involved in lipid synthesis (e.g., fatty acid synthase, acetyl-CoA carboxylase, perilipin). Third, TNF-α enhances lipolysis by increasing cAMP levels, and through other poorly defined mechanisms (Zhang *et al.* 2002, Cawthorn and Sethi 2008). The combined effects of TNF-α greatly impair triglyceride storage in adipose tissue. Decreased storage of triglycerides, combined with enhanced lipolysis in adipocytes, elevates circulating FFA levels.

The lipotoxicity effects of excess FFA exacerbate insulin-resistant phenotypes, including excessive hepatic glucose production and VLDL secretion and decreased glucose uptake and metabolism in muscle (Ruan and Lodish 2003).

Adipose tissue synthesizes and secretes a wide array of adipokines, including leptin, adiponectin, resistin, omentin, visfatin, RBP4, vaspin, lipocalin 2 and CTRPs (MacDougald and Burant 2007, Wong *et al.* 2008, 2009). The expression of most adipokines is dysregulated in obesity and diabetes. Elevated TNF-α levels in adipose tissue increase the synthesis and secretion of pro-inflammatory cytokines that contribute to macrophage infiltration, further exacerbating adipose tissue inflammation (Hotamisligil 2006). Concomitantly, TNF-α decreases the production of the insulin-sensitizing adipokine adiponectin (Ruan and Lodish 2003, Arner 2005). Together, these changes impair the endocrine function of adipose tissue, indirectly contributing to decreased systemic insulin sensitivity.

Surprisingly, despite potent pleiotropic metabolic effects, short-term TNF-α blockade by anti-TNF-α antibody or sTNFR1-IgG does not restore insulin sensitivity in humans (Ofei *et al.* 1994). Indeed, increased insulin resistance occurs in rheumatoid arthritis patients treated with the TNF-α blockers etanercept and infliximab (Seriolo *et al.* 2008). The benefits of targeting TNF-α or its receptors in treating metabolic diseases, therefore, remain unclear.

CONCLUSION

Given the incidence of obesity, diabetes and other metabolic diseases, a nuanced understanding of the roles of TNF-α in these and related disorders may provide novel therapeutic approaches for treatment.

Table I   Key points about TNF-α

- Insulin resistance is a major contributor to the pathogenesis of diabetes.
- Adipose tissue, a storage depot for fat, also acts as an endocrine organ and is involved in regulating whole-body metabolism.
- Some adipokines, such as TNF-α and IL-6, also serve as cytokines; therefore, metabolic dysregulation often affects the immune system.

Summary Points

- TNF-α is a physiologically important cytokine and adipokine with pleiotropic effects on multiple cell and tissue types.
- TNF-α down-regulates genes involved in the uptake and storage of free fatty acids and glucose in adipocytes.
- TNF-α decreases insulin sensitivity by down-regulating IRS-1 expression and interfering with the insulin signaling pathway.

- TNF-α inhibits adipogenesis via down-regulation of key transcription factors PPAR-γ and C/EBPα.
- TNF-α decreases the uptake, synthesis and storage of triglycerides and promotes lipolysis in adipocytes.
- TNF-α increases the expression of pro-inflammatory cytokines and decreases the production of the insulin-sensitizing adipokine adiponectin.

Abbreviations

aP2	:	adipocyte protein 2; fatty acid binding protein 4 (FABP4)
ADAM17	:	ADAM metallopeptidase domain 17; TNF-α converting enzyme (TACE)
C/EBPα	:	CCAAT/enhancer-binding protein alpha
CTRPs	:	C1q/TNF-related proteins
Erk1/2	:	extracellular regulated kinase 1/2
FAT	:	fatty acid translocase
FATP	:	fatty acid transport protein
GLUT4	:	glucose transporter type 4
IKK complex, IKK2	:	inhibitor κB kinase complex, inhibitor κB kinase, subunit 2
IL	:	interleukins
IR	:	insulin receptor
IRS-1	:	insulin receptor substrate 1
JNK1- c-	:	Jun N-terminal kinase 1
LPL	:	lipoprotein lipase
MAPK	:	mitogen-activated protein kinase
NF-κB	:	nuclear Factor-kappa B
PPAR-γ	:	peroxisome proliferator-activated receptor gamma
RBP4	:	retinol binding protein 4
S6K1	:	ribosomal protein S6 kinase 1; p70-S6K1
TNF-α	:	tumor necrosis factor alpha; also known as cachexin or cachectin
TNFR1 and TNFR2	:	TNF-α receptor I and TNF-α receptor II
VLDL	:	very low density lipoprotein

Definition of Terms

Autocrine: Secreted factor that acts on the cells that produce it.

Dyslipidemia: Abnormal concentrations of lipids or lipoproteins in the blood.

Lipotoxicity: The pathological effects of elevated fat levels in blood or tissues.

Paracrine: Effects of hormones that act locally, usually on nearby cells.

Pleiotropic: Producing more than one effect.

References

Arner, P. 2005. Insulin resistance in type 2 diabetes–role of the adipokines. Curr. Mol. Med. 5: 333-339.

Black, R.A. *et al.* and D.P. Cerretti. 1997. A metalloproteinase disintegrin that releases tumour-necrosis factor-alpha from cells. Nature 385: 729-733.

Boura-Halfon, S. and Y. Zick. 2009. Phosphorylation of IRS proteins, insulin action, and insulin resistance. Am. J. Physiol. Endocrinol. Metab. 296: E581-91.

Cawthorn, W.P. and J.K. Sethi. 2008. TNF-α and adipocyte biology. FEBS Lett. 582: 117-131.

Eck, M.J. and S.R. Sprang. 1989. The structure of tumor necrosis factor-α at 2.6 Å resolution: implications for receptor binding. J. Biol. Chem. 264: 17595-17605.

Farmer, S.R. 2006. Transcriptional control of adipocyte formation. Cell Metab. 4: 263-273.

Gao, Z. and D. Hwang, F. Bataille, M. Lefevre, D. York, M.J. Quon, and J. Ye. 2002. Serine phosphorylation of insulin receptor substrate 1 by inhibitor κB kinase complex. J. Biol. Chem. 277: 48115-48121.

Hotamisligil, G.S. and P. Arner, J.F. Caro, R.L. Atkinson, and B.M. Spiegelman. 1995. Increased adipose tissue expression of tumor necrosis factor-α in human obesity and insulin resistance. J. Clin. Invest. 95: 2409-2415.

Hotamisligil, G.S. 1999. The role of TNFα and TNF receptors in obesity and insulin resistance. J. Intern. Med. 245: 621-625.

Hotamisligil, G.S. and N.S. Shargill, and B.M. Spiegelman. 1993. Adipose expression of tumor necrosis factor-α: direct role in obesity-linked insulin resistance. Science 259: 87-91.

Hotamisligil, G.S. 2006. Inflammation and metabolic disorders. Nature 444: 860-867.

Kishore, U. and C. Gaboriaud, P. Waters, A.K. Shrive, T.J. Greenhough, K.B.M. Reid, R.B. Sim, and G.J. Arlaud. 2004. C1q and tumor necrosis factor superfamily: modularity and versatility. Trends Immunol. 25: 551-561.

MacDougald, O.A. and C.F. Burant. 2007. The rapidly expanding family of adipokines. Cell Metab. 6: 159-161.

Moller, D.E. 2000. Potential role of TNF-alpha in the pathogenesis of insulin resistance and type 2 diabetes. Trends Endocrinol. Metab. 11: 212-217.

Ofei, F. and S. Hurel, J. Newkirk, M. Sopwith, and R. Taylor. 1996. Effects of an engineered human anti-TNF-α antibody (CDP571) on insulin sensitivity and glycemic control in patients with NIDDM. Diabetes 45: 881-885.

Rosen, E.D. and B.M. Spiegelman. 2006. Adipocytes as regulators of energy balance and glucose homeostasis. Nature 444: 847-853.

Ruan, H. and H.F. Lodish. 2003. Insulin resistance in adipose tissue: direct and indirect effects of tumor necrosis factor-α. Cytokine Growth Fact. Rev. 14: 447-455.

Ruan, H. and H.F. Lodish. 2004. Regulation of insulin sensitivity by adipose tissue-derived hormones and inflammatory cytokines. Curr. Opin. Lipidol. 15: 297-302.

Ruan, H. and H.J. Pownall, and H.F. Lodish. 2003. Troglitazone antagonizes TNF-α-induced reprogramming of adipocyte gene expression by inhibiting the transcriptional regulatory functions of NF-κB. J. Biol. Chem. 278: 28181-28192.

Seriolo, B. and C. Ferrone, and M. Cutolo. 2008. Longterm anti-tumor necrosis factor-α treatment in patients with refractory rheumatoid arthritis: relationship between insulin resistance and disease activity. J. Rheumatol. 35: 355-357.

Sethi, J.K. and G.S. Hotamisligil. 1999. The role of TNFα in adipocyte metabolism. Semin. Cell Dev. Biol. 10: 19-29.

Shapiro, L. and P.E. Scherer. 1998. The crystal structure of a complement-1q family protein suggests an evolutionary link to tumor necrosis factor. Curr. Biol. 8: 335-338.

Tang, P. and M.-C. Hung, and J. Klostergaard. 1996. Human pro-tumor necrosis factor is a homotrimer. Biochemistry 35: 8216-8225.

Tartaglia, L.A. and D.V. Goeddel. 1992. Two TNF receptors. Immunol. Today 13: 151-153.

White, M.F. 2002. IRS proteins and the common path to diabetes. Am. J. Physiol. Endocrinol. Metab. 283: E413-22.

Wong, G.W. and S.A. Krawczyk, C. Kitidis-Mitrokostas, T. Revett, R. Gimeno, and H.F. Lodish. 2008. Molecular, biochemical and functional characterizations of C1q/TNF family members: adipose-tissue-selective expression patterns, regulation by PPAR-γ agonist, cysteine-mediated oligomerizations, combinatorial associations and metabolic functions. Biochem. J. 416: 161-177.

Wong, G.W. and S.A. Krawczyk, C. Kitidis-Mitrokostas, G. Ge, E. Spooner, C. Hug, R. Gimeno, and H.F. Lodish. 2009. Identification and characterization of CTRP9, a novel secreted glycoprotein, from adipose tissue that reduces serum glucose in mice and forms heterotrimers with adiponectin. FASEB J. 23: 241-258.

Yuan, M. and N. Konstantopoulos, J. Lee, L. Hansen, Z.W. Li, M. Karin, and S.E. Shoelson. 2001. Reversal of obesity- and diet-induced insulin resistance with salicylates or targeted disruption of Ikkbeta. Science 293: 1673-1677.

Zhang, H.H. and M. Halbleib, F. Ahmad, V.C. Manganiello, and A.S. Greenberg. 2002. Tumor necrosis factor-α stimulates lipolysis in differentiated human adipocytes through activation of extracellular signal-related kinase and elevation of intracellular cAMP. Diabetes 51: 2929-2935.

Zhang, J. and Z. Gao, J. Yin, M.J. Quon, and J. YeJ. 2008. S6K directly phosphorylates IRS-1 on Ser-270 to promote insulin resistance in response to TNF-α signaling through IKK2. J. Biol. Chem. 283: 35375-35382.

CHAPTER 16

Transforming Growth Factor as an Adipokine

Yutaka Yata[1,*] and Terumi Takahara[2]

[1]Department of Gastroenterology, Saiseikai Maebashi Hospital, 564-1 Kamishindenmachi, Maebashi-city, Gunma, Japan 371-0821
[2]Third Department of Internal Medicine, University of Toyama, 2630 Sugitani, Toyama, Japan 930-0194

ABSTRACT

Transforming growth factor (TGF) beta and its superfamily, macrophage inhibitory cytokine (MIC) 1, are produced in adipose tissue. Adipocytes also produce thrombospondin (TSP) 1, an activator of TGF-beta1. TGF-beta regulates the gene expression related to the adipose tissue metabolism and plays an important role as an adipokine in the pathogenesis of many diseases.

INTRODUCTION

Transforming growth factor (TGF) beta is a key cytokine at inflammatory site. Adipose tissue produces TGF-beta and its superfamily macrophage inhibitory cytokine (MIC) 1. In this review we focus on the roles of TGF-beta and its superfamily in the metabolism and pathogenesis related to adipose tissue.

TRANSFORMING GROWTH FACTOR (TGF) BETA

TGF-beta1 is known to be produced in adipose tissue (adipocytes and nonfat cells in adipose tissue) (Fain *et al.* 2005) (Fig. 1). First of all, it regulates several gene

---
*Corresponding author

expressions including plasminogen activator inhibitor (PAI) 1, which is associated with insulin resistance characterized by fat accumulation (Alessi *et al.* 2000) (Fig. 2). Circulating TGF-beta1 or adipose tissue TGF-beta1 content are significantly correlated with human body mass index (BMI), and TGF-beta1 antigen is also correlated with PAI-1 antigen (Fig. 3) (Fain *et al.* 2005, Alessi *et al.* 2000).

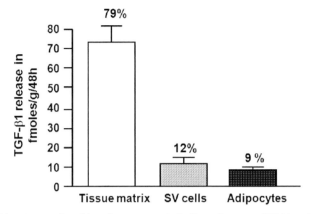

Fig. 1 TGF-beta1 is produced by adipose tissue including adipocytes. TGF-beta1 was mainly released by nonfat cells including stromal-vascular (SV) cells of human adipose tissue, and its release by adipocyte was less than 10% of that by the nonfat cells of adipose tissue.

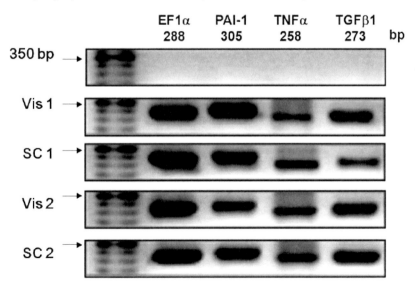

Fig. 2 The mRNA expression of PAI-1, TNF-α, and TGF-beta1 in adipose tissues. RT-PCR revealed the mRNA expressions of PAI-1, TNF-α, and TGF-beta1 in visceral (Vis) and subcutaneous (SC) adipose tissues from two individuals (1 and 2). EF1 α is a control as a housekeeping gene.

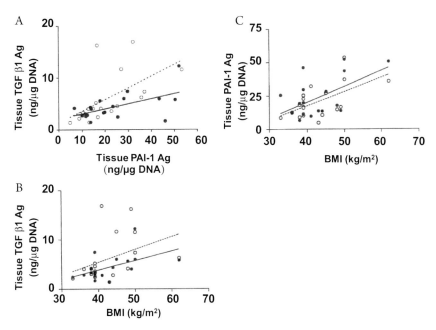

Fig. 3 Correlation between tissue TGF-beta1 antigen, tissue PAI-1 antigen and BMI of humans. Adipose tissue TGF-beta1 antigen was significantly correlated with tissue PAI-1 antigen (A) and BMI of humans (B). PAI-1 antigen was also correlated with BMI of humans (C).

Second, TGF-beta is known to suppress adipocye differentiation (Ignotz and Massagué 1985). TGF-beta inhibits human adipose tissue development and reduces the activity of a lipogenic key enzyme in newly formed fat cells (Petruschke *et al.* 1994) (Fig. 4). Over-expression of TGF-beta1 in adipose tissue of transgenic animals induces severe reduction of adipose tissue masses, which results in the failure of adipocyte differentiation (Clouthier *et al.* 1997). In addition, TGF-beta treatment induced strong inhibition of PPAR γ transactivation activity and decreased the level of mature adipocyte marker aP2 (Hong *et al.* 2007). TGF-beta1 also suppresses leptin expression in human adipose tissue (Gottschling-Zeller *et al.* 1999). Enhanced expression of TGF-beta1 in human obesity is supposed to be linked to its ability to inhibit the inflammatory cytokines such as IL-8 whose release is enhanced in obesity (Fain *et al.* 2005). These findings suggest that TGF-beta is associated with the regulation of adipocyte differentiation and secretion of the other adipokines.

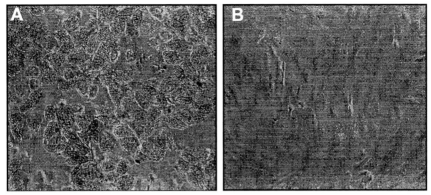

Fig. 4 TGF-beta suppresses adipocyte differentiation. Photomicrographs of cultured human adipocyte precursor cells on day 16 in the absence (A) or presence of 100 pmol/l TGF beta for 24 h (B). In control, approximately 70% of the cells were filled with lipid inclusions (A), whereas lipid accumulation was not detectable in the presence of TGF-beta (B).

MACROPHAGE INHIBITORY CYTOKINE I (MIC-I)

MIC-1 is a member of TGF-beta superfamily and is a newly reported adipokine. It is expressed in adipose tissue and secreted from adipocytes (Ding *et al.* 2009). In mouse adipose tissue, MIC-1 mRNA was detected in the major adipose depots (epididymal, perirenal, and subcutaneous) and evident in both isolated mature adipocytes and stromal-vascular cells. In addition, MIC-1 mRNA and protein secretion were evident in both human preadipocytes and differentiated adipocytes. They were also observed in human subcutaneous and visceral fat. The recombinant MIC-1 increased adiponectin secretion in differentiated human adipocytes. MIC-1 mRNA levels were positively correlated with adiponectin mRNA and negatively associated with BMI and body fat mass in human subjects (Fig. 5). Thus, MIC-1 may have a paracrine role in the modulation of adipose tissue function and body fat mass (Ding *et al.* 2009).

THROMBOSPONDIN (TSP-I)

TSP-1 is an adipokine that is highly expressed in obese subject and strongly correlated with adipose inflammation. TSP-1 is important as a major activator of TGF-beta by converting latent TGF-beta procytokine to biologically active form (Crawford *et al.* 1998). TSP-1 is expressed in adipocytes, platelets, megakaryocytes, and macrophages (Varma *et al.* 2008). TGF-beta1 also enhanced TSP-1 expression in cultured cells obtained from vascular smooth muscle or lung epithelium (RayChaudhury *et al.* 1994). TGF-beta together with TSP-1 plays important roles in pathogenesis of angiogenesis and fibrosis in many organs including liver (El-Youssef *et al.* 1999).

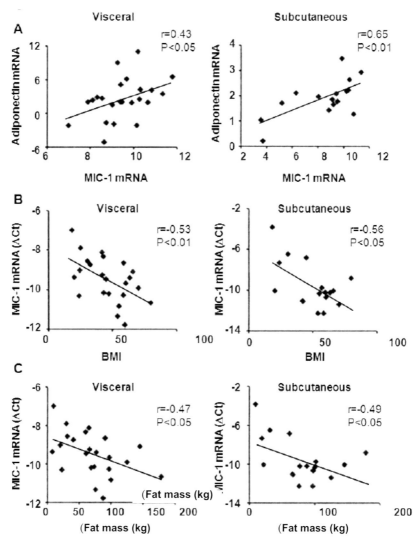

Fig. 5 MIC-1 gene and protein expression in human adipose tissue. MIC-1 mRNA levels were positively correlated with adiponectin mRNA in both human visceral and subcutaneous fat (A). MIC-1 mRNA levels were negatively associated with BMI (B) and body fat mass (C) in both visceral and subcutaneous fat.

Summary Points

- TGF-beta and its superfamily MIC-1 are produced in adipose tissue as adipokines.
- TSP-1, major activator of TGF-beta, is also produced by adipocytes.
- TGF-beta suppresses adipocyte differentiation.
- MIC-1 increases adiponectin secretion.
- TGF-beta regulates several expressions related to adipose tissue metabolism such as PAI-1, leptin and PPAR γ.
- TGF-beta plays an important role as an adipokine in pathogenesis of angiogenesis and fibrosis in many organs.

Abbreviations

BAT	:	brown adipose tissue
BMI	:	body mass index
MIC	:	macrophage inhibitory cytokine
PAI	:	plasminogen activator inhibitor
PPAR	:	peroxisome proliferator activated receptor
TGF	:	transforming growth factor
TSP	:	thrombospondin

Definition of Terms

BMI: One of the measurements of obesity. BMI equals a person's weight in kilograms divided by height in meters squared (BMI = kg/m^2).

PAI-1: The principal inhibitor of tissue plasminogen activator (tPA), urokinase (uPA), the activators of plasminogen, and fibrinolysis.

PPARγ: A member of the nuclear receptor superfamily of ligand-dependent transcription factors that is predominantly expressed in adipose tissue, adrenal gland and spleen.

TGF-beta: A multifunctional cytokine that plays important roles in immunity, cancer, heart disease, diabetes, and fibrosis. It inhibits growth of normal epithelial cells, promotes tumor angiogenesis and metastasis, and acts as a fibrogenic cytokine.

TSP-1: Interacts with blood coagulation and anticoagulant factors. It is involved in cell adhesion, platelet aggregation, cell proliferation, angiogenesis, tumor metastasis, and tissue repair.

References

Alessi, M.C. and D. Bastelica, P. Morange, B. Berthet, I. Leduc, M. Verdier, O. Geel, and I. Juhan-Vague. 2000. Plasminogen activator inhibitor 1, transforming growth factor-beta1, and BMI are closely associated in human adipose tissue during morbid obesity. Diabetes 49: 1374-1380.

Clouthier, D.E. and S.A. Comerford, and R.E. Hammer. 1997. Hepatic fibrosis, glomerulosclerosis, and a lipodystrophy-like syndrome in PEPCK-TGF-beta1 transgenic mice. J. Clin. Invest. 100: 2697-2713.

Crawford, S.E. and V. Stellmach, J.E. Murphy-Ullrich, S.M. Ribeiro, J. Lawler, R.O. Hynes, G.P. Boivin, and N. Bouck. 1998. Thrombospondin-1 is a major activator of TGF-beta1 in vivo. Cell 93: 1159-1170.

Ding, Q. and T. Mracek, P. Gonzalez-Muniesa, K. Kos, J. Wilding, P. Trayhurn, and C. Bing. 2009. Identification of macrophage inhibitory cytokine-1 in adipose tissue and its secretion as an adipokine by human adipocytes. Endocrinology 150: 1688-1696.

El-Youssef, M. and Y. Mu, L. Huang, V. Stellmach, and S.E. Crawford. 1999. Increased expression of transforming growth factor-beta1 and thrombospondin-1 in congenital hepatic fibrosis: possible role of the hepatic stellate cell. J. Pediatr. Gastroenterol. Nutr. 28: 386-392.

Fain, J.N. and D.S. Tichansky, and A.K. Madan. 2005. Transforming growth factor beta1 release by human adipose tissue is enhanced in obesity. Metabolism 54: 1546-1551.

Gottschling-Zeller, H. and M. Birgel, D. Scriba, W.F. Blum, and H. Hauner. 1999. Depot-specific release of leptin from subcutaneous and omental adipocytes in suspension culture: effect of tumor necrosis factor-alpha and transforming growth factor-beta1. Eur. J. Endocrinol. 141: 436-442.

Hong, K.M. and J.A. Belperio, M.P. Keane, M.D. Burdick, and R.M. Strieter. 2007. Differentiation of human circulating fibrocytes as mediated by transforming growth factor-beta and peroxisome proliferator-activated receptor gamma. J. Biol. Chem. 282: 22910-22920.

Ignotz, R.A. and J. Massagué. 1985.Type beta transforming growth factor controls the adipogenic differentiation of 3T3 fibroblasts. Proc. Natl. Acad. Sci. USA 82: 8530-8534.

Petruschke, T. and K. Röhrig, and H. Hauner. 1994. Transforming growth factor beta (TGF-beta) inhibits the differentiation of human adipocyte precursor cells in primary culture. Int. J. Obes. Relat. Metab. Disord. 18: 532-536.

RayChaudhury, A. and W.A. Frazier, and P.A. D'Amore. 1994. Comparison of normal and tumorigenic endothelial cells: differences in thrombospondin production and responses to transforming growth factor-beta. J. Cell Sci. 107: 39-46.

Varma, V. and A. Yao-Borengasser, A.M. Bodles, N. Rasouli, B. Phanavanh, G.T. Nolen, E.M. Kern, R. Nagarajan, H.J. Spencer III, M.J. Lee, S.K. Fried, R.E. McGehee Jr, C.A. Peterson, and P.A. Kern. 2008. Thrombospondin-1 is an adipokine associated with obesity, adipose inflammation, and insulin resistance. Diabetes 57: 432-439.

CHAPTER 17

Vaspin as an Adipokine: a Link between Obesity and Metabolic Alterations

Peter Kovacs,[1] Matthias Blüher[2] and Gabriela Aust[3],[*]

[1]Interdisciplinary Centre for Clinical Research
[2]Department of Internal Medicine
[3]Department of Surgery, Research Laboratories, University of Leipzig, Faculty of
Medicine, Germany

ABSTRACT

Vaspin (visceral adipose tissue-derived serpin) is a member of the broadly distributed serpins, a protein superfamily of serine protease inhibitors of ~500 genes, and is identical to serpin A12. The target serine protease of vaspin is still unknown.

In the context of adipokines, vaspin was first described to be expressed in visceral white adipose tissue of a rat model of obesity and type 2 diabetes mellitus (T2DM). However, adipose tissue may not be the only source of vaspin in both rodents and human; other tissues including the skin express high levels of vaspin mRNA. This explains the observation that in several individuals circulating vaspin is found despite undetectable vaspin mRNA in adipose tissue. In healthy individuals, serum vaspin correlates with body mass index (BMI), showing a U-shaped distribution with highest levels in overweight subjects suggesting a complex relationship between circulating vaspin and fat mass.

Application of vaspin in animal models indicates that it lowers glucose levels and influences insulin sensitivity, suggesting that vaspin may play a role in the pathophysiology of T2DM. Thus, identification of the proteases inhibited by vaspin may lead to the development of novel strategies in the treatment of diabetes and insulin resistance.

[*]Corresponding author

INTRODUCTION

Vaspin as an adipokine has attracted interest in obesity research after the observation that its expression in fat is related to a deterioration of metabolic parameters in a rat model of obesity. Here, we summarize the current, still incomplete knowledge of vaspin by focusing on the biochemistry, cellular sources, regulation, potential function and pathophysiolgy of this adipokine with specific emphasis on metabolic aspects.

IDENTIFICATION OF VASPIN AS AN ADIPOKINE

Vaspin was found to be expressed in visceral white, but not subdermal and brown adipose tissue of the Otsuka Long-Evans Tokushima Fatty (OLETF) rat, a model of obesity and T2DM, at the age when obesity and insulin plasma concentrations reached a peak (Hida *et al.* 2005). Furthermore, vaspin serum levels could be normalized by insulin or pioglitazone treatment. Administration of vaspin to obese mice improved glucose tolerance and insulin sensitivity and led to the reversal of altered expression of candidate genes for insulin resistance (Hida *et al.* 2005). Vaspin was therefore suggested to be a compensatory molecule in the pathogenesis of metabolic syndrome.

BIOCHEMISTRY AND CELLULAR SOURCES OF VASPIN

Biochemistry

The serpin superfamily is divided into 16 clades ranging in size from 3 to 77 members based on their phylogenic relationships. Vaspin is identical to serpin A12, a clade A serpin. In their native, active state, serpins are monomeric proteins consisting of a variably glycosylated ~350-500 amino acid polypeptide chain.

Serpins inhibit serine proteases by a unique suicide mechanism. They contain an exposed reactive centre loop (RCL) that is presented to the target protease as a pseudosubstrate. The amino acid sequence of the RCL determines which serin protease will be inhibited by the serpin. Binding of the protease to the RCL induces conformation changes of the serpin, which thus deforms the reactive centre of the protease and inactivates it. The target protease of vaspin has not been identified yet. It has been shown that recombinant human vaspin failed to inhibit protease activity of trypsin and other known common proteases (Hida *et al.* 2005).

Cellular Sources of Vaspin

According to the tissue expression pattern of vaspin mRNA in mouse and man (BioGPS, http://biogps.gnf; GeneNote, http://bioinfo2.weizmann.ac.il/cgi-bin/

genenote), the expression levels in almost all tissues including white fat are rather low. Lean human subjects (BMI< 25 kg/m^2) had undetectable vaspin mRNA in visceral and subcutaneous fat (Klöting *et al.* 2006). The frequency of subjects with detectable vaspin mRNA expression in visceral white fat increased from overweight (BMI ≥ 25-30 kg/m^2) to obese (BMI > 30 kg/m^2) individuals. In a comparison of mRNA distribution for 60 proteins in fat vs. the non-fat cells in human omental adipose tissue, vaspin mRNA was found to be enriched in non-fat cells in addition to vaspin mRNA positive preadipocytes and differentiated adipocytes.

Skin is the only tissue with prominent vaspin mRNA levels, in both mouse and human. Changes in the proteolytic balance between proteases and their inhibitors as serpins disrupt the integrity and protective function of the skin and contribute to inflammatory skin disease (Meyer-Hoffert 2009).

PHYSIOLOGY OF VASPIN

Serum Level of Vaspin

Compared to levels of other adipokines, circulating vaspin levels are rather low and range between 0.22 and 0.42 ng/ml (Table 1). Because lean subjects, negative for vaspin mRNA in white fat, show detectable vaspin serum levels, tissues other than adipose tissue may be the main source of circulating vaspin.

Age

In healthy lean individuals (n = 164), vaspin serum concentration significantly correlates with age (r = 0.33, p < 0.05), whereas no relationship between circulating vaspin and age was found in subgroups of overweight, obese subjects and individuals with T2DM (unpublished data). As demonstrated in animal models, age might predict serum vaspin levels (Gonzalez *et al.* 2009).

Gender

In most studies, circulating vaspin levels were higher in females than in males at least in lean and healthy individuals (Youn *et al.* 2007, Seeger *et al.* 2008, Loeffelholz *et al.* 2009). In particular, studies that included patients with different chronic diseases failed to observe sexual dimorphism in serum vaspin concentrations (Aust et al. 2009, Handisurya et al. 2009, Ye et al. 2009).

It has been postulated that gender-related differences are due to distinct fat distribution or estrogen-mediated induction of vaspin in females (Hida *et al.* 2005, Seeger *et al.* 2008). However, no significant differences in serum vaspin were found in pre- and post-menopausal women (Handisurya 2009 4837/id /d). Consistently, white adipose tissue vaspin mRNA levels were unchanged in gonadectomized rats

Table 1 Correlation of clinical and anthropometric characteristics with vaspin, adiponectin, and leptin serum concentrations

		vaspin	leptin	adiponectin
reference sera level		0.22-0.42 ng/ml		8.1-19.5 µg/ml
	male		1.2-8.9 ng/ml	
	female		8-24 ng/ml	
Correlation to				
sex		male < female?	male < female	male < female
age		positive?	positive	positive
BMI		U-shaped (Youn *et al.* 2007, Aust *et al.* 2009) positive (Suleymanoglu *et al.* 2009, Handisurya *et al.* 2009, Cho *et al.* 2009)	positive	negative
lipids	total cholesterol	no		
	triglycerides	positive (Suleymanoglu *et al.* 2009)		negative
	low-density lipoprotein (LDL) cholesterol	no		
	high-density lipoprotein (HDL) cholesterol	no		positive
T2DM		unclear		
HOMA of insulin resistance		positive (Youn *et al.* 2007, Suleymanoglu *et al.* 2009, Handisurya *et al.* 2009, Cho *et al.* 2009) no (Loeffelholz *et al.* 2009)		
Leptin		positive (Aust *et al.* 2009)		
inflammation	C-reactive protein	positive? (Aust *et al.* 2009)		negative
	leukocyte blood count	no		negative
atherosclerosis	statin treatment	lowering (Aust *et al.* 2009)		
	stenosis	no		

(Gonzalez *et al.* 2009). Pregnant rats also did not show changes in adipose tissue vaspin mRNA despite marked changes in gonadal function. These data do not support a potential role of estrogens in the regulation of vaspin.

Exercise

Physical training significantly increases circulating vaspin, which may mediate the improvement of insulin resistance after exercise (Youn *et al.* 2007). Recently, we determined serum vaspin in 40 healthy young men who were randomly assigned to either antioxidant treatment (vitamin C and E) or no supplementation during a standardized 4 wk physical training program (unpublished data). Vaspin was decreased by exercise-induced oxidative stress, but not by exercise-associated improvement in insulin sensitivity (unpublished data).

Diet-induced Weight Loss

In the recent DIRECT study in which a 2 yr dietary weight loss program was evaluated, serum vaspin significantly decreased in response to moderate weight loss following different hypocaloric diets (unpublished data).

PATHOPHYSIOLOGY

Obesity

Obesity was associated with increased serum vaspin in young Korean men (Cho *et al.* 2009). Vaspin levels were positively correlated with BMI as well as triglycerides, fasting insulin and insulin resistance in pubertal obese children (Suleymanoglu *et al.* 2009). On the other hand, several studies failed to show a simple correlation between serum vaspin levels and BMI (Aust *et al.* 2009).

Youn *et al.* found this correlation only in male subjects of normal glucose tolerance (NGT) (2007). In healthy females, lowest serum vaspin levels were measured in lean subjects and highest in overweight subjects, whereas the levels in obese individuals were not different from those in lean ones. Such a U-shaped distribution of serum vaspin was also observed in carotid stenosis patients (Aust *et al.* 2009).

Further evidence for a link between vaspin and obesity came from recent studies investigating circulating vaspin after gastric surgery to reduce body weight. Following Roux-en-Y gastric bypass (RYGB) surgery, reduction in vaspin correlated with the reduction of BMI, circulating leptin, and insulin levels and with the amelioration of insulin sensitivity (Handisurya 2009 4837/id).

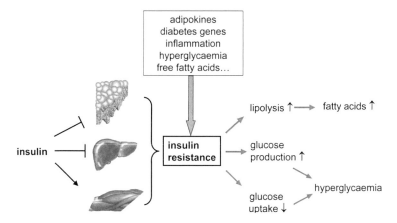

Fig. 1 Key features of insulin resistance. Insulin resistance affects insulin action in major target tissues (adipose tissue, lipolysis; liver, glucose production; skeletal muscle, glucose uptake). The resulting increased circulating concentrations of fatty acids and glucose lead to a deterioration in both insulin secretion and insulin resistance.

Vaspin: a Link between Obesity and Metabolic Alterations?

In contrast to obese subjects, vaspin mRNA is not detectable in either visceral or subcutaneous fat of lean NGT individuals. Thus, induction of vaspin mRNA in adipose tissue could represent a compensatory mechanism associated with obesity, severe insulin resistance, and T2DM (Klöting *et al.* 2006). Furthermore, caloric restriction resulted in decreased white adipose tissue vaspin mRNA in rats, which was reversed by leptin administration, suggesting that vaspin is strongly regulated by nutritional status (Gonzalez *et al.* 2009). Taken together, these studies demonstrate that vaspin might be a novel candidate linking human obesity to its related metabolic disorders and suggest that rather than affecting obesity directly, vaspin exerts insulin-sensitizing effects in the state of obesity.

Supporting the latter notion, the association between percentage change of vaspin and homeostasis model assessment (HOMA) of insulin resistance remained significant even after adjustment for RYGB-induced changes in BMI (Handisurya 2009 4837/id). Moreover, correlation between changes in vaspin and C-peptide suggests a link between vaspin and RYGB-induced attenuation of pancreatic insulin secretion. However, no association between serum vaspin and insulin resistance was found in non-diabetic subjects, indicating no major role of vaspin in the regulation of insulin sensitivity in healthy individuals (Loeffelholz *et al.* 2009).

The vaspin gene represents a plausible candidate gene possibly affecting the risk of T2DM. Kempf *et al.* (2008) genotyped 25 single nucleotide polymorphisms (SNPs) in vaspin in 2759 subjects of the population-based, cross-sectional German study (KORA) and assessed the association with diabetes and obesity. One SNP, rs2236242, was significantly associated with T2DM.

Interestingly, vaspin influences insulin-induced glucose uptake *in vivo*, but not *in vitro*. Vaspin probably modulates insulin action only in the presence of its target proteases, which are most likely triggering altered insulin sensitivity. Therefore, identification of vaspin´s target protease is the major challenge for future studies related to vaspin. Unraveling the proteases might lead to the development of novel anti-diabetic therapy, which may improve insulin sensitivity in patients with T2DM.

Summary Points

- Vaspin is a member of the broadly distributed serpins, a protein superfamily of serine proteinase inhibitors.
- The target serine protease of vaspin is still unknown.
- Adipose tissue may not be the only source of vaspin in normal human; other tissues such as the skin express higher levels of vaspin mRNA. Serum vaspin is also detectable in subjects negative for vaspin mRNA in visceral or subcutaneous fat.
- Vaspin does not seem to be directly linked to obesity, but may potentially link obesity to its related metabolic alterations.
- The insulin-sensitizing effect of vaspin is only observed *in vivo*, indicating the relevance of the still unidentified target protease in the pathophysiology of insulin resistance.

Abbreviations

BMI	:	body mass index
HOMA	:	homeostasis model assessment
NGT	:	normal glucose tolerance
OGTT	:	oral glucose tolerance test
OLETF rat	:	Otsuka Long-Evans Tokushima Fatty rat
RYGB	:	Roux-en-Y gastric bypass
T2DM	:	type 2 diabetes mellitus

Definition of Terms

Insulin: A hormone produced by the beta-cells in the pancreas, it enables cells to absorb glucose and turn it into energy.

Insulin resistance: Failure of insulin-sensitive cells as muscle and fat to respond properly to insulin.

Type 2 diabetes mellitus (T2DM): Disorder that results from insulin resistance, altered insulin secretion or combination of both, in which glucose accumulates in the blood and leads to clinical complications.

Normal glucose tolerance (NGT): Normal tolerance of glucose according to the oral glucose tolerance test (OGTT), which determines how quickly glucose is cleared from the blood.

Homeostasis model assessment (HOMA): A method to quantify insulin resistance and beta-cell function.

References

Aust, G. and O. Richter, S. Rohm, C. Kerner, J. Hauss, N. Klöting, K. Ruschke, B.S. Youn, and M. Blüher. 2009. Vaspin serum concentrations in patients with carotid stenosis. Atherosclerosis 204: 262-266.

Cho, J.K. and T.K. Han, and H.S. Kang. 2009. Combined effects of body mass index and cardio/respiratory fitness on serum vaspin concentrations in Korean young men. Eur. J. Appl. Physiol. Epub.

Gonzalez, C.R. and J.E. Caminos, M.J. Vazquez, M.F. Garces, L.A. Cepeda, A. Angel, A.C. Gonzalez, M.E. Garcia-Rendueles, S. Sangiao-Alvarellos, M. Lopez, S.B. Bravo, R. Nogueiras, and C. Dieguez. 2009. Regulation of visceral adipose tissue-derived serine protease inhibitor by nutritional status, metformin, gender and pituitary factors in rat white adipose tissue. J. Physiol. 587: 3741-3750.

Handisurya, A. and M. Riedl, G. Vila, C. Maier, M. Clodi, T. Prikoszovich, B. Ludvik, G. Prager, A. Luger, and A. Kautzky-Willer. 2010. Serum vaspin concentrations in relation to insulin sensitivity following RYGB-Induced weight loss. Obes. Surg. 20: 198–203.

Hida, K. and J. Wada, J. Eguchi, H. Zhang, M. Baba, A. Seida, I. Hashimoto, T. Okada, A. Yasuhara, A. Nakatsuka, K. Shikata, S. Hourai, J. Futami, E. Watanabe, Y. Matsuki, R. Hiramatsu, S. Akagi, H. Makino, and Y. S. Kanwar. 2005. Visceral adipose tissue-derived serine protease inhibitor: a unique insulin-sensitizing adipocytokine in obesity. Proc. Natl. Acad. Sci. USA 102: 10610-10615.

Kempf, K. and B. Rose, T. Illig, W. Rathmann, K. Strassburger, B. Thorand, C. Meisinger, H.E. Wichmann, C. Herder, and C. Vollmert. 2008. Vaspin (serpina12) genotypes and risk of type 2 diabetes: results from the MONICA/KORA studies. Exp. Clin. Endocrinol. Diabetes. Epub.

Klöting, N. and J. Berndt, S. Kralisch, P. Kovacs, M. Fasshauer, M.R. Schon, M. Stumvoll, and M. Bluher. 2006. Vaspin gene expression in human adipose tissue: association with obesity and type 2 diabetes. Biochem. Biophys. Res. Commun. 339: 430-436.

Loeffelholz, C. and M. Moehlig, A.M. Arafat, F. Isken, J. Spranger, K. Mai, H.S. Randeva, A. Pfeiffer, and M. Weickert. 2009. Circulating vaspin is unrelated to insulin sensitivity in a cohort of non-diabetic humans. Eur. J. Endocrinol. Epub.

Meyer-Hoffert, U. 2009. Reddish, scaly, and itchy: how proteases and their inhibitors contribute to inflammatory skin diseases. Arch. Immunol. Ther. Exp. (Warsz.). 57: 345-354.

Seeger, J. and M. Ziegelmeier, A. Bachmann, U. Lossner, J. Kratzsch, M. Bluher, M. Stumvoll, and M. Fasshauer. 2008. Serum levels of the adipokine vaspin in relation to metabolic and renal parameters. J. Clin. Endocrinol. Metab. 93: 247-251.

Suleymanoglu, S. and E. Tascilar, O. Pirgon, S. Tapan, C. Meral, and A. Abaci. 2009. Vaspin and its correlation with insulin sensitivity indices in obese children. Diabetes Res. Clin. Pract. 84: 325-328.

Ye, Y., X. Hou, X. Pan, J. Lu, and W. Jia. 2009. Serum vaspin level in relation to postprandial plasma glucose concentration in subjects with diabetes. Chin. Med. J. 122: 2530–2533.

Youn, B.S. and N. Klöting, J. Kratzsch, J.W. Park, N. Lee, E.S. Song, K. Ruschke, A. Oberbach, M. Fasshauer, M. Stumvoll, and M. Blüher. 2007. Serum vaspin concentrations in human obesity and type 2 diabetes. Diabetes 57: 372-377.

CHAPTER 18

Visfatin as an Adipokine

Ariadne Malamitsi-Puchner[1,*] and Despina D. Briana[1]

[1]Neonatal Division, 2nd Department of Obstetrics and Gynecology, Athens University Medical School, Athens, Greece

ABSTRACT

Visceral fat accumulation has been shown to play crucial roles in the development of obesity-related disorders. Additionally, insulin resistance in visceral fat obesity is thought to be one of the key abnormalities related to metabolic disorders. Given these clinical findings, adipocyte functions have been intensively investigated in the past 10 years, and adipocytes have been revealed to act as endocrine cells, secreting various bioactive substances termed adipokines. Current evidence suggests that these secretory products are implicated in modulation of appetite, insulin sensitivity, energy expenditure, inflammation and immunity. One of the most recent proteins shown to be highly expressed in adipose tissue is visfatin, originally identified as pre-B-cell colony-enhancing factor (PBEF). Visfatin appears to be preferentially produced by the visceral adipose tissue and has insulin-mimetic actions. Therefore, visfatin may be involved in the development of obesity-related diseases. In addition, recent data indicate that visfatin may exert various cardiovascular effects and may be implicated in inflammatory disease states beyond insulin resistance, such as acute lung injury and inflammatory bowel disease. Some of these observed actions of visfatin indicate that this secreted protein may be an interesting therapeutic target. Several studies, however, suggest that our understanding of visfatin is still speculative.

INTRODUCTION

Visceral obesity is associated with insulin resistance, often leading to the development of type 2 diabetes mellitus (T2DM) and metabolic syndrome

---

*Corresponding author

(Matsuzawa 2006). Furthermore, although evidence so far is limited, obesity might be associated with inflammatory and immune-mediated disorders (Tilg and Moschen 2008).

One of the most recent proteins highly expressed in AT is visfatin, originally identified as pre-B-cell colony-enhancing factor (PBEF). PBEF was originally cloned during a research for novel cytokine-like molecules secreted from human peripheral blood lymphocytes (Samal *et al.* 1994). PBEF is a growth factor for early B lymphocytes and is primarily expressed in bone marrow, liver, and muscle (Samal *et al.* 1994). Furthermore, PBEF has been linked to a variety of cellular processes including the following: (1) it acts as a biomarker of acute lung injury; (2) it is up-regulated in infected fetal membranes; (3) it inhibits neutrophil apoptosis in experimental inflammation and clinical sepsis; and (4) it participates in the maturation of vascular smooth muscle cells (Stephens and Vidal-Puig 2006).

The recent discovery that PBEF, now also termed visfatin, is highly expressed in visceral fat and that its circulating levels correlate with obesity is of great interest to many researchers (Fukuhara *et al.* 2005). The exogenous administration of visfatin reduced serum glucose levels and improved insulin sensitivity in both healthy and obese diabetic mice (Fukuhara *et al.* 2005). It was thus proposed that visfatin may act at the level of the insulin receptor (Fukuhara *et al.* 2005); however, no follow-up studies confirmed these original observations (Stephens and Vidal-Puig 2006). Instead, a growing body of evidence now suggests that visfatin is involved in inflammation and innate immunity (Moschen *et al.* 2007) and may exert a variety of cardiovascular effects, including endothelial dysfunction, angiogenesis, and acute cardioprotection (Hausenloy 2009). Interestingly, it was hypothesized that visfatin may not only provide a potential new target for acute cardioprotection, but it may also act as an anti-diabetic agent, thereby offering a potentially novel drug target for the diabetic patient (Hausenloy 2009).

This chapter considers the main roles of visfatin in pathophysiology with particular attention to its role in obesity-associated disorders and inflammation.

VISFATIN IN OBESITY AND OBESITY-ASSOCIATED DISEASES

PBEF was also termed visfatin when the levels of this protein were shown to be higher in visceral AT than in subcutaneous AT in mice and humans (Fukuhara *et al.* 2005). Circulating visfatin levels strongly correlated with the amount of visceral fat in both humans and mice. Moreover, recombinant visfatin directly binds to the insulin receptor, leading to enhanced glucose uptake *in vitro* and *in vivo*. This suggests a possible role for visfatin production as a compensatory response in diet- or obesity-induced insulin resistance (Fukuhara *et al.* 2005). This original observation was supported by studies demonstrating that visfatin levels are increased in T2DM, regardless of administration of hypoglycemic medication or the presence of obesity (Chen *et al.* 2005, Dogru *et al.* 2007). Accordingly, elevated visfatin concentrations in morbidly obese subjects were reduced after weight loss, probably relating to changes in insulin resistance over time (Haider *et al.* 2006a). Moreover, up-regulation of visfatin is evident in obese children (Haider *et al.* 2006b) and in patients with polycystic ovary syndrome (Tan 2006). Additionally, Haider *et al.* showed that elevated visfatin levels in patients with T1DM can be lowered by regular physical exercise (Haider *et al.* 2006c). The authors concluded that elevated visfatin levels may be a consequence of intermittent or continued hyperglycemia or lack of physiological insulin exposure and may represent a feedback mechanism of glucose homeostasis. Indeed, an acute regulation of visfatin by glucose and insulin was demonstrated in healthy subjects and in isolated adipocytes (Haider et al 2006d).

In support of the above studies, it has been suggested that drugs such as rosiglitazone and fenofibrate may prevent T2DM by regulating visfatin and tumor necrosis factor α (TNF-α) (Choi *et al.* 2005). Furthermore, basal visfatin release was enhanced *in vitro* and in healthy subjects by rosiglitazone treatment, which may contribute to its anti-diabetic pharmacological action (Haider *et al.* 2006e). This effect was acutely reduced by elevation of free fatty acids, which suggests that visfatin bioactivity may be modulated by food intake.

However, the actual role that visfatin plays in the pathophysiology of metabolic syndrome is presently unclear, as demonstrated by the fact that clinical studies measuring plasma visfatin levels in humans have yielded conflicting results, which partly may relate to the methodology used and the circadian variation in visfatin secretion (Stephens and Vidal-Puig 2006, Sethi and Vidal-Puig 2005). In this respect, subsequent reports demonstrated lack of correlation between plasma visfatin levels and visceral fat mass (Berndt *et al.* 2005). In fact, no correlations were observed between plasma visfatin concentrations and various parameters of insulin sensitivity (Berndt *et al.* 2005, de Luis *et al.* 2009, Chang *et al.* 2010). However, it was suggested that visfatin is a pro-inflammatory marker of adipose tissue, associated with systemic insulin resistance and hyperlipidemia (Chang *et al.* 2010). Additional reports in animals and humans also observed that visfatin gene expression was not associated with most parameters of the metabolic syndrome (Kloting *et al.* 2005, de Luis *et al.* 2009). In line, Dogru *et al.* 2007 demonstrated that visfatin levels did not correlate with body mass index in subjects with T2DM, impaired glucose tolerance and controls (Dogru *et al.* 2007). More recent data also indicate that visfatin is not related to glucose metabolism and perhaps is related to a low grade of inflammation (de Luis *et al.* 2009).

Larger-scale studies are clearly required in order to ascertain the pathophysiological significance of plasma visfatin levels in diabetic and obese patients. It remains to be established whether visfatin production is a compensatory response to tissue-specific insulin resistance or more simply a marker of tissue-specific inflammatory-cytokine action.

PRO-INFLAMMATORY AND IMMUNO-MODULATING PROPERTIES OF VISFATIN

Over the last decade, much evidence has emerged that obesity is closely linked to systemic inflammation (Wellen and Hotamisligil 2005). Accordingly, visfatin was recently shown to exert pro-inflammatory activities, since it dose-dependently up-regulated the production of the pro-inflammatory cytokines IL-1β, IL-6 and TNF-α in human monocytes (Moschen 2007). Several studies suggest that visfatin is primarily a pro-inflammatory cytokine, as its serum/plasma levels are increased in various inflammatory disorders, such as inflammatory bowel disease and chronic obstructive pulmonary disease (Moschen 2007). Moreover, the expression of visfatin was increased in the macrophages of unstable atherosclerotic plaques, in the synovial tissue of patients with rheumatoid arthritis, and in the neutrophils of septic patients (Moschen 2007).

Overall, there is a growing body of evidence, challenging the role of visfatin as an adipokine and supporting its involvement in inflammation and innate immunity.

VISFATIN IN PREGNANCY-ASSOCIATED INSULIN RESISTANCE

Visfatin transcript and protein are present in fetal membranes and the human placenta (Ognjanovic and Bryant-Greenwood 2002). Visfatin concentrations were either comparable between non-pregnant women and women in the third trimester of pregnancy, suggesting that the placenta may not contribute to maternal circulating visfatin during pregnancy (Hu *et al.* 2008), or gradually increased, probably compensating for the gradual increase in insulin resistance (Telejko *et al.* 2009). Interestingly, serum visfatin concentrations in the first trimester of pregnancy may predict insulin sensitivity in the second trimester, but this close association later disappears, possibly because of an increase in visfatin secretion by the placenta (Mastorakos *et al.* 2007).

On the other hand, while some investigators reported lower maternal visfatin concentrations in gestational diabetes mellitus, others demonstrated increased visfatin levels with a worsening degree of maternal glucose intolerance in the third trimester of pregnancy (reviewed in Briana and Malamitsi-Puchner 2009). Similarly, maternal serum visfatin levels were elevated or decreased in preeclampsia, regardless of the severity of the disease or maternal body mass index (reviewed in Briana and Malamitsi-Puchner 2009). Finally, a recent study from our group demonstrated higher maternal visfatin concentrations at term in the IUGR state, suggesting that visfatin may be a novel marker up-regulated in this pregnancy disorder (Malamitsi-Puchner *et al.* 2007).

Taken together, there are huge discrepancies in reports of circulating visfatin in pregnancy, possibly due to large variations in serum visfatin data or due to a paracrine/autocrine action of visfatin. It may be speculated, however, that, since visfatin improves glucose tolerance through insulin-mimetic effects, up-regulation of maternal visfatin concentrations in insulin resistance–associated pregnancy complications may be part of a physiological feedback mechanism, improving insulin signaling.

Table I Key features of visfatin

- Visfatin is identical with pre-B-cell colony-enhancing factor.
- Visfatin is highly expressed in adipose tissue and up-regulated in obesity.
- Visfatin may mimic the glucose-lowering effect of insulin.
- Visfatin may be an adipokine with pro-inflammatory and immunomodulating properties.
- Visfatin may be responsible for a number of different cardiovascular effects.
- Visfatin is produced by the human placenta and is differentially regulated in pregnancy-associated disorders.
- Visfatin may be a useful drug target for diabetes and cardiovascular disease.

Summary Points

- The preferential accumulation of visceral fat is a strong and independent predictor for the development of metabolic syndrome.
- The adipokine visfatin may facilitate the accumulation of fat in the intra-abdominal depot and may be implicated in the development of insulin resistance.
- Visfatin is able to mimic the anti-hyperglycemic effects of insulin.
- The origin of circulating visfatin is not yet defined.
- Visfatin may exert several cardiovascular effects.
- Visfatin may be a new marker of inflammation.

Abbreviations

AT   :   adipose tissue
PBEF   :   pre-B-cell colony-enhancing factor
TNF-α   :   tumor necrosis factor-α
T2DM   :   type 2 diabetes mellitus

Definition of Terms

Adipokines: Soluble mediators that are mainly, but not exclusively, produced by adipocytes and exert their biological function in an autocrine, paracrine or systemic manner.

Adipose tissue: A complex endocrine and immune organ involved in biological processes, such as insulin sensitivity, appetite, endocrine functions, inflammation and immunity.

Metabolic syndrome: A variety of common disorders, including hyperglycemia, hyperlipidemia, hypertension and visceral obesity, which predispose individuals to the development of cardiovascular disease.

Pre-B-cell colony-enhancing factor (PBEF): A growth factor for early B lymphocytes.

Visceral obesity: The accumulation of adipose tissue inside the abdominal cavity.

References

Berndt, J. 2005. Plasma visfatin concentrations and fat depot-specific mRNA expression in humans. Diabetes 54: 2911-2916.

Briana, D.D. and A. Malamitsi-Puchner. 2009. Reviews: adipocytokines in normal and complicated pregnancies. Reprod. Sci. 16: 921-937.

Chang, Y.C. *et al.* 2010. The relationship of visfatin/pre-B-cell colony-enhancing factor/ nicotinamide phosphoribosyltransferase in adipose tissue with inflammation, insulin resistance, and plasma lipids. Metabolism 59: 93-99.

Chen, M.P. *et al.* 2005. Elevated plasma level of visfatin/pre-B-cell colony enhancing-factor in patients with type 2 diabetes mellitus. J. Clin. Endocrinol. Metab. 18: 295-299.

Choi, K.C. *et al.* 2005. Effect of PPAR alpha and gamma on the expression of visfatin, adiponectin and TNF alpha in visceral fat of OLETF rats. Biochem. Biophys. Res. Commun. 336: 747-753.

de Luis, D.A. *et al.* 2009. Relation of visfatin to cardiovascular risk factors and adipocytokines in patients with impaired fasting glucose. Nutrition (in press).

Dogru, T. *et al.* 2007. Plasma visfatin levels in patients with newly diagnosed and untreated type 2 diabetes mellitus and impaired glucose tolerance. Diabetes Res. Clin. Pract. 76: 24-29.

Fukuhara, A. *et al.* 2005. Visfatin: a protein secreted by visceral fat that mimics the effects of insulin. Science 307: 426-430.

Haider, D.G. *et al.* 2006a. Increased plasma visfatin concentrations in morbidly obese subjects are reduced after gastric banding. J. Clin. Endocrinol. Metab. 91: 1578-1581.

Haider, D.G. *et al.* 2006b. The adipokine visfatin is markedly elevated in obese children. J. Pediatr. Gastroenterol. Nutr. 43: 548-549.

Haider, D.G. *et al.* 2006c. Exercise training lowers plasma visfatin concentrations in patients with type 1 diabetes. J. Clin. Endocrinol. Metab. 91: 4702-4704.

Haider, D.G. *et al.* 2006d. The release of the adipocytokine visfatin is regulated by glucose and insulin. Diabetologia 49: 1909-1914.

Haider, D.G. *et al.* 2006e. Free fatty acids normalize a rosiglitazone-induced visfatin release. Am. J. Physiol. Endocrinol. Metab. 291: E885-E890.

Hausenloy, D.J. 2009. Drug discovery possibilities from visfatin cardioprotection? Curr. Opin. Pharmacol. 9: 202-207.

Hu, W. *et al.* 2008. Serum visfatin levels in late pregnancy and pre-eclampsia. Acta. Obstet. Gynecol. Scand. 87: 413-418.

Kloting, N. and I. Kloting. 2005. Visfatin: gene expression in isolated adipocytes and sequence analysis in obese WOKW rats compared with lean control rats. Biochem. Biophys. Res. Commun. 332: 1070-1072.

Malamitsi-Puchner, A. 2007. Perinatal circulating visfatin levels in intrauterine growth restriction. Pediatrics 119: E1314-1318.

Mastorakos, G. *et al.* 2007. The role of adipocytokines in insulin resistance in normal pregnancy: visfatin concentrations in early pregnancy predict insulin sensitivity. Clin. Chem. 53: 1477-1483.

Matsuzawa, Y. 2006. The metabolic syndrome and adipocytokines. FEBS Lett. 580: 2917-2921.

Moschen, A.R. and A. Kaser, B. Enrich, B. Mosheimer, M. Theurl, H. Niederregger, and H. Tilg. 2007. Visfatin, an adipocytokine with proinflammatory and immunomodulating properties. J. Immunol. 178: 1748-1758.

Ognjanovic, S. and G.D. Bryant-Greenwood. 2002. Pre-B-cell colony-enhancing factor, a novel cytokine of human fetal membranes. Am. J. Obstet. Gynecol. 187: 1051-1058.

Samal, B. *et al.* 1994. Cloning and characterization of the cDNA encoding a novel human pre-B-cell colony-enhancing factor. Mol. Cell. Biol. 14: 1431-1437.

Sethi, J. and A. Vidal-Puig. 2005. Visfatin: the missing link between intra-abdominal obesity and diabetes? Trends Mol. Med. 11: 344-347.

Stephens, J.M. and A.J. Vidal-Puig. 2006. An update on visfatin/pre-B cell colony-enhancing factor, an ubiquitously expressed, illusive cytokine that is regulated in obesity. Curr. Opin. Lipidol. 17: 128-131.

Tan, B.K. *et al.* 2006. Increased visfatin messenger ribonucleic acid and protein levels in adipose tissue and adipocytes in women with polycystic ovary syndrome: parallel increase in plasma visfatin. J. Clin. Endocrinol. Metab. 91: 5022-5028.

Telejko, B. *et al.* 2009. Visfatin in gestational diabetes: Serum level and mRNA expression in fat and placental tissue. Diab. Res. Clin. Pract. 84: 68-75.

Tilg, H. and A.R. Moschen. 2008. Role of adiponectin and PBEF/visfatin as regulators of inflammation: involvement in obesity-associated diseases. Clin. Sci. 114: 275-288.

Wellen, K.E. and G.S. Hotamisligil. 2005. Inflammation, stress, and diabetes. J. Clin. Invest. 115: 1111-1119.

Section II
General Cellular Aspects

CHAPTER 19

Thermal Stress and Adipokines Expression

Umberto Bernabucci,[1,*] Patrizia Morera[1] and Loredana Basiricò[1]

[1]Department of Animal Science, Università della Tuscia-Viterbo, Via San C. De Lellis, s.n.c., 01100, Viterbo, Italy

ABSTRACT

Thermal stress induces profound alterations in cell physiology. These alterations are constituted by an extensive reencoding of gene expression and biochemical adaptive responses, characterized by the impairment of major cellular functions and by an adaptive reprogramming of the cell metabolism. Among cells, adipocytes are of particular interest, since they fulfil the dual role of energy storage and thermal insulation. Because of its proximity to the skin surface, the temperature of subcutaneous adipose tissue may be strongly influenced by environmental temperature. Heat and cold stress strongly affect adipocyte metabolism. Recent studies report effects of thermal stress (heat and cold) on biology of some adipokines. Heat stress exposure has been found to induce an up-regulation of leptin and down-regulation of adiponectin. In contrast, cold exposure is responsible for a down-regulation of leptin and, to some extent, for the up-regulation of adiponectin. This may suggest a different role of those adipokines in the acclimation of adipose tissue to thermal stress. To date, the biological and physiological meanings of adipokine responses to heat or cold exposure are still to be clarified. Further studies are encouraged for better understanding the mechanisms through which thermal stress modifies adipokine biology, and the physiological and biological meanings of adipokine changes.

*Corresponding author

INTRODUCTION

Upon thermal stress (TS), cell physiology is profoundly altered. Alterations are constituted by an extensive reencoding of gene expression and biochemical adaptive responses, characterized by the impairment of major cellular functions and by an adaptive reprogramming of the cell metabolism (Fujita 1999). The extent of the alterations depends on the severity of the stress and may lead to cell death (Lacetera *et al.* 2009). Biologically, the ability to survive and adapt to TS appears to be a fundamental requirement of cell life, as cell stress responses are ubiquitous among both eukaryotes and prokaryotes (Lindquist 1986). Cells from all organisms respond to physiologically relevant variations in temperature, by rapidly increasing the expression and the synthesis of a selected group of proteins, the heat shock proteins (HSP). The resultant increase and accumulation of HSP give the stressed cell added protection, allowing it to survive. In medicine, evidence is mounting that the ability to survive and adapt to severe systemic physiological stress is critically dependent on the ability of cells to mount an appropriate compensatory stress response (Sonna *et al.* 2002).

Two forms of TS, heat and cold, that show important similarities as well as critical differences in cellular response, can be identified. The differences include the specific profile of gene expression, the temporal sequence of expression and the cellular activity altered (Sonna *et al.* 2002).

In this article, we review our current knowledge on the effects of TS on gene expression and secretion of some adipokines and speculated how TS of adipocytes may therefore help to shed some light on the modulation and biological actions of the adipose-derived proteins and to characterize their metabolic roles in the human stress response.

THERMAL STRESS AND ADIPOCYTES

It is widely accepted that changes in gene expression are an integral part of the cellular response to TS. Although HSP are perhaps the best-studied examples of genes whose expression is affected by heat and cold shock, it has become apparent that TS also leads to induction of a substantial number of genes not traditionally considered to be HSP. Some of these genes are affected by a wide variety of stressors and probably represent a non-specific cellular response to stress, whereas others may eventually be found to be specific to certain type of cells and stress.

In recent years a growing number of reports have been published on the effects of TS, on pathological and normal mammalian cells and on intact animal (Sonna *et al.* 2002). Among cells markedly affected by exposure to TS, adipocytes are of particular interest, since they fulfil the dual role of energy storage and thermal insulation. As a consequence of its insulating function, and its proximity to the skin surface, the temperature of subcutaneous adipose tissue is strongly influenced by environmental temperature and varies over a wide range, particularly during cold exposure (Zeyl *et al.* 2004).

Recently Jiang *et al.* (2007) observed that heat stress directly stimulates adipocyte lipolysis in rat, which may be a cellular basis for elevated circulating fatty acids under hyperthermal conditions. In addition, Ezure and Amano (2009) found that heat stimulation reduces adipogenesis in 3T3-L1 pre-adipocytes by decreasing the gene expression of early adipogenesis.

Cold exposure is well known to increase energy expenditure and thermogenesis by stimulating adipocytes to increase the release of free fatty acids (FFA) to be used as energy supplements and consumptive thermogenesis (Yoda *et al.* 2001). The thermogenic response of brown adipocytes to cold exposure is mediated in part by adrenergic stimulation and thyroid hormone (Yoda *et al.* 2001). Granneman and coworkers (2003) reported that white adipocytes also contribute to thermogenesis. Therefore, in both adipose tissues, cold exposure leads to an immediate fatty acid release from adipose tissue to provide fuel for thermoregulatory heat production. This observation supports the hypothesis that white adipocytes act primarily as a net FFA provider, whereas brown adipocytes act as a net FFA user.

Emerging evidence shows that changes in environmental temperature may affect adipokine gene expression and secretion. Few data are available on the effects of TS on adipokines. In the following paragraphs we report results from our and other studies showing as adipokines can be affected by TS.

EFFECTS OF HEAT STRESS ON ADIPOKINES EXPRESSION

Very few data have been reported on the effect of extreme temperatures on food intake in human subjects (Westerterp-Plantenga 1999). Conversely, a considerable amount of evidence indicates that hot climatic conditions can modify appetite, energy intake and metabolism in animals (Bernabucci *et al.* 2010). Earlier studies demonstrated that leptin and adiponectin, expressed predominantly by adipocytes, are crucial in the regulation of energy balance and carbohydrate/lipid metabolism in humans and animals (Trayurn *et al.* 1995), suggesting a modulating role of these adipokynes in the metabolic adaptive responses observed during changes in environmental temperature. The effect of cold exposure on adipokine gene expression was extensively studied in human, mice and rats, as reported in the next section. On the contrary, to date, there is very little information available concerning the interactions between heat stress and adipokine expression and secretion in other mammalian and non-mammalian species. The only studies available concern the effect of heat stress on leptin gene expression and secretion in birds and farm animals and, in our *in vitro* studies, in mouse. For example, in broiler chickens, Dridi *et al.* (2008) investigated the influence of chronic heat exposure (32°C for 10 d) on hepatic leptin gene expression. The authors observed that heat exposure significantly up-regulated hepatic leptin mRNA levels (+46%) compared with thermoneutrality (22°C). The up-regulation of hepatic leptin gene expression was accompanied also by an increase in plasma leptin levels (+34%), indicating that leptin may be regulated at transcriptional level.

The effect of heat stress on plasma concentration of leptin was also studied in pigeons (Azraqi 2008). During the 40 d experiment, pigeons were kept either thermoneutral (22°C for 24 h/d) or under heat stress conditions (34°C for 8 h/d and 22°C for the rest of the day). Heat stress caused parallel increase in the concentration of leptin and metabolites such as glucose, cholesterol and triglycerides, and oxidative markers. In lactating sows, Renaudeau *et al.* (2003) investigated the effect of elevated temperatures on plasma leptin concentration. Sows were exposed to 28°C between d 8 and 14 of lactation, and to 20°C between d 15 and 21. A significant decrease of feed intake in heat-stressed sows, accompanied by an increase in plasma leptin concentration, was observed. In contrast, Collin (2000) reported a greater tendency for lower plasma leptin concentration in growing pigs fed ad libitum and exposed to 33°C compared with subjects exposed to 23°C.

Recently we investigated the effect of heat stress on gene expression and secretion of leptin in 3T3-L1 adipocytes (Bernabucci *et al.* 2009). 3T3-L1 adipocytes were incubated at 37°C (control temperature), 39°C and 41°C for 24 h. Temperature treatments were adopted to mime mild or severe hyperthermia typical of hot season. To value the kinetic of heat shock response, some culture plates of adipocytes were exposed at 41°C for 2 h, and then at 37°C for 24 h. The samples were collected immediately before initiation and end of TS and at 2, 4, 8, 16, and 24 h after recovering at 37°C. Compared with control temperature (37°C), leptin mRNA and protein levels at 39°C (mild-hyperthermia) remained unchanged (Fig. 1a, b). When temperature was raised to 41°C (severe hyperthermia) leptin was increased compared with 37°C (Fig. 1a, b). Kinetics of leptin mRNA and protein secretion (Fig. 2) studied after 2 h heat shock at 41°C and after 2, 8, 16 and 24 h exposure at 37°C showed an increase of leptin during the 2 h heat shock. Leptin mRNA continued to increase until 2 h after heat shock and then started to decrease, reaching the basal levels 8 h after heat shock. Leptin secretion increased after 2 h exposure and remained higher until the end of the experiment.

Several studies reported that human or laboratory animals lymphocytes express genes for leptin and its receptor. In our study (Lacetera *et al.* 2009), we tested the effects of TS on gene expression of leptin and the long form of its receptor (Ob-Rb) in lymphocyte of dairy cows (Fig. 3a, b). In this *in vitro* study, peripheral blood mononuclear cells (PBMC) were cultured under 39°C continuously for 65 h or three 13 h cycles at 40, 41 or 42°C, respectively. The temperature of 39°C mimicked normal temperature conditions; 40, 41 and 42°C mimicked conditions of hyperthermia alternating with normal temperature. Compared with 39°C, levels of mRNA for leptin or Ob-Rb were lower in PBMC cultured under 40, 41 and 42°C, respectively.

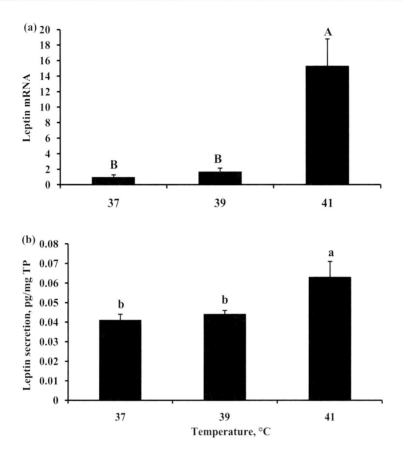

Fig. 1 Temperature-dependent leptin mRNA (a) and leptin protein (b) in 3T3-L1 adipocytes. Severe heat shock induced up-regulation of leptin gene expression and protein secretion. Results are Lsmeans ± SEM of triplicate determinations. [a,b]P<0.05, [A,B]P<0.01. Reprinted from Bernabucci *et al.* 2009, with permission.

The molecular mechanism underlying the increase of leptin under hot conditions is still unknown. However, a comprehensive evaluation of data suggests that up-regulation of leptin expression, with a consequent increase in circulating leptin, by heat shock is probably one of the mechanisms involved in thermoregulatory processes acting to limit body hyperthermia by a central action that is responsible for the decrease of feed intake, energy metabolism and body fat. Moreover, adipocytes express leptin receptors allowing leptin to act directly on adipocytes to regulate energy and metabolism. This "short-loop" leptinergic system is independent of the far more complex hypothalamic "long-loop" energy regulation. Another possible mechanism that may explain the up-regulation of leptin in severe heat-shocked cells is the role of leptin in inducing cell apoptosis.

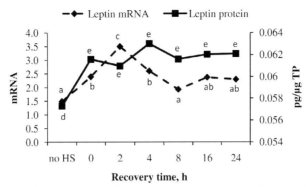

Fig. 2 Kinetics of leptin mRNA and protein secretion response to heat shock. Two hours' exposure of 3T3-L1 cells to heat shock induced up-regulation of leptin mRNA and protein secretion. Results are Lsmeans ± SEM of triplicate determinations. No HS = no heat shock. [a, b, c]P<0.05 between hours for leptin mRNA, [d, e]P<0.05 between hours for leptin protein. Reprinted from Bernabucci *et al.* 2009, with permission.

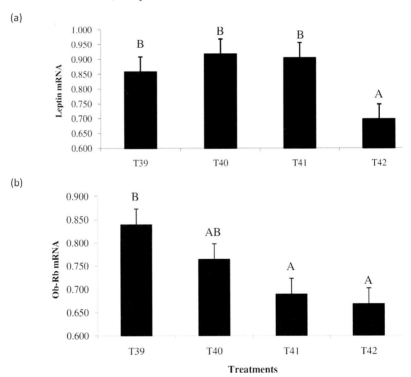

Fig. 3 Levels of mRNA for leptin (a) and ObRb receptor (b) in concanavalin A-stimulated peripheral blood mononuclear cells (PBMC). Exposure of PBMC to heat shock down-regulated leptin and ObRb mRNA. Results are Lsmeans ± SEM. [A, B]P<0.01. Reprinted from Lacetera *et al.* 2009, with permission.

Lymphocytes provide a small contribution to the pool of circulating leptin and are likely to be subjected to the effects of leptin mainly through an autocrine mechanism. The reduced expression of leptin and leptin receptor by heat shock may limit the sensitivity and biological response of lymphocytes to circulating leptin (leptin resistance). In this regard, we can speculate that heat shock–associated down-regulation of leptin and Ob-Rb gene expression in lymphocytes may represent one of the homeostatic mechanisms through which heat-stressed animals attenuate less crucial functions such as immunity in the short term.

Only one study has been done to evaluate the effects of heat shock on adiponectin expression and secretion in 3T3-L1 adipocytes (Bernabucci *et al.* 2009). The experimental protocol followed the temperature treatments and exposure time reported before in the *in vitro* study on leptin. Results of this study showed that heat shock affected the two adipokines differently. In fact, unlike the case with leptin, the exposure of adipocytes to severe hyperthermia (41°C) caused down-regulation of adiponectin (Fig. 4a,b). In particular, compared with control temperature (37°C),

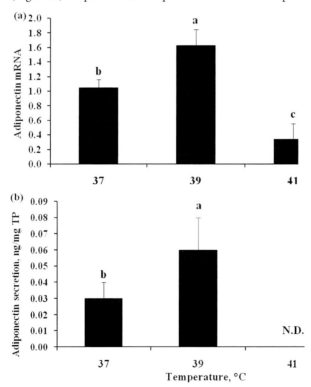

Fig. 4 Temperature-dependent adiponectin mRNA (a) and adiponectin protein (b) in 3T3-L1 adipocytes. Severe heat shock induced down-regulation of adiponectin gene expression and protein secretion. Results are Lsmeans ± SEM of triplicate determinations. [a, b, c]P<0.05. Reprinted from Bernabucci *et al.* 2009, with permission.

adiponectin mRNA and secretion levels increased at 39°C (mild hyperthermia). When temperature was raised to 41°C (severe hyperthermia), adiponectin was significantly decreased from levels at 37°C. The kinetics of adiponectin mRNA and protein secretion studied after 2 h heat shock at 41°C and after 2, 8, 16 and 24 h exposure at 37°C showed a decrease of adiponectin mRNA and secretion after a 2 h period of heat shock at 41°C (Fig. 5). After that time, adiponectin mRNA and secretion started to recover after just 2 h or 4 h of recovery, respectively.

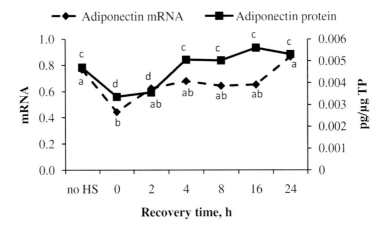

Fig. 5 Kinetics of adiponectin mRNA and protein secretion response to heat shock. Two hours' exposure of 3T3-L1 cells to heat shock induced down-regulation of adiponectin mRNA and protein secretion. Results are Lsmeans ± SEM of triplicate determinations. No HS = no heat shock. [a, b, c, d]$P<0.05$ between hours for leptin mRNA, [d, e]$P<0.05$ between hours for leptin protein. Reprinted from Bernabucci *et al.* 2009, with permission.

Despite the physiological significance of adiponectin, the mechanism regulating the decrease of adiponectin by heat shock is still not understood. Besides the increase of leptin expression as a consequence of a possible adaptive response to heat shock, down-regulation of adiponectin might be explained as a cell heat shock response that was accompanied by a reduction in protein synthesis, favoring the induction of heat shock response over the ongoing gene program.

EFFECTS OF COLD STRESS ON ADIPOKINES EXPRESSION

As mentioned earlier, at the adipose tissue level, cold exposure is accompanied by an increased noradrenaline yield that activates the β-adrenergic receptors and, in turn, raises the mobilization of non-esterified fatty acids from the white adipose tissue (Puerta *et al.* 2002). In 3T3-L1 adipocytes, it has also been observed that stimulation of the β-adrenergic receptors regulates gene expression and levels of some adipokines, suggesting a role of the sympathetic system in the modulation of these adipokines (Imbeault *et al.* 2009).

To our knowledge, the majority of experiments performed on the effects of cold shock on adipokine expression concern above all leptin and adiponectin and only one resistin.

There are several *in vivo* and *in vitro* studies (Moinat *et al.* 1995, Trayurn *et al.* 1995, Puerta *et al.* 2002, Zeyl *et al.* 2004) reporting the effects of cold stress on leptin expression and production. For example, it has been shown that cold exposure down-regulated leptin gene expression and reduced plasma leptin levels in mammals, and this effect has been explained as an adaptive mechanism to cold environments. Trayurn *et al.* (1995) found that acute cold-air exposure (4°C for 2-18 h) induced suppression of leptin gene expression in white adipose tissue in mice, whereas Moinat *et al.* (1995) observed significant reductions of leptin mRNA expression in the brown adipose tissue of rats exposed to 6°C for 24 h. Similar cold exposures (4°C for 24 h) have elicited a 25% reduction in circulating leptin in lean Zucker rats, but not in obese rats (Hardie et al 1996). Puerta *et al.* (2002) have examined the effects of acute cold exposure (18 h at 6°C) on the expression of the gene encoding leptin in both white and brown fat in rats. These authors observed the down-regulation of leptin mRNA in white and brown adipose tissue.

In humans, Ricci *et al.* (2000) observed significant decrease of plasma leptin concentration (-14%, -17%, -22%) during exposure to cold environment (6.3°C: 30, 60 and 90 min, respectively). Peino *et al.* (2000) reported local temperature effects on leptin secretion from human subcutaneous adipose tissue *in vitro*, showing that low temperatures directly reduced leptin secretion rate (-41% and -68% at 34.5°C and 32°C, respectively, compared with 37°C). More recently, Zeyl *et al.* (2004), by *in vivo* and *in vitro* studies in human, demonstrated cold-induced reductions in circulating leptin concentration *in vivo* (-14% and -22% after 25 and 90 min at 18°C, respectively) and *in vitro* (-3.7-fold from 37°C to 27°C). The exact mechanism by which cold exposure reduces leptin is not fully understood. To date, it has not been elucidated whether cold acts directly or indirectly upon the adipose tissue. Moreover, its physiological meaning is not yet clear.

In vivo the inhibitory effect of low temperatures on adipose tissue may be explained as the result of an increase in the adrenergic tone induced by exposure to cold temperatures, which would in turn act through the b-3 receptors present in the adipose tissue inducing a reduction in the expression of the leptin mRNA (Trayhurn *et al.* 1995). In addition, *in vitro* study indicated the possibility that environmental temperature may exert direct local effects on leptin secretion from subcutaneous adipose tissues, and that these local effects may make a substantial supplementary contribution to the cold-induced reduction in circulating leptin *in vivo* (Zeyl *et al.* 2004). Collectively, these observations show that cold exposure reduces circulating leptin levels in rodents and in humans, suggesting that this hormone may participate in the adaptive mechanisms triggered by variations in environmental temperature to modify energy expenditure and appetite and to determine relative adiposity.

There are few and conflicting results of cold effects on adiponectin production in white adipose tissue of rodents. Puerta *et al.* (2002) reported no effect of cold exposure (18 h at 6°C) on gene expression in white adipose tissue (epididymal) and serum concentration of adiponectin in rats. Conversely, Yoda *et al.* (2001) reported up-regulation of gene expression of high molecular weight adiponectin isoform [namely gelatin-binding protein of 28 kDa (GBP28)] in epididymal white adipose tissue and higher serum concentration of adiponectin in mice exposed to cold environment (24 h at 4°C). More recently, Imai *et al.* (2006) showed that cold exposure (24 h at 4°C) reduced serum and mRNA levels (subcutaneous, epididymal, and mesenteric fat tissues) of adiponectin in mice without any change in body mass.

Cold exposure is one of the few stimuli identified so far that affects adiponectin levels in a short period of time in humans. Imbeault *et al.* (2009) recently explored the effect of cold exposure on adiponectin plasma levels in humans. Plasma levels of this protein were analyzed in the samples from glycogen-depleted and glycogen-loaded men exposed to a temperature of 10°C for 2 h. These results showed that adiponectin levels increased significantly after 90 min of cold exposure in both glycogen-depleted and glycogen-loaded men, and changes in circulating adiponectin tended to be associated with changes in plasma glucose oxidation rates but not with changes in lipid oxidation rates during cold exposure. In healthy young men, it has been demonstrated that adiponectin levels significantly increase during a 2 h period of cold (10°C) exposure (Imbeault *et al.* 2009). This rise in adiponectin levels observed during shivering is inhibited by glucose ingestion but not by diets varying in carbohydrate content.

The mechanisms underlying the rise in adiponectin levels during cold exposure are still to be identified. The involvement of the sympathetic nervous system in adiponectin regulation is unclear and remains controversial (Puerta *et al.* 2002). The inhibitory effect of β-adrenergic stimulation on adiponectin messenger RNA levels has also been demonstrated in mice *in vivo* and in human adipose tissue explants (Imbeault *et al.* 2009). Cold exposure would be accompanied by a significant decrease in adiponectin levels, but a number of studies in human and in rodents showed that cold increases adiponectin levels. Therefore, the sympathetic system is not a key regulator of adiponectin expression to cold shock.

Yoda *et al.* (2001) demonstrated a role of thyroid hormones in the up-regulation of adiponectin expression in mice exposed to cold. The relationship between thyroid hormones and adiponectin may be either direct, through the stimulation of hormone synthesis, or indirect, through improvement in insulin sensitivity during cold exposure (Yoda *et al.* 2001).

The physiological meaning of the modulation in adiponectin levels during cold exposure remains to be explored. It could be postulated that adiponectin regulates body temperature and basal metabolic rate in response to changing environmental conditions. In fact, adiponectin is known to have structural similarities with hibernation-associated plasma proteins: HP-27, HP-25, and HP-20 in chipmunks (Yoda *et al.* 2001). It is possible to conclude that adiponectin has a role in adaptive thermogenesis under cold exposure.

Little published information is available about the effect of TS on gene expression and secretion of resistin. Puerta *et al.* (2002) observed that acute cold exposure (18 h at 6°C) did not affect gene expression of resistin in rat white adipose tissue and there is little effect on rat brown adipose tissue despite the cold-induced stimulation of sympathetic activity and fatty acid flux. It is currently unclear whether the sympathetic system regulates synthesis of the resistin in white adipose tissue. Different studies indicated inhibition (Shojima *et al.* 2002), stimulation (Martinez at al. 2001) or no effect of catecholamines (Haugen *et al.* 2001) on transcription of this gene. The absence of effects of acute cold exposure on resistin mRNA levels in white adipose tissue suggests that the sympathetic system does not play a significant role in the regulation of the transcription of this gene as a response to cold. Moreover, the modest effect of acute cold on resistin mRNA would suggest that resistin, involved in insulin sensitivity, adipocyte proliferation and nutrient sensing, is not implicated in the extensive metabolic acclimation to cold, at least in terms of early responses.

FINAL REMARKS

The biological and physiological meanings of adipokine responses to heat or cold exposure are still to be clarified. Results available in literature clearly show that heat and cold stress have opposite effects on leptin and adiponectin expression and secretion. This may suggest a different role of these adipokines in the acclimation of adipose tissue to TS. Finally, further studies are encouraged to verify whether the

Table 1   Key features of thermal stress and adipokines

- Thermal stress (heat or cold) is a physical factor that induces profound alterations in cell physiology.
- As a consequence of its proximity to the skin surface, the temperature of subcutaneous adipose tissue is strongly influenced by environmental temperature.
- Heat and cold stress affect gene expression and secretion of leptin and adiponectin.
- Changes of leptin and adiponectin might have a role in adaptive response to thermal stress (heat or cold) conditions.
- These adaptive responses may be involved in mechanisms implicated in the modification of carbohydrate and lipid metabolism.
- The impairment of adipokine biology induced by thermal stress may be involved in the exacerbation of medical conditions in subjects with metabolic diseases when exposed to heat or cold environment.

TS-induced impairment of adipokines biology may be a cofactor for aggravation of the clinical status observed in patients suffering from metabolic diseases and exposed to heat or cold.

Summary Points

- Thermal stresses (heat and cold) profoundly alter cell physiology by the impairment of major cellular functions and by an adaptive reprogramming of the cell metabolism.
- Cells from all organisms respond similarly to thermal stress by rapidly increasing the expression and synthesis of the heat shock proteins that protect the stressed cell.
- Adipokines, secreted by adipose tissue, are well-known factors involved in the regulation of energy intake and energy expenditure that are markedly modified by thermal stresses.
- Heat stress is responsible for up-regulation of leptin and down-regulation of adiponectin.
- In contrast, cold stress down-regulates leptin and, to some extent, up-regulates adiponectin.
- Results available in literature show that heat and cold stress have opposite effects on leptin and adiponectin expression and secretion.

Abbreviations

FFA	:	free fatty acids
GBP28	:	gelatin-binding protein of 28 kDa
HSP	:	heat shock proteins
Ob-Rb	:	long form leptin receptor
PBMC	:	peripheral blood mononuclear cells
TS	:	thermal stress

Definition of Terms

Adipose tissue: The major depot for energy storage and an active endocrine organ secreting a variety of proteins that influence metabolism.

Cold stress: A condition that occurs when body heat is lost from being in a cold environment faster than it can be replaced. When the body temperature drops below the normal 37°C to around 35°C, the organism begins to show symptoms.

Heat shock proteins: Intracellular proteins present in every species, from bacteria to humans. These proteins protect nascent proteins, halt protein de-folding and have many other functions. Their generation may alter the threshold for thermal injury.

Heat stress: An acute condition that occurs when the body produces or absorbs more heat than it can dissipate. It is usually due to excessive exposure to heat.

Thermal stress: Any change in the thermal relation between a temperature regulator and its environment which, if uncompensated by temperature regulation, would result in hyper- or hypothermia.

References

Al-Azraqi, A.A. 2008. Pattern of leptin secretion and oxidative markers in heat-stressed pigeons. Int. J. Poultry Sci. 7: 1174-1176.

Bernabucci, U. and L. Basiricò, P. Morera, N. Lacetera, B. Ronchi, and A. Nardone. 2009. Heat shock modulates adipokines expression in 3T3-L1 adipocytes. J. Mol. Endocrinol. 42: 139-147

Bernabucci, U. and N. Lacetera, L.H. Baumgard, R.P. Rhoads, B. Ronchi, and A Nardone. 2010. Metabolic and hormonal acclimation to heat stress in domesticated ruminants. Animal 4(7): 1667-1183.

Collin, A. 2000. Effets de la température ambiante élevée sur le métabolisme énergétique du porcelet. Thèse de Doctorat. INRA-Unité mixte de recherches sur le veau et le porc. Ecole Supérieure Agronomique de Rennes, France.

Dridi, S. and S. Temim, M. Derouet, S. Tesseraud, and M. Taouis 2008. Acute cold- and chronic heat-exposure upregulate hepatic leptin and muscle uncoupling protein (UCP) gene expression in broiler chickens. J. Exper. Zool. 309A: 381-388.

Ezure, T. and S. Amano. 2009. Heat stimulation reduces early adipogenesis in 3T3-L1 preadipocytes. Endocrinology 35: 402-408.

Fujita, J. 1999. Cold shock response in mammalian cells. J. Mol. Microbiol. Biotechnol. 1: 243-255.

Granneman, J.G. and M. Burnazi, Z. Zhu, and L.A. Schwamb. 2003. White adipose tissue contributes to UCP1-independent thermogenesis. Am. J. Physiol. Endocrinol. Metab. 285: E1230-E1236.

Hardie, L.J. and D.V. Rayner, S. Holmes, and P. Trayhurn. 1996. Circulating leptin levels are modulated by fasting, cold exposure and insulin administration in lean but not Zucker (fa/fa) rats as measured by ELISA. Biochem. Biophys. Res. Commun. 223: 660-665.

Haugen, F. and A. Jørgensen, C.A. Drevon and P. Trayhurn. 2001. Inhibition by insulin of resistin gene expression in 3T3-L1 adipocytes. FEBS Lett. 507: 105-108.

Imai, J. and H. Katagiri, T. Yamada, Y. Ishigaki, T. Ogihara, and K. Uno. 2006. Cold exposure suppresses serum adiponectin levels through sympathetic nerve activation in mice. Obesity 14: 1132-1141.

Imbeault, P. and I. Dépault, and F. Haman. 2009. Cold exposure increases adiponectin levels in men. Metab. Clin. Exp. 58: 552-559.

Jiang, H. and J. He, S. Pu, C. Tang, and G. Xu. 2007. Heat shock protein 70 is translocated to lipid droplets in rat adipocytes upon heat stimulation. Biochim. Biophys. Acta. 1771: 66-74.

Lacetera, N. and U. Bernabucci, L. Basiricò, P. Morera, and A. Nardone. 2009. Heat shock impairs DNA synthesis and down-regulates gene expression for leptin and Ob-Rb receptor in concanavalin A-stimulated bovine peripheral blood mononuclear cells. Vet. Immunol. Immunopathol. 127: 190-194.

Lindquist, S. 1986. The heat-shock response. Annu. Rev. Biochem. 55: 1151-1191.

Martínez, J.A. and J. Margareto, A. Marti, F.I. Milagro, E. Larrarte, and M.J. Moreno Aliaga. 2001. Resistin overexpression is induced by a beta3 adrenergic agonist in diet-related overweightness. J. Physiol. Biochem. 57: 287-288.

Moinat, M. and C. Deng, P. Muzzin, F. Assimacopoulos-Jeannet, J. Seydoux, A.G. Dulloo, and J.P. Giacobino. 1995. Modulation of obese gene expression in rat brown and white adipose tissues. FEBS Lett. 373: 131-134.

Peino, R. and V. Pineiro, O. Gualillo, C. Menendez, J. Brenlla, X. Casabiell, C. Dieguez, and F.F. Casanueva. 2000. Cold exposure inhibits leptin secretion *in vitro* by a direct and non-specific action on adipose tissue. Eur. J. Endocrinol. 142: 195-199.

Puerta, M. and M. Abelenda, M. Rocha, and P. Trayhum. 2002. Effect of acute cold exposure on the expression of the adiponectin, resistin and leptin genes in rat white and brown adipose tissues. Horm. Metab. Res. 34: 629-634.

Renaudeau, D. and J. Noblet, and J. Y. Dourmad. 2003. Effect of ambient temperature on mammary gland metabolism in lactating sows. J. Anim. Sci. 81: 217-231.

Ricci, M.R., and S.K. Fried, and K.D. Mittleman. 2000. Acute cold exposure decreases plasma leptin in women. Metabolism 49: 421-423.

Shojima, N. and H. Sakoda, T. Ogihara, M. Fujishiro, H. Katagiri, M. Anai, Y. Onishi, H. Ono, K. Inukai, M. Abe, Y. Fukushima, M. Kikuchi, Y. Oka, and T. Asano. 2002. Humoral regulation of resistin expression in 3T3-L1 and mouse adipose cells. Diabetes 51: 1737-1744.

Sonna, L.A. and J. Fujita, S.L. Gaffin, and C.M. Lilly. 2002. Invited review: Effects of heat and cold stress on mammalian gene expression. J. Appl. Physiol. 92: 1725-1742.

Trayhurn, P. and J.S. Duncan, and D.V. Rayner. 1995. Acute cold-induced suppression of ob (obese) gene expression in white adipose tissue of mice: mediation by the sympathetic system. Biochem. J. 311: 729-733.

Westerterp-Plantenga, M.S. 1999. Effects of extreme environments on food intake in human subjects. Proc. Nutr. Soc. 58: 791-798.

Yoda, M. and Y. Nakano, T. Tobe, S. Shioda, N.H. Choi-Miura, and M. Tomita. 2001. Characterization of mouse GBP28 and its induction by exposure to cold. Int. J. Obes. Relat. Metab. Disord. 25: 75-83.

Zeyl, A. and J.M. Stocks, N.A.S. Taylor, and A.B. Jenkins. 2004. Interactions between temperature and human leptin physiology *in vivo* and *in vitro*. Eur. J. Appl. Physiol. 92: 571-578.

Central Nervous System Roles for Adiponectin in Neuroendocrine and Autonomic Function

Ted D. Hoyda,[1] **Willis K. Samson**[2] **and Alastair V. Ferguson**[1,*]

[1]Department of Physiology, Queen's University, Kingston, Ontario, Canada
K7L 3N6
[2]Department of Pharmacological and Physiological Science, Saint Louis
University, St. Louis, Missouri, USA

ABSTRACT

Adiponectin (ADP) is a fat cell–derived peptide that acts as an insulin-sensitizing hormone, the circulating concentrations of which are inversely correlated with adipose tissue mass. ADP concentrations are also reduced in obesity-related diseases such as insulin resistance and the metabolic syndrome. In this chapter we review recent evidence demonstrating that in addition to its peripheral effects, ADP also has potentially important actions in the brain that appear to complement those peripheral actions, thus contributing to the integrated regulation of energy balance. *In situ* hybridization studies have identified adiponectin receptors in a number of important autonomic control centers in hypothalamus and medulla. The cellular actions of ADP on identified cell groups in a number of these nuclei have also been identified by studies combining patch clamp recordings with single cell RT-PCR techniques to identify chemical phenotypes of recorded neurons. Such cellular actions correlate well with demonstrated central actions of ADP on endocrine, metabolic and cardiovascular function. Collectively these studies are laying the groundwork for establishing the integrated roles of ADP in the brain, and in particular understanding how such central actions potentially contribute to the integrated physiological roles of this adipokine in the regulation of energy homeostasis.

*Corresponding author

INTRODUCTION

Obesity is associated with development of insulin resistance, diabetes, cardiovascular disease and hypertension, but the basis of these associations is presently unclear. Since the discovery of leptin in the early 1990s, our view of adipose tissue has changed dramatically. Once thought of as strictly an energy repository, fat cells have emerged as an active player in energy homeostasis, producing a variety of signaling molecules (adipokines) that appear to play essential roles in metabolic, immune, cardiovascular, and endocrine regulation. Adiponectin (ADP) is a fat cell-derived peptide that acts as an insulin-sensitizing hormone (Scherer et al. 1995, Kadowaki and Yamauchi 2005), the circulating concentrations of which are *inversely correlated with adipose tissue mass*. ADP concentrations are also reduced in obesity-related diseases such as insulin resistance and the metabolic syndrome. Treatment with ADP lowers hepatic gluconeogenesis, serum glucose and ameliorates insulin resistance in mice (Combs *et al.* 2001, Yamauchi *et al.* 2001, Kadowaki and Yamauchi 2005). Transgenic mice over-expressing the globular portion of ADP demonstrate increased fatty acid clearance and insulin sensitivity (Combs *et al.* 2004). Conversely, ADP knockout mice demonstrate obesity, impaired insulin sensitivity, impaired fatty acid clearance, and hypertension (Kubota *et al.* 2002). Human genetic studies indicate that alterations in ADP signaling may lead to susceptibility to obesity, insulin resistance, diabetes and hypertension (Kadowaki and Yamauchi 2005). Collectively these observations suggest important roles for ADP in maintaining a variety of normal metabolic functions.

ADIPONECTIN ACTIONS IN THE CENTRAL NERVOUS SYSTEM

The first evidence that the brain is a major target for adiponectin came in 2004 with the demonstration that a single injection of globular adiponectin (gAd) into the lateral cerebral ventricles of C57Bl/6J mice is associated with a sustained decrease in body weight (for up to 4 d) *without* a concomitant decrease in food intake, suggesting an increase in total energy expenditure (Qi *et al.* 2004). The physiological effects of central adiponectin on weight loss are dose dependent and observed at a concentration approximately 1/1000 the effective concentration of adiponectin in the periphery (Qi *et al.* 2004). Levels of O_2 consumption, brown adipose tissue UCP-1 mRNA, and hypothalamic corticotropin releasing hormone (CRH) mRNA are all increased following intracerebroventricular (icv) injections of the peptide (Qi *et al.* 2004), while examination of Fos-positive cells (a marker of neuronal activation) in the hypothalamus following icv *or* iv injections of gAd indicates that the paraventricular nucleus of the hypothalamus (PVN) may be one of the primary sites of neuron activation (Qi *et al.* 2004). Interestingly, the arcuate nucleus (ARC), a major source of orexigenic and anorexigenic peptides including neuropeptide Y (NPY), agouti related peptide (AgRP), pre-opiomelanocortin (POMC), and cocaine and amphetamine related transcript (CART), is not activated by either route of adiponectin administration (icv or iv), as illustrated by the lack of Fos activation (Qi *et al.* 2004). Importantly, in leptin-deficient obese mice, adiponectin acts in the brain to reduce plasma levels of glucose by 71%, insulin by 52%, triglycerides by 17% and total cholesterol by 29%, revealing a crucial role for the CNS in mediating adiponectin-dependent control of glucose homeostasis and metabolic function independent of the other major adipokine, leptin (Qi *et al.* 2004).

ADIPONECTIN TRANSPORT ACROSS THE BLOOD-BRAIN BARRIER

While the large, hydrophilic adiponectin molecule is unlikely to cross the blood-brain barrier (Spranger *et al.* 2006), the peptide is present in the cerebral spinal fluid (CSF) at 1-4% the concentration of circulating levels (Qi *et al.* 2004) predominantly in trimeric and low molecular weight (LMW) hexameric complexes. IV injection of adiponectin linearly increases cerebroventricular concentration of the peptide, while the same vascular increase is seen following icv injections of adiponectin (Kubota *et al.* 2007, Qi *et al.* 2004). Importantly, either iv or icv injections made into *adipo*$^{-/-}$ mice demonstrate the same increases seen in wild-type mice supporting the presence of a bi-directional, non-saturating central transport mechanism for the peptide, although the precise nature of such transport has yet to be described.

RECEPTOR EXPRESSION IN THE CENTRAL NERVOUS SYSTEM

Mapping the central expression of adiponectin receptor mRNA has been advanced through the use of RT-PCR (both from tissue and single neurons) and *in situ* hybridization techniques. Our lab (see Fig. 1) and others show mRNA expression for AdipoR1 and AdipoR2 in the area postrema (AP) (Fry *et al.* 2006), nucleus tractus solitarius (NTS) (Hoyda *et al.* 2009b), subfornical organ (Hindmarch *et al.* 2008), PVN (Kubota *et al.* 2007, Hoyda *et al.* 2007, Coope *et al.* 2008, Guillod-Maximin *et al.* 2009) and ARC (Kubota *et al.* 2007). Within the PVN, NTS and AP, data obtained from combining single cell RT-PCR with electrophysiological recordings demonstrate that most neurons that respond to adiponectin with changes in membrane excitability also express mRNA for either AdipoR1, AdipoR2 or both of the receptors in the same cell (Fry *et al.* 2006, Hoyda *et al.* 2007, 2009a, b). These experiments demonstrate that cells in these nuclei have the capacity to produce functional receptors the activation of which presumably mediates adiponectin's central actions on energy homeostasis and feeding behavior. *In situ*

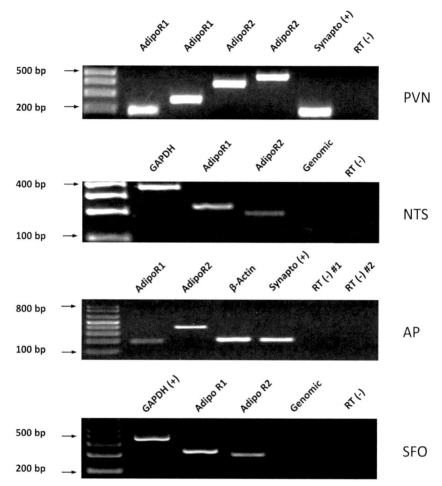

Fig. 1 Adiponectin receptor expression in central nervous system sites. These agarose gels show RT-PCR analysis of paraventricular nucleus (PVN), nucleus tractus solitarius (NTS), area postrema (AP) and subfornical organ (SFO) cDNA for AdipoR1 and AdipoR2. PCR products for GADPH/synaptotagmin/β-Actin are observed (positive control) while no products were observed for genomic cDNA (Genomic) or in the RT- lanes, in which template has been omitted from the cDNA synthesis reaction (adapted from Fry *et al.* 2006, Hoyda *et al.* 2007, 2009b, Alim and Ferguson unpublished observation, with permission).

hybridization studies using anti-sense probes for the two receptors demonstrate the expression of AdipoR1 and AdipoR2 mRNA in the PVN and the ARC and immunohistochemical analysis suggests that AdipoR1 co-localizes with leptin receptor in the ARC (Kubota *et al.* 2007). Kubota *et al.* (2007) show iv injections of adiponectin stimulating food intake through AMP-activated protein kinase (AMPK) and acetyl-CoA carboxylase activation in the ARC, a phenomenon dependent upon AdipoR1 activation. This activation of AMPK is directly opposed to, and may override, the ability of leptin to inhibit AMPK in the ARC, thereby inhibiting feeding (Ahima 2005).

Recent histochemical studies using the appropriate controls demonstrate ubiquitous AdipoR1 and specific AdipoR2 expression in the brain (Coope *et al.* 2008). Importantly, these studies show the presence of the receptors in the PVN and ARC nuclei (Coope *et al.* 2008), confirming our initial reports of mRNA expression in the PVN (Hoyda *et al.* 2007). Examination of individual cellular expression patterns for the two receptors presents interesting observations. The two main cell groups in the ARC, NPY/AgRP and POMC/CART neurons, both express the two adiponectin receptors, as measured by fluorescent immunohistochemical detection in either GFP-NPY or GFP-POMC transgenic mice (Guillod-Maximin *et al.* 2009). Because the experiments were performed separately, it is unknown if these cell types co-localize both receptors in every cell, although our molecular phenotyping of receptor mRNA in the PVN (Hoyda *et al.* 2007, 2009a) suggests that co-localization of both receptors may be a feature of ARC neurons as well. It is important to note also that not all of these neurons express AdipoR1 or R2 and that receptors are expressed in ARC *astrocytes* as well (Guillod-Maximin *et al.* 2009, Hoyda *et al.* 2009a). Precedent therefore exists for the possibility that astrocytes in the PVN and other areas important in metabolic control play important roles in adiponectin signaling.

ACTIVATION OF THE "STRESS AXIS"

Proper coordination of the stress axis is vital to producing appropriate responses to external physiological stressors. Control of this integrated physiological response is important to emotional well-being, performance, executive function and proper social interaction culminating in an increased chance of survival. Disruptions in the stress axis can lead to pathological changes in growth and development, thyroid function, reproduction, metabolic and gastrointestinal function, immune function and the development of some psychiatric disorders. A number of different studies provide evidence supporting the concept that adiponectin plays important roles in controlling the stress axis (Qi *et al.* 2004, Hoyda *et al.* 2009a). CRH produced in neuroendocrine cells of the PVN is the principle hypothalamic regulator of the pituitary-adrenal axis as the CRH released from these neurons at the median eminence is the primary controller of the secretion of ACTH from anterior pituitary corticotropes (Bale and Vale 2004). CRH is also a major anorexigenic peptide whose secretion is paradoxically *stimulated* by NPY, the major orexigenic peptide in the CNS.

Adiponectin Modulation of CRH Neurons

Key Points: _Adiponectin actions on the HPA axis_
- ICV adiponectin causes *c-fos* activation in PVN
- ICV adiponectin increases CRH mRNA in PVN
- Adiponectin depolarizes PVN CRH neurons *in vitro*
- ICV adiponectin increases plasma ACTH concentrations

Qi *et al.* report that icv injections of adiponectin are accompanied by increases in PVN c-fos immunoreactivity, an activated autonomic nervous system, demonstrated by increases in VO_2 usage and thermogenesis, as well as an increase in CRH mRNA within the PVN, culminating in positive energy expenditure and decreased body weight (Qi *et al.* 2004). In accordance with these findings, adiponectin has depolarizing effects on neuroendocrine CRH producing PVN neurons (see Fig. 2), while icv injections of adiponectin increases plasma adrenocorticotropin (ACTH) concentrations in conscious freely moving animals (Hoyda *et al.* 2009a, b), effects that would all contribute to a generalized activation of the stress axis. In addition to our results demonstrating direct depolarizing actions of adiponectin on PVN CRH neurons, recordings from NTS NPY neurons indicate that this medullary nucleus may be an additional site where an adiponectin-sensitive mechanism for modulating food intake and energy homeostasis exists (Hoyda *et al.* 2009b). Single cell RT-PCR analysis combined with electrophysiological recordings reveal a population of neurons expressing NPY only (separate from neurons co-expressing NPY/GAD67), which are depolarized by adiponectin. Immunohistochemical evidence suggests that these NPY neurons project to the medial parvocellular region of the PVN, the anatomical location of CRH expressing neurons involved in the hypothalamo-pituitary-adrenal (HPA) axis, where they act to regulate hypothalamic peptide release. Combined, our experiments suggest two mechanisms for increasing CRH secretion; directly at PVN CRH neurons and through influencing the excitability of putative PVN projecting NPY NTS neurons. Indeed, the ability of centrally administered NPY to activate the HPA axis depends upon interaction with NPY Y5 receptors in the PVN (Kakui and Kitamura 2007).

Since the effects of adiponectin on neuronal excitability as well as its transport into the CSF are dose dependent (Hoyda *et al.* 2007, Kubota *et al.* 2007), we can speculate on the activity level of the stress axis in pathological extremes of body adiposity. In the obese state, serum adiponectin is low (Hu *et al.* 1996) suggesting limited transport to CNS sites behind the blood-brain barrier and therefore depressed HPA axis activation and further orexigenic feeding behavior due to diminished concentrations of serum CRH and ACTH. However, many obese people are also hypercortisolemic, suggesting that their increased appetite may

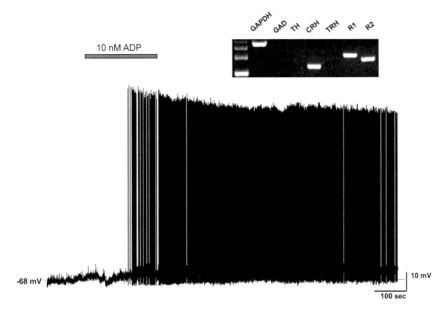

Fig. 2 Adiponectin depolarizes CRH neurons in the paraventricular nucleus of the hypothalamus. This electrophysiological trace shows a current clamp recording from an electrophysiologically identified neuroendocrine neuron identified post hoc using single cell RT-PCR (gel inset) as a CRH neuron that expresses both AdipoR1 (R1) and AdipoR2 (R2). This cell was negative for GAD (GABA marker), tyrosine dehydroxylase (TH, dopamine marker), and TRH. The red bar shows the time of bath application of 10 nM adiponectin and this CRH neuron responded with a small depolarization (8 mV) and a robust increase in action potential frequency (adapted from Hoyda *et al.* 2009a, with permission).

Color image of this figure appears in the color plate section at the end of the book.

be the result of increased cortisol feedback to inhibit central release of CRH. In contrast, in the anorexic state, adiponectin levels are drastically elevated (Delporte *et al.* 2003) and presumably increased in the CSF. This would lead to an overactive HPA axis potentiating anorexigenic behavior (Key Points). This begs the question whether extreme concentrations of adiponectin are the cause of such diseases or simply co-incident with them, further aggravating the pathologies associated with the extremes of energy homeostasis and feeding behavior. Clearly there may be important consequences to unregulated adiponectin signaling in the CNS, suggesting opportunities for therapeutic intervention.

THE THYROID AXIS

Thyrotropin releasing hormone (TRH), a tri-peptide synthesized from a larger precursor protein within cells of the CNS, is widely distributed throughout the CNS in areas such as the hippocampus, brainstem and hypothalamus, where it is highly expressed in the parvocellular regions of the PVN. Neuroendocrine PVN neurons that express TRH are known as "hypophysiotrophic TRH neurons" and release TRH at median eminence into the perivascular space of the pituitary portal system, which carries TRH to its targets in the anterior pituitary. TRH confers its effects on metabolism primarily through its ability to increase release of thyroid stimulating hormone (TSH), thus playing a critical role in controlling the thyroid gland.

TRH increases thermogenesis, locomotor activity and autonomic function. Central injections of TRH are associated with reduced feeding in normal, fasting and stressed animals, indicating this peptide plays important roles as an anorexigenic modulator of food intake and energy homeostasis. Both neuroendocrine and preautonomic (PA) TRH neurons receive monosynaptic connections from α-MSH/CART and NPY/AgRP neurons of the ARC, cells that are involved in mediating the effects of leptin on food intake (Schwartz *et al.* 2000). Glutamatergic synapses innervate TRH neurons, further supporting the integrative nature of this peptide in CNS circuitry regulating energy homeostasis.

Adiponectin Modulation of TRH Neurons

Increases in TRH secretion at the median eminence are associated with activation of the thyroid axis, which increases energy expenditure. Given that icv injections of adiponectin also increase energy expenditure (Qi *et al.* 2004), it is surprising that adiponectin *does not* affect plasma concentrations of TSH following central adiponectin injections and has no effect on the membrane potential of TRH PVN neurons responsible for median eminence release of TRH (Hoyda *et al.* 2009a).

Putative hindbrain projecting PA TRH neurons, however, are selectively depolarized by adiponectin (Hoyda *et al.* 2009a). This subpopulation of PVN TRH neurons have been suggested to project to autonomic preganglionic neurons in the dorsal vagal complex and spinal cord, and to transmit information regarding core body temperature to the periphery. Adiponectin acting through PA TRH neurons may therefore represent an important connection coupling adiposity stores to central thermoregulation.

THE NEUROHYPOPHYSIAL SYSTEM

Oxytocin (OT) is a nonapeptide synthesized in neurons distributed throughout the CNS with the highest densities seen in the PVN and supraoptic nucleus (SON). OT-positive neurons are found in both magnocellular (MNC) and parvocellular regions of the PVN, where they are involved in different aspects of endocrine, autonomic and metabolic regulation. OT plays important roles in energy homeostasis through its actions on consumptive behavior, as illustrated by many studies showing that either ip or icv injections of OT inhibit both feeding and drinking in rats. In the presence of OT, meal duration times are decreased and latency to first meal following starvation is increased (Arletti *et al.* 1990). Parvocellular PA neurons expressing OT innervate the NTS, a site of OT receptor expression and area involved in the regulation of feeding. This pathway has been shown to play a critical role in regulating meal size (Blevins *et al.* 2003), which further implicates OT as a major regulator of energy homeostasis and a therefore a target of signals involved in mediating this function. Additionally, CSF OT levels are decreased in patients with anorexia nervosa (Demitrack *et al.* 1990), a metabolic state associated with high concentrations of serum adiponectin (Misra *et al.* 2007).

Vasopressin (VP) is also a nonapeptide synthesized throughout the brain with the highest expression in the SON and the PVN. Immunohistochemical staining demonstrates that VP, along with its associated neurophysin, is present in both parvocellular and MNC regions of the PVN. These studies also provide evidence that VP-synthesizing MNC neurons exist as a distinct population of cells from those of OT-synthesizing MNC neurons, a phenomenon that similarly holds true in the SON. Recent analysis of mRNA obtained from single MNC neurons suggests, however, that single SON MNC neurons express both VP and OT mRNA and may produce both peptides. MNC-derived VP is packaged into secretory granules and transported down axons to the pars nervosa of the posterior pituitary, where it is stored and released upon appropriate stimuli.

MNC VP plays a vital role in regulating body fluid homeostasis through its actions in the kidneys. Hyperosmotic or hypovolemic stimuli induce VP release from the posterior pituitary, which acts on V2 receptors expressed in the distal convoluted tubules and apical surface of collecting ducts of the kidney, increasing the expression of aquaporin-2 and thus permeability to water. The VP system acts as a defense against disruptions in body fluid homeostasis and therefore plays an important role in regulating energy homeostasis and metabolic function.

Adiponectin Modulation of Oxytocin and Vasopressin Neurons

Activation of adiponectin receptors has direct consequences on the membrane properties of both OT and VP neurohypophysial neurons in the PVN, selectively hyperpolarizing OT neurons (see Fig. 3), exhibiting no effect on those expressing OT and VP and either depolarizing or hyperpolarizing VP neurons, effects that are likely a consequence of the distinct adiponectin receptor and/or ion channel expression profiles of these groups of neurons.

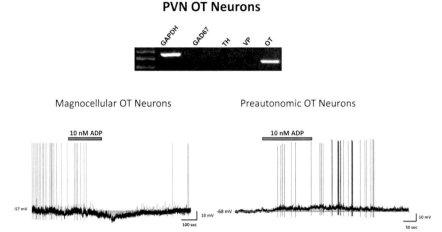

Fig. 3 Adiponectin has opposite effects on different subpopulations of paraventricular nucleus oxytocin neurons. This figure shows examples of recordings from two PVN neurons identified as OT cells using post hoc single cell RT-PCR techniques as shown in the agarose gel presented at the top of the figure. Using electrophysiological fingerprinting we can further subdivide these cells into magnocellular (expression of a dominant transient potassium current, projecting to posterior pituitary), or preautonomic (expression of a low threshold calcium current, projecting to caudal autonomic centers). Intriguingly, while magnocellular OT neurons respond to bath administration of 10 nmol adiponectin with hyperpolarizations (as illustrated in the current clamp recording on the left side), preautonomic OT neurons are depolarized by adiponectin (adapted from Hoyda *et al.* 2007, 2009a, with permission).

Color image of this figure appears in the color plate section at the end of the book.

Intriguingly, adiponectin exerts opposite effects on preautonomic OT neurons, which as illustrated in Fig. 3 are activated by the adipokine, an action that presumably leads to the release of OT from axon terminals of these neurons in the NTS. This effect may promote the sensitization of NTS neurons to satiety cues such as CCK, advancing meal termination (Blevins *et al.* 2003), and supports the concept that adiponectin may act centrally to control feeding behavior and weight loss (Qi *et al.* 2004).

ADIPONECTIN AND THE CONTROL OF FOOD INTAKE

The expression of adiponectin receptors in the ARC and PVN and the roles they play in mediating the effect of peptides on food intake and energy homeostasis remain controversial. Two major studies suggest opposite effects of adiponectin on feeding behavior and energy expenditure in different species of rodent (Kubota *et al.* 2007, Qi *et al.* 2004). Differences in injection protocols, sites of peptide action and the feeding status of species used in these studies may explain some of these disparities. In mice, icv injections of gAd decrease body weight by increasing energy expenditure and enhancing autonomic function without a concomitant decrease in food intake (Qi *et al.* 2004). These effects are dependent upon an activated PVN and are accompanied by an increase in CRH mRNA expression in that nucleus (Qi *et al.* 2004). Coope *et al.* report in addition that icv injections of high concentrations of adiponectin cause an anorexigenic response in rats and illustrate the importance of AdipoR1 signaling in the hypothalamus as oligonucleotide anti-sense primers for the receptor mRNA abolish the anorexic effect of the peptide (Coope *et al.* 2008). Adiponectin also induces activation of the insulin/leptin-like signaling pathways IRS1/2, ERK, Akt, FOXO1, JAK2 and STAT3 in the ARC (Coope *et al.* 2008). Conversely, a study performed by Kadowaki's group demonstrates that iv injections of adiponectin *increase* food intake in mice, an effect dependent upon AdipoR1-induced activation of AMPK in the ARC (Kubota *et al.* 2007). Close examination of the protocols used in these studies reveals two major differences that may account for the discrepancies in their data. First, as opposed to most studies examining the effects of neuropeptides on feeding behavior, the studies of Kubota *et al.* (2007) have examined the effects of injections into *fed* rather than fasting mice (Kubota *et al.* 2007). The importance of pre-injection feeding status to the central effects of adiponectin (and any other metabolic signal) becomes clear when one considers the many other signals that together contribute to the regulation of feeding. Post-prandial changes in other feeding and satiety peptides such as amylin, ghrelin, leptin and insulin, if not controlled, will clearly contribute to the final outcome (food intake or not). Second, the route of peptide administration is also likely to play a role in the response to adiponectin. Kubota *et al.* examined the feeding behavior of mice receiving iv injections of adiponectin rather than icv injections as performed by the other two studies (Kubota *et al.* 2007, Qi *et al.* 2004, Coope *et al.* 2008). This difference may have profound effects on animal behavior as sites of action and response time could be markedly different between the two protocols. Kubota *et al.* implicate the ARC as the primary site of action for iv adiponectin, whereas Ahima's group suggests the PVN is a critical mediator of the effects of icv adiponectin. Despite the inconsistencies, these studies do highlight potential roles for adiponectin in the regulation of feeding, while at the same time they reveal the complexities of understanding the integrated role of this adipokine in regulating feeding behavior and energy homeostasis. Kadowaki's group proposes that adiponectin acts in a sophisticated feedback loop through peripheral and central function as a "starvation" signal (Kadowaki *et al.* 2008). In times of limited energy storage, adiposity is low while serum adiponectin is elevated, engaging the ARC to promote feeding. This intake behavior over time increases adiposity stores and limits adiponectin in the serum, effectively removing the orexigenic drive (Kadowaki *et al.* 2008).

EFFECTS ON SYMPATHETIC NERVOUS ACTIVITY AND AUTONOMIC FUNCTION

An emerging literature suggests that adiponectin also exerts control over cardiovascular function by modulating the sympathetic nervous system (SNS) as a result of actions at hypothalamic and brainstem centers controlling autonomic function (Wang and Scherer 2008). This effect may be particularly important in view of the well-established association between the dysregulation of the SNS and obesity, as well as co-morbidities including hypertension (Wang and Scherer 2008). A reciprocal connection between SNS activation and adiponectin is illustrated both by the reduced serum adiponectin levels following SNS activation using cold stimulus (Imai *et al.* 2006) and by the observations that iv injections of adiponectin decrease blood pressure and sympathetic nerve activity (Tanida *et al.* 2007), although the CNS site at which ADP acts to elicit such effects has yet to be identified. It has, however, been suggested that adiponectin may act in a number of autonomic control nuclei including the area postrema (Fry *et al.* 2006) and the NTS (Hoyda *et al.* 2009a) (see Fig. 4). The latter observation in particular is quite intriguing in view of the additional demonstration of effects of adiponectin on NPY neurons in this region. Although NPY is one of the most widely expressed neuropeptides in the brain, its high expression in the ARC nucleus and its role in feeding induction has dominated interest in the central actions of this peptide. NPY immunoreactivity exists in known cardiovascular regulatory locations within the NTS as these are sites of synaptic termination of baroreceptor afferents from the aortic arch and carotid sinus.

Microinjection of NPY into the NTS induces hypotension and facilitates the aortic baroreflex , an effect that appears to be mediated by the Y_2 receptor as only a Y_2-selective truncated form of NPY (13-36) could elicit the same hypotensive effects, which are not blocked by a Y_1-selective antagonist (Barraco *et al.* 1991). Further support for a role for NPY in cardiovascular regulation comes from the observation that in streptozotocin-induced diabetic rats (which develop hypertension), intra-NTS microinjections of NPY have a far less significant hypotensive effect than in wild-type rats (Dunbar *et al.* 1992). NPY also controls peripheral insulin secretion through actions at NTS neurons (Dunbar *et al.* 1992) and it has been suggested that the pathogenesis of hypertension in diabetic patients may be due to a decrease of intra-NTS efficacy of NPY signaling. These studies clearly indicate important CNS actions for adiponectin and point to the need for a clearer understanding of the mechanisms by which this adipokine controls specific networks and cellular subtypes in the brain to exert control over integrated autonomic function (Table 1).

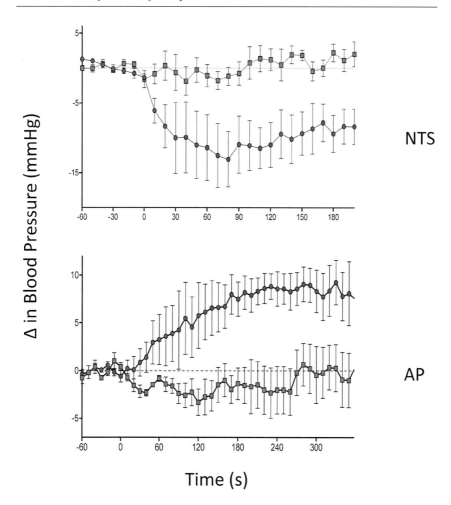

Fig. 4 Local administration of adiponectin into area postrema and nucleus tractus solitarius of anaesthetized rats influences blood pressure. Normalized mean blood pressure (upper) traces showing the different cardiovascular responses to adiponectin microinjection (0.5 µl at time = 0) into the medial nucleus tractus solitarius (NTS-50 fmol, upper graph blue circles) and the area postrema (AP-1 pmol, lower graph blue circles). Importantly, despite the close anatomical proximity of these two regions in the medulla, microinjections of adiponectin exert opposite effects on blood pressure. In both graphs the red square indicates the effects of control microinjections (vehicle or inactive peptide) into the same region (adapted from Fry *et al.* 2006, Hoyda *et al.* 2009b, with permission).

Color image of this figure appears in the color plate section at the end of the book.

Table I   Key questions for future study

- How does adiponectin access the brain?
  - o Does circulating ADP get into the CNS?
  - o Is ADP produced in the CNS?
- How does adiponectin influence neuronal excitability?
  - o Does ADP modulate specific ion channels?
  - o Is AMPK critical to actions in CNS?
- Does adiponectin act in other CNS regions?
  - o What are ADP effects on NPY/POMC cells in the arcuate?
  - o Does ADP act in motivational brain centers?

CONCLUSION

Clearly the last decade has seen an explosion of research in the adiponectin field, particularly since the cloning of the adiponectin receptors and the development of reliable detection assays for the different forms of the peptide. Investigation of the central effects of this adipokine is still at an early stage, ready to emerge as new studies contribute to the literature. Data from different groups will undoubtedly allow the generation of clearer integrative models describing the central actions of this adipokine. In this chapter we have described data highlighting the complexity of such central actions of adiponectin by showing differential effects on the excitability of different subpopulations of neurons in the PVN and NTS neurons, but the complexity is perhaps best highlighted by our demonstration of opposite effects of ADP on magnocellular and preautonomic OT neurons in the PVN. These effects on single cells have functional consequences for the animal including cardiovascular, neuroendocrine and behavioral effects. These experiments continue in the stream of "foundational" studies, beginning to place a few pieces of the puzzle together, while new studies will need to address the means by which adiponectin gains access to the CNS, and whether in fact it is produced within the brain. If the latter is shown to be true, we will then need to identify ADP-producing cells, mechanisms controlling synthesis and release of this adipokine, and we will also need to come to grips with understanding whether central "neurotransmitter" actions contribute to the coordinated metabolic control functions of ADP in the periphery.

Summary Points

- Adiponectin (ADP) is a fat cell–derived peptide, circulating concentrations of which are inversely correlated with adipose tissue mass.
- ADP concentrations are reduced in obesity-related diseases such as insulin resistance and in metabolic syndrome.
- Adiponectin acts in the CNS to influence energy homeostasis.
- Adiponectin receptors are localized in a number of important autonomic control centers in hypothalamus and medulla.

- Cellular actions of ADP on circumventricular organ neurons (subfornical organ and area postrema) suggest mechanisms through which this adipokine can access and influence central autonomic function.
- Adiponectin also influences neuroendocrine CRH neurons in the paraventricular nucleus (PVN), actions that are correlated with increased circulating concentrations of ACTH.
- Adiponectin also influences CNS neurons in PVN and the nucleus tractus solitarius known to be involved in the regulation of autonomic function.
- Collectively the literature now suggests essential integrative roles for ADP in the brain in the regulation of autonomic function.

Abbreviations

ACTH	:	adrenocorticotropin hormone
ADP	:	adiponectin
AgRP	:	agouti related peptide
AMPK	:	AMP activated protein kinase
AP	:	area postrema
ARC	:	arcuate nucleus
CART	:	cocaine and amphetamine related transcript
CNS	:	central nervous system
CRH	:	corticotropin releasing hormone
CSF	:	cerebral spinal fluid
gAd	:	globular adiponectin
HPA	:	hypothalamo-pituitary-adrenal
MNC	:	magnocellular
NPY	:	neuropeptide Y
NTS	:	nucleus tractus solitarius
OT	:	oxytocin
PA	:	preautonomic
POMC	:	pro-opiomelanocortin
PVN	:	paraventricular nucleus
RT-PCR	:	reverse transcription polymerase chain reaction
SNS	:	sympathetic nervous system
SON	:	supraoptic nucleus
TRH	:	thyrotropin releasing hormone
TSH	:	thyroid stimulating hormone
VP	:	vasopressin

Definition of Terms

Blood-brain barrier: The barrier to large/lipophobic particle movement from extracellular fluid of the CNS to plasma resulting from glial cells and tight junctions between vascular endothelial cells.

Circumventricular organs: Specialized regions of the central nervous system lacking the normal blood-brain barrier.

Corticotrophin releasing hormone: Hypothalamic peptide synthesized in paraventricular nucleus neurons and released at the median eminence to control ACTH release from the anterior pituitary.

Nucleus tractus solitarius: Medullary autonomic control center for integration of autonomic function.

Patch clamp electrophysiology: Recordings of single cell membrane potential or current.

Paraventricular nucleus: Hypothalamic autonomic control center for integration of neuroendocrine and autonomic function.

Single cell RT-PCR: Method for analysis of gene expression in single neurons following patch clamp analysis of functional phenotype.

References

Ahima, R.S. 2005. Central actions of adipocyte hormones. Trends Endocrinol. Metab. 16: 307-313.

Arletti, R. and A Benelli, and A. Bertolini. 1990. Oxytocin inhibits food and fluid intake in rats. Physiol. Behav. 48: 825-830.

Bale, T.L. and W.W. Vale. 2004. CRF and CRF receptors: role in stress responsivity and other behaviors. Annu. Rev. Pharmacol. Toxicol. 44: 525-557.

Barraco, R.A. and E. Ergene, J.C. Dunbar, Y.L. Ganduri, and G.F. Anderson. 1991. Y2 receptors for neuropeptide Y in the nucleus of the solitary tract mediate depressor responses. Peptides 12: 691-698.

Blevins, J.E. and T.J. Eakin, J.A. Murphy, M.W. Schwartz, and D.G. Baskin. 2003. Oxytocin innervation of caudal brainstem nuclei activated by cholecystokinin. Brain Res. 993: 30-41.

Combs, T.P. and A.H. Berg, S. Obici, P.E. Scherer, and L. Rossetti. 2001. Endogenous glucose production is inhibited by the adipose-derived protein Acrp30. J. Clin. Invest. 108: 1875-1881.

Combs, T.P. and U.B. Pajvani, A.H. Berg, Y. Lin, L.A. Jelicks, M. Laplante, A.R. Nawrocki, M.W. Rajala, A.F. Parlow, L. Cheeseboro, Y.Y. Ding, R.G. Russell, D. Lindemann, A. Hartley, G.R. Baker, S. Obici, Y. Deshaies, M. Ludgate, L. Rossetti, and P.E. Scherer. 2004. A transgenic mouse with a deletion in the collagenous domain of adiponectin displays elevated circulating adiponectin and improved insulin sensitivity. Endocrinology 145: 367-383.

Coope, A. and M. Milanski, E.P. Araujo, M. Tambascia, M.J. Saad, B. Geloneze, and L.A. Velloso. 2008. AdipoR1 mediates the anorexigenic and insulin/leptin-like actions of adiponectin in the hypothalamus. FEBS Lett. 582: 1471-1476.

Delporte, M.L. and S.M. Brichard, M.P. Hermans, C. Beguin, and M. Lambert. 2003. Hyperadiponectinaemia in anorexia nervosa. Clin. Endocrinol. (Oxf.) 58: 22-29.

Demitrack, M.A. and M.D. Lesem, S.J. Listwak, H.A. Brandt, D.C. Jimerson, and P.W. Gold. 1990. CSF oxytocin in anorexia nervosa and bulimia nervosa: clinical and pathophysiologic considerations. Am. J. Psychiatry 147: 882-86.

Dunbar, J.C. and E. Ergene, G.F. Anderson, and R.A. Barraco. 1992. Decreased cardiorespiratory effects of neuropeptide Y in the nucleus tractus solitarius in diabetes. Am. J. Physiol. 262: R865-R871.

Fry, M. and P.M. Smith, T.D. Hoyda, M. Duncan, R.S. Ahima, K.A. Sharkey, and A.V. Ferguson. 2006. Area postrema neurons are modulated by the adipocyte hormone adiponectin. J. Neurosci. 26: 9695-9702.

Guillod-Maximin, E. and A.F. Roy, C.M. Vacher, A. Aubourg, V. Bailleux, A. Lorsignol, L. Penicaud, M. Parquet, and M. Taouis. 2009. Adiponectin receptors are expressed in hypothalamus and colocalized with proopiomelanocortin and neuropeptide Y in rodent arcuate neurons. J. Endocrinol. 200: 93-105.

Hindmarch, C. and M. Fry, S.T. Yao, P.M. Smith, D. Murphy, and A.V. Ferguson. 2008. Microarray analysis of the transcriptome of the subfornical organ in the rat: regulation by fluid and food deprivation. Am. J. Physiol. Regul. Integr. Comp. Physiol. 295: R1914-R1920.

Hoyda, T.D. and M. Fry, R.S. Ahima, and A.V. Ferguson. 2007. Adiponectin selectively inhibits oxytocin neurons of the paraventricular nucleus of the hypothalamus. J. Physiol. 585: 805-816.

Hoyda, T.D. and W.K. Samson, and A.V. Ferguson. 2009a. Adiponectin depolarizes parvocellular paraventricular nucleus neurons controlling neuroendocrine and autonomic function. Endocrinology 150: 832-840.

Hoyda, T.D. and P.M. Smith, and A.V. Ferguson. 2009b. Adiponectin acts in the nucleus of the solitary tract to decrease blood pressure by modulating the excitability of neuropeptide Y neurons. Brain Res. 1256: 76-84.

Hu, E. and P. Liang, and B.M. Spiegelman. 1996. AdipoQ is a novel adipose-specific gene dysregulated in obesity. J. Biol. Chem. 271: 10697-10703.

Imai, J. and H. Katagiri, T. Yamada, Y. Ishigaki, T. Ogihara, K. Uno, Y. Hasegawa, J. Gao, H. Ishihara, H. Sasano, and Y. Oka. 2006. Cold exposure suppresses serum adiponectin levels through sympathetic nerve activation in mice. Obesity (Silver Spring) 14: 1132-1141.

Kadowaki, T. and T. Yamauchi. 2005. Adiponectin and adiponectin receptors. Endocr. Rev. 26: 439-451.

Kadowaki, T. and T. Yamauchi, and N. Kubota. 2008. The physiological and pathophysiological role of adiponectin and adiponectin receptors in the peripheral tissues and CNS. FEBS Lett. 582: 74-80.

Kakui, N. and K. Kitamura. 2007. Direct evidence that stimulation of neuropeptide Y Y5 receptor activates hypothalamo-pituitary-adrenal axis in conscious rats via both corticotropin-releasing factor- and arginine vasopressin-dependent pathway. Endocrinology 148: 2854-2862.

Kubota, N. and Y. Terauchi, T. Yamauchi, T. Kubota, M. Moroi, J. Matsui, K. Eto, T. Yamashita, J. Kamon, H. Satoh, W. Yano, P. Froguel, R. Nagai, S. Kimura, T. Kadowaki, and T. Noda. 2002. Disruption of adiponectin causes insulin resistance and neointimal formation. J. Biol. Chem. 277: 25863-25866.

Kubota, N. and W. Yano, T. Kubota, T. Yamauchi, S. Itoh, H. Kumagai, H. Kozono, I. Takamoto, S. Okamoto, T. Shiuchi, R. Suzuki, H. Satoh, A. Tsuchida, M. Moroi, K. Sugi, T. Noda, H. Ebinuma, Y. Ueta, T. Kondo, E. Araki, O. Ezaki, R. Nagai, K. Tobe, Y. Terauchi, K. Ueki, Y. Minokoshi, and T. Kadowaki 2007. Adiponectin stimulates AMP-activated protein kinase in the hypothalamus and increases food intake. Cell Metab. 6: 55-68.

Misra, M. and K.K. Miller, J. Cord, R. Prabhakaran, D.B. Herzog, M. Goldstein, D.K. Katzman, and A. Klibanski. 2007. Relationships between serum adipokines, insulin levels, and bone density in girls with anorexia nervosa. J. Clin. Endocrinol. Metab. 92: 2046-2052.

Qi, Y. and N. Takahashi, S.M. Hileman, H.R. Patel, A.H. Berg, U.B. Pajvani, P.E. Scherer, and R.S. Ahima. 2004. Adiponectin acts in the brain to decrease body weight. Nat. Med. 10: 524-529.

Scherer, P.E. and S. Williams, M. Fogliano, G. Baldini, and H.F. Lodish. 1995. A novel serum protein similar to C1q, produced exclusively in adipocytes. J. Biol. Chem. 270: 26746-26749.

Schwartz, M.W. and S.C. Woods, D. Porte Jr., R.J. Seeley, and D.G. Baskin. 2000. Central nervous system control of food intake. Nature 404: 661-671.

Spranger, J. and S. Verma, I. Gohring, T. Bobbert, J. Seifert, A.L. Sindler, A. Pfeiffer, S.M. Hileman, M. Tschop, and W.A. Banks. 2006. Adiponectin does not cross the blood-brain barrier but modifies cytokine expression of brain endothelial cells. Diabetes 55: 141-147.

Tanida, M. and J. Shen, Y. Horii, M. Matsuda, S. Kihara, T. Funahashi, I. Shimomura, H. Sawai, Y. Fukuda, Y. Matsuzawa, and K. Nagai. 2007. Effects of adiponectin on the renal sympathetic nerve activity and blood pressure in rats. Exp. Biol. Med. (Maywood) 232: 390-397.

Wang, Z.V. and P.E. Scherer. 2008. Adiponectin, cardiovascular function, and hypertension. Hypertension 51: 8-14.

Yamauchi, T. and J. Kamon, H. Waki, Y. Terauchi, N. Kubota, K. Hara, Y. Mori, T. Ide, K. Murakami, N. Tsuboyama-Kasaoka, O. Ezaki, Y. Akanuma, O. Gavrilova, C. Vinson, M.L. Reitman, H. Kagechika, K. Shudo, M. Yoda, Y. Nakano, K. Tobe, R. Nagai, S. Kimura, M. Tomita, P. Froguel, and T. Kadowaki. 2001. The fat-derived hormone adiponectin reverses insulin resistance associated with both lipoatrophy and obesity. Nat. Med. 7: 941-946.

CHAPTER 21

The Astroglial Leptin Receptors and Obesity

Weihong Pan,[1,*] **Hung Hsuchou,**[1] **Yi He**[1] **and Abba J. Kastin**[1]

[1]Blood-Brain Barrier Group, Pennington Biomedical Research Center, Baton Rouge, LA, USA

ABSTRACT

Leptin plays an essential role in the neuroendocrine regulation of obesity. It is now well established that leptin enters the brain from blood by a saturable transport system. Although it is known that perivascular astrocytes play important roles in nutrient transport across the blood-brain barrier (BBB) in general, it is not clear how astrocytes affect leptin transport specifically. The significance of astrocytic-endothelial interactions in leptin transport is emphasized by two novel findings in our laboratory: (1) astrocytes show region-specific expression of leptin receptors (ObR); (2) mouse models of adult-onset obesity are associated with astrogliosis and up-regulation of astrocytic ObR in the hypothalamus. These key findings challenge the neuronal dogma in the leptin field. It is possible that the glial ObR is more important in energy homeostasis once obesity is full-blown. Here we review early evidence that astrocytic ObR is involved in leptin transport, leptin signaling, obesity, and neonatal development. The results indicate that astrocytic ObR shows an opposing effect on neuronal ObR and functions as a negative regulator in energy balance. Astrocytes may determine the overall CNS actions of leptin not only by modulating the amount of leptin relayed to neurons, but also by generation of astrocytic factors. In this way, astrocytes should be considered essential functional components of the BBB that regulate leptin transport. The novel role of astrocytic ObR is providing new directions in the regulation of neuroendocrine disturbances related to obesity.

---

*Corresponding author

INTRODUCTION

Many forms of obesity are associated with increased blood leptin levels proportional to the mass of adipose tissue. Circulating leptin can cross the blood-brain barrier (BBB) and blood-cerebrospinal fluid barrier (Banks *et al.* 1996, 2003, Zlokovic *et al.* 2000, Pan and Kastin 2007a). The permeation of leptin into the hypothalamus provides a negative feedback control of feeding behavior and energy balance. Leptin that reaches the hippocampus and cerebellum may be associated with trophic support, memory consolidation, and other central nervous system (CNS) functions (Shanley *et al.* 2001, Oomura *et al.* 2006, O'Malley *et al.* 2007, Harvey 2007a, b). In conjunction with actions at circumventricular organs, retrograde axonal transport, and relay by indirect signals, the BBB provides a major route mediating the communication between leptin (and many other adipokines) and the CNS.

The major structural component of the BBB is the endothelial cell in capillaries and postcapillary venules. Endothelial cells are joined by tight junctions and underlined by a continuous basement membrane. There is high metabolic activity with increased ATP and enzymes within the endothelial cells, and there are fewer vesicles. On the other side of the basement membrane, astrocyte endfeet are closely associated with the vascular structure, and further reinforce the tightness of the non-fenestrated microvessels. While most of the studies on the functions of leptin in the CNS focus on either the cerebral endothelia composing the BBB or the effector neurons, very few studies have addressed the expression and functions of the glial leptin system. However, it is now clear that both astrocytes and microglial cells express leptin receptors (ObR). Here, we mainly review the regulatory changes and functions of astrocytic ObR in the interactions between blood-borne leptin and the CNS.

LEPTIN, BBB AND ASTROCYTES IN THE A^{vy} MICE

The agouti variable yellow (A^{vy}) mouse provides a unique model to study the defects of melanocortin receptor (especially MC4R) signaling and metabolic syndromes related to adult-onset obesity. It was discovered in the early 1960s (Dickes 1962). The spontaneous mutation in the A^{vy} mice involves the insertion of a retrotransposon at the promoter region of the gene encoding agouti signaling protein. The resultant ectopic over-expression of agouti-related protein (AgRP) antagonizes melanocortin receptors. In the skin, dysfunction of MC1R signaling increases pheomelanin synthesis and leads to subapical yellow bands of hair. In the hypothalamus, dysfunction of MC4R signaling results in hyperphagia and obesity. This produces the two prominent phenotypic features of an agouti-colored coat and obesity (Fig. 1). As a result, A^{vy} mice have been used for various obesity and metabolic studies in the past few decades (Wolff and Whittaker 2005).

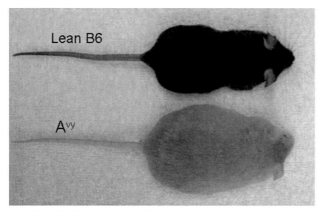

Fig. 1   Two-month-old adult-onset obese A^{vy} mouse and its lean 2-mo-old B6 control.
Color image of this figure appears in the color plate section at the end of the book.

There are only a few studies examining the permeation of ingestive peptides across the BBB in the A^{vy} mouse. AgRP is a candidate as a reverse antagonist to MC4R and inhibitor of leptin-induced suppression of food intake and body weight (Ebihara *et al.* 1999). However, AgRP does not cross most parts of the BBB by a saturable transport system (Kastin *et al.* 2000, Pan *et al.* 2005). By contrast, the C-terminal mahogany peptide crosses the BBB by a specific system (Kastin and Akerstrom 2000). In A^{vy} mice, the brain uptake of mahogany peptide shows age-related changes; in 1-mo-old (m/o) A^{vy} mice it is higher than in 7 m/o A^{vy} mice, 12 m/o A^{vy} mice, or B6 mice studied at any of these three ages (Pan and Kastin 2007b). Since mahogany protects against obesity (Dinulescu *et al.* 1998, Nagle *et al.* 1999), the increased permeation of mahogany peptide preceding the fat surge might serve to delay the onset of obesity in the A^{vy} mice, which are susceptible to strong epigenetic influences (Pan and Kastin 2007b).

The A^{vy} and ob/ob mice share an obesity phenotype but have different underlying etiologies (Wolff 1997). We propose that leptin transport across the BBB in A^{vy} mice not only shows adaptive changes related to the dynamics of their blood leptin concentrations, but also plays an active role in "repairing" the neuroendocrine circuitry induced by excess AgRP in the brain. Indeed, there is a decrease in the apparent influx rate of ^{125}I-leptin crossing the BBB after intravenous injection in the A^{vy} mice at ages 2 mo and 8 mo compared with the lean B6 controls (Fig. 2). This difference is not observed by *in situ* brain perfusion studies, indicating that it is mainly caused by circulating factors, such as elevated leptin levels or soluble receptors, interacting with leptin. More interestingly, the expression of the ObRa subtype, known to mediate most of the leptin transport, differs in astrocytes between the A^{vy} and B6 groups but not in the cerebral microvessels composing the BBB. Immunofluorescent staining unexpectedly revealed that many of the ObR (+) cells are astrocytes, and that the A^{vy} mice show a significantly greater increase of ObR(+) astrocytes in the hypothalamus than the B6 mice (Pan *et al.* 2008a). Thus, although leptin permeation from the circulation is slower in the A^{vy} mice, the increase in the overall ObR expression in astrocytes and the increased ObRb mRNA in microvessels suggest a heightened CNS sensitivity to circulating leptin in A^{vy} mice.

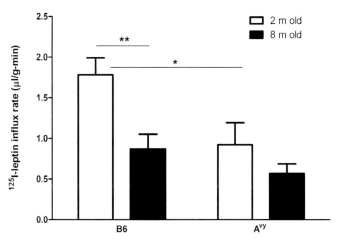

Fig. 2 Permeation of leptin into the brain of 2- and 8-mo-old A^{vy} and B 6 mice. The fast influx in the 2-mo-old normal mouse decreases with age and is higher than at either age in the obese A^{vy} mouse. Modified from Pan *et al.* (2008a).

ASTROCYTIC ObR AND LEPTIN SIGNALING

Primary astrocytes obtained from 7-d-old neonates were purified and passaged to 6 cm dishes. Confluent cells were serum-starved for 12 h and stimulated with PBS (the 0 min control) or leptin (30 nM) in serum-free DMEM for 10, 30, or 60 min. The cells were lysed and the signaling activation was determined by western blotting. Leptin induced a time-dependent activation of pERK1/2, pAkt, and pSTAT3, particularly at the serine^{727} residue. At 1 h, pAkt and pSTAT3-Y^{705} signals were decreased. By contrast, pAMPK, *c-Fos*, and the housekeeping gene β-actin did not show apparent changes. The results show that astrocytes activate a specific subset of cellular signaling elements by phosphorylation (Fig. 3). This pattern differs from what we have observed in SH-SY5Y neurons (He *et al.* 2009) and indicates the specificity of astrocytic leptin signaling.

Primary astrocytes can also respond to leptin by an increase of intracellular calcium signals. The induction is quite rapid, occurring within 100 sec, and lasting about 1 min. There is desensitization of the response, shown by the reduced amplitude of the calcium signals with repetitive application of leptin (Hsuchou *et al.* 2009). It is conceivable that astrocytes in different brain and spinal cord regions show variable levels of ObR subtype expression and coupling of intracellular signaling pathways, thus exhibiting differential responses.

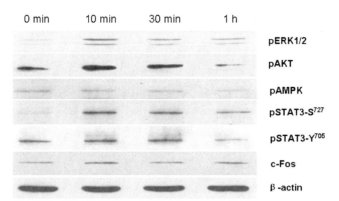

	0 min	10 min	30 min	1 h	
					pERK1/2
					pAKT
					pAMPK
					pSTAT3-S^{727}
					pSTAT3-Y^{705}
					c-Fos
					β -actin

Fig. 3 Western blot of protein changes in astrocytes after treatment with leptin. Leptin time-dependently increased pERK1/2, pAkt, and pSTAT3, particularly at the serine^{727} residue, while pAkt and pSTAT3-Y^{705} signals were decreased at 1 h.

How does astrocytic leptin signaling affect neuronal functions? The astrocyte is well positioned to regulate synaptic transmission and neurovascular coupling. Activation of astrocytes with resulting calcium oscillations induces the accumulation of arachidonic acid and the release of the gliotransmitters glutamate, D-serine, and ATP. Inflammatory mediators (in particular prostaglandin E2 and epoxyeicosatrienoic acids in astrocytic endfeet) in turn control the cerebral microcirculation (Haydon and Carmignoto 2006). Insults to the CNS induce reactive astrogliosis, marked by hypertrophy, proliferation, and increased expression of the intermediate elements GFAP, vimentin, and nestin (Pekny et al. 2007). There is also induction of specific markers such as M22, Ml, 6.17, J1-31, S100β, growth factors, cytokines, recognition molecules, gangliosides, enzymes, proto-oncogenes, and even selective neuronal markers (Ridet et al. 1997). The reactive astrocytes appear to play dual roles in neuroregeneration, initially being neuroprotective, and later inhibiting axonal reconnection by contributing to the chemical barrier and mechanical scar (Faulkner et al. 2004, Xiang et al. 2005).

After intracerebroventricular (icv) injection of leptin that is conjugated with an Alexa-488 fluorescent dye, we followed its distribution at different time intervals. It appears that Alexa488-leptin is mainly taken up by neurons during the time course studied. In Fig. 4, the mice were perfuse-fixed 0 min, 20 min, 60 min, and 3 h after Alexa568-leptin icv (0.5 μg/μl in 2 μl of PBS). The 0 time group involved insertion of the cannula without initiation of the syringe pump, and the mice were perfused 5 min later. In the lean control mice, Alexa568-leptin reached neurons in the hippocampus by 60 min. In mice with diet-induced obesity (DIO), prominent fluorescence was seen in the choroid plexus and part of the hilus at 20 min, but was no longer present in the hippocampus at 60 min and 3 h. Overall, the results show that obesity modulates the distribution of leptin in the hippocampus.

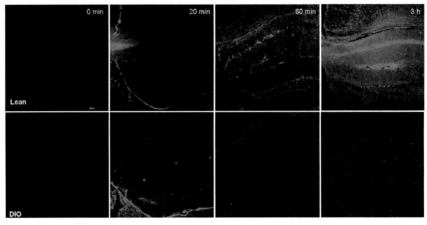

Fig. 4 Distribution of fluorescently labeled leptin in the hippocampus at various times after icv administration. In the lean control (top), Alexa568-leptin reached neurons in the hippocampus by 60 min. In the obese mouse (bottom), prominent fluorescence was seen in the choroid plexus and part of the hilus at 20 min, but was no longer present in the hippocampus at 60 min and 3 h.

Color image of this figure appears in the color plate section at the end of the book.

The role of astrocytic activities in the distribution of Alexa568-leptin was determined by use of fluorocitrate pretreatment before icv injection of Alexa568-leptin. In the presence of this metabolic inhibitor of astrocytic activity (Clarke 1991, Swanson and Graham 1994), there were more Alexa568-leptin (+) cells in the arcuate nucleus of the hypothalamus (Fig. 5). This finding in rats was replicated

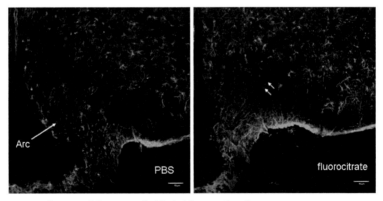

Fig. 5 Distribution of fluorescently labeled leptin after fluorocitrate pretreatment preceding icv injection of Alexa568-leptin. In the presence of this metabolic inhibitor of astrocytic activity, there was an increased amount of Alexa568-leptin (+) cells in the arcuate nucleus (arrows) of the hypothalamus as compared with the diluent-treated controls.

Color image of this figure appears in the color plate section at the end of the book.

in the obese A^{vy} mice, as fluorocitrate treatment increased neuronal leptin uptake after icv injection and increased leptin-induced pSTAT3 signaling (Pan *et al.* 2010). In the obese A^{vy} mice, there was a higher level of GFAP expression in the hypothalamus (Fig. 6). This may at least partly explain the delayed and decreased distribution of leptin after icv injection.

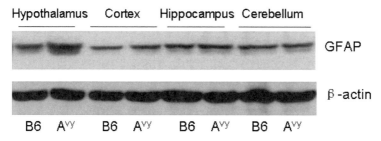

Fig. 6 Western blot of protein changes in obese A^{vy} and lean B6 control mice, showing a higher level of GFAP expression in the hypothalamus of the A^{vy} mouse.

Obesity also modulates leptin signaling by pSTAT3 activation. The basal activation of pSTAT3 in the DIO mice showed a low-level, microglia-like distribution in the hypothalamus. At 20 min after leptin icv, the activation of pSTAT3 in the lean B6 mice was seen in focal areas in the CA1 of the hippocampus, in the α2 tanycytes lining the third ventricle, and in the adjacent arcuate nucleus. The fluorescent intensity of pSTAT3 was much diminished in the corresponding regions of the DIO mouse (Fig. 7). However, *c-Fos* immunohistochemistry suggests that there was a more sustained and higher level of cellular activation in obesity. Thus, DIO not only reduced the extent of pSTAT3 activation in response to leptin icv, it also shifted cellular subtypes in the pattern of activation.

CELLULAR STUDIES OF THE INFLUENCE OF ASTROCYTIC OBR ON LEPTIN TRANSPORT

To determine whether astrocytes affect leptin transport, we used hCMEC/D3 cerebral endothelial cells as an *in vitro* model of the BBB. The endothelia were grown to a confluent monolayer on Transwell inserts to measure the apical-to-basolateral flux of leptin over time. The presence of C6 astrocytes caused a significant reduction of leptin flux across hCMEC/D3, partly because of its induction of tight junction proteins and facilitation of their intracellular distribution. In the presence of astrocytes, the endothelial barrier was reinforced, as shown by a decrease of paracellular permeation of sodium fluoroscein and albumin (Hsuchou *et al.* 2010). However, C6 cells over-expressing different subtypes of leptin receptors (ObR) have differential effects on leptin permeation (Fig. 8). Specifically, over-expression of the long isoform ObRb completely reverses the effect of C6 cells.

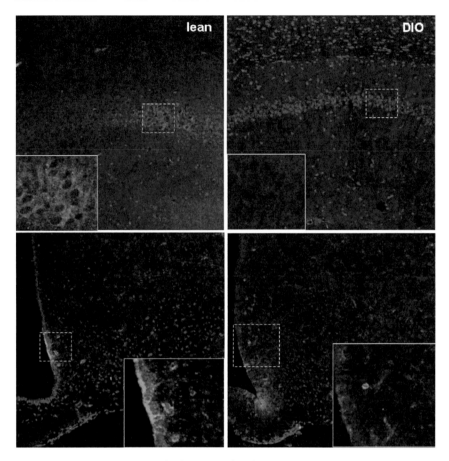

Fig. 7   pSTAT3 signaling in lean (left) and DIO (right) mice 20 min after leptin administered icv. The activation of pSTAT3 in the lean B6 mice, seen in focal areas in the CA1 of the hippocampus, in the α2 tanycytes lining the third ventricle, and in the adjacent arcuate nucleus (boxes) was much less in the corresponding regions of the DIO mouse.

Color image of this figure appears in the color plate section at the end of the book.

Over-expression of the soluble isoform ObRe also shows a partial effect. This suggests that the effects of astrocytes in decreasing leptin transport are mainly caused by ObRa, and that ObRb and ObRe have antagonistic effects. Thus, the reduction of permeation is a combination of decreased leptin transport and decreased paracellular permeability.

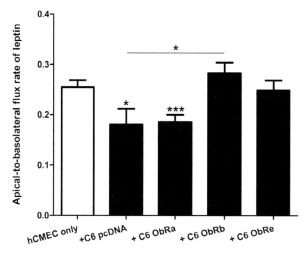

Fig. 8 Flux rates of leptin across the hCMEC cerebral endothelial cells co-cultured with C6 astrocytic cells over-expressing different subtypes of leptin receptors (ObR). Over-expression of the long isoform ObRb completely reverses the effect of C6 cells in decreasing leptin flux. Over-expression of the soluble isoform ObRe also shows a partial effect, whereas over-expression of ObRa, the major transporting isoform, is without effect. This suggests that the effects of astrocytes in decreasing leptin transport are mainly caused by ObRa, and that ObRb and ObRe have antagonistic effects. Modified from Hsuchou *et al.* (2010).

LEPTIN AND NEONATAL DEVELOPMENT

In the hypothalamus, experimental evidence has led to a proposed model in which anorexigenic signals, including Janus kinases and Signal Transducer and Activator for Transcription (STAT), activated by leptin converge on melanocortin receptor 4 neurons (Uotani *et al.* 1999, Nijenhuis *et al.* 2001, Bouret and Simerly 2007). In cerebral microvessels from neonates, ObRa, ObRb, ObRc, and ObRe mRNA were all higher than in adults, but ObRd was not detectable. The hypothalamus showed similar age-related changes except for ObRb, which was higher in adults. The homologous receptor gp130 did not show significant age-related changes in either region. Despite the increase of leptin receptors, leptin permeation across the BBB after intravenous injection was less in the neonates. *In situ* brain perfusion with blood-free buffer showed no significant difference in the brain uptake of leptin between neonates and adults, indicating an antagonistic role of leptin-binding proteins in the circulation, especially the soluble receptor ObRe (Pan *et al.* 2008b). Thus, the developmental changes observed for leptin receptors unexpectedly failed to correlate with the entry of leptin into brain, indicating different functions of the receptors in neonates and adults. Further analysis of the developmental changes of astrocytic ObR subtypes and distribution would likely provide novel information on the role of astrocytes in neuroendocrine control of obesity.

Summary Points

- Leptin enters the brain from blood by a saturable transport system.
- Astrocytic ObR is involved in leptin transport.
- Astrocytic ObR is involved in leptin signaling.
- Astrocytic ObR is involved in obesity.
- The cellular models showed differential role of various ObR subtypes in the astrocytic effects on leptin transport across endothelial cell monolayer.
- Leptin is involved in neonatal development.

Abbreviations

AgRP	:	agouti-related protein
BBB	:	blood-brain barrier
CNS	:	central nervous system
DIO	:	diet-induced obesity
icv	:	intracerebroventricular
ObR	:	leptin receptor

Acknowledgement

Grant support was provided by NIH (DK54880 and NS62291).

References

Banks, W.A. 2003. Is obesity a disease of the blood-brain barrier? Physiological, pathological, and evolutionary considerations. Curr. Pharm. Des. 9: 801-809.

Banks, W.A. and A.J. Kastin, W. Huang, J.B. Jaspan, and L.M. Maness 1996. Leptin enters the brain by a saturable system independent of insulin. Peptides 17: 305-311.

Bouret, S.G. and R.B. Simerly. 2007. Development of leptin-sensitive circuits. J. Neuroendocrinol. 19: 575-582.

Clarke, D.D. 1991. Fluoroacetate and fluorocitrate: mechanism of action. Neurochem. Res. 16: 1055-1058.

Dickes, M.M. 1962. A new viable yellow mutation in the house mouse. J. Hered. 53: 84-86.

Dinulescu, D.M. and W. Fan, B.A. Boston, K. McCall, M.L. Lamoreux, K.J. Moore, J. Montagno, and R.D. Cone. 1998. Mahogany (mg) stimulates feeding and increases basal metabolic rate independent of its suppression of agouti. Proc. Natl. Acad. Sci. USA 95: 12707-12712.

Ebihara, K. and Y. Ogawa, G. Katsuura, Y. Numata, H. Masuzaki, N. Satoh, M. Tamaki, T. Yoshioka, M. Hayase, N. Matsuoka, M. Aizawa-Abe, Y. Yoshimasa, and K. Nakao. 1999. Involvement of agouti-related protein, an endogenous antagonist of hypothalamic melanocortin receptor, in leptin action. Diabetes 48: 2028-2033.

Faulkner, J.R. and J.E. Herrmann, M.J. Woo, K.E. Tansey, N.B. Doan, and M.V. Sofroniew. 2004. Reactive astrocytes protect tissue and preserve function after spinal cord injury. J. Neurosci. 24: 2143-2155.

Harvey, J. 2007a. Leptin regulation of neuronal excitability and cognitive function. Curr. Opin. Pharmacol. 7: 643-647.

Harvey, J. 2007b. Leptin: a diverse regulator of neuronal function. J. Neurochem. 100: 307-313.

Haydon, P.G. and G. Carmignoto. 2006. Astrocyte control of synaptic transmission and neurovascular coupling. Physiol. Rev. 86: 1009-1031.

He, Y. and A.J. Kastin, H. Hsuchou, and W. Pan. 2009. The Cdk5/p35 kinases modulate leptin-induced STAT3 signaling. J. Mol. Neurosci. 39: 49-58.

Hsuchou, H. and Y. He, A.J. Kastin, H. Tu, E.N. Markadakis, R.C. Rogers, P.B. Fossier, and W. Pan. 2009. Obesity induces functional astrocytic leptin receptors in hypothalamus. Brain 132: 889-902.

Hsuchou, H. and A.J. Kastin, H. Tu, N.J. Abbott, P.O. Couraud, and W. Pan. 2010. Role of astrocytic leptin receptor subtypes on leptin permeation across hCMEC/D3 endothelial cells. J. Neurochem.,115: 1288-98.

Kastin, A.J. and V. Akerstrom. 2000. Mahogany (1377-1428) enters brain by a saturable transport system. J. Pharmacol. Exp. Ther. 294: 633-636.

Kastin, A.J. and V. Akerstrom, and L. Hackler. 2000. Agouti-related protein(83-132) aggregates and crosses the blood-brain barrier slowly. Metabolism 49: 1444-1448.

Nagle, D.L. and S.H. McGrail, J. Vitale, E.A. Woolf, B.J. Dussault Jr., L. DiRocco, L. Holmgren, J. Montagno, P. Bork, D. Huszar, V. Fairchild-Huntress, P. Ge, J. Keilty, C. Ebeling, L. Baldini, J. Gilchrist, P. Burn, G.A. Carlson, and K.J. Moore. 1999. The mahogany protein is a receptor involved in suppression of obesity. Nature 398: 148-152.

Nijenhuis, W.A. and J. Oosterom, and R.A. Adan. 2001. AgRP(83-132) acts as an inverse agonist on the human-melanocortin-4 receptor. Mol. Endocrinol. 15: 164-171.

O'Malley, D. and N. MacDonald, S. Mizielinska, C.N. Connolly, A.J. Irving, and J. Harvey. 2007. Leptin promotes rapid dynamic changes in hippocampal dendritic morphology. Mol. Cell. Neurosci. 35: 559-572.

Oomura, Y. and N. Hori, T. Shiraishi, K. Fukunaga, H. Takeda, M. Tsuji, T. Matsumiya, M. Ishibashi, S. Aou, X.L. Li, D. Kohno, K. Uramura, H. Sougawa, T. Yada, M.J. Wayner, and K. Sasaki 2006. Leptin facilitates learning and memory performance and enhances hippocampal CA1 long-term potentiation and CaMK II phosphorylation in rats. Peptides 27: 2738-2749.

Pan, W. and A.J. Kastin. 2007a. Adipokines and the blood-brain barrier. Peptides 28: 1317-1330.

Pan, W. and A.J. Kastin. 2007b. Mahogany, blood-brain barrier, and fat mass surge in $A(^{VY})$ mice. Int. J. Obesity 31: 1030-1032.

Pan, W. and A.J. Kastin, Y. Yu, C.M. Cain, T. Fairburn, A.M. Stutz, C. Morrison, and G. Argyropoulos. 2005. Selective tissue uptake of agouti-related protein(82-131) and its modulation by fasting. Endocrinology 146: 5533-5539.

Pan, W. and H. Hsuchou, Y. He, A. Sakharkar, C. Cain, C. Yu, and A.J. Kastin. 2008a. Astrocyte leptin receptor (ObR) and leptin transport in adult-onset obese mice. Endocrinology 149: 2798-2806.

Pan, W. and H. Hsuchou, H. Tu, and A.J. Kastin. 2008b. Developmental changes of leptin receptors in cerebral microvessels: unexpected relation to leptin transport. Endocrinology 149: 877-885.

Pan, W. and H. Hsuchou, S.G. Bouret, and A.J. Kastin. 2011. Astrocytes modulate distribution and neuronal signaling of leptin in the hypothalamus of obese A(vy) mice. J. Mol. Neurosci. 43: 478-84.

Pekny, M. and U. Wilhelmsson, Y.R. Bogestal, and M. Pekna. 2007. The role of astrocytes and complement system in neural plasticity. Int. Rev. Neurobiol. 82: 95-111.

Ridet, J.L. and S.K. Malhotra, A. Privat, and F.H. Gage. 1997. Reactive astrocytes: cellular and molecular cues to biological function. Trends Neurosci. 20: 570-577.

Shanley, L.J. and A.J. Irving, and J. Harvey. 2001. Leptin enhances NMDA receptor function and modulates hippocampal synaptic plasticity. J. Neurosci. 21: RC186.

Swanson, R.A. and S.H. Graham. 1994. Fluorocitrate and fluoroacetate effects on astrocyte metabolism *in vitro*. Brain Res. 664: 94-100.

Uotani, S. and C. Bjorbaek, J. Tornoe, and J.S. Flier. 1999. Functional properties of leptin receptor isoforms: internalization and degradation of leptin and ligand-induced receptor downregulation. Diabetes 48: 279-286.

Wolff, G. 1997. Obesity as a pleiotropic effect of gene action. J. Nutr. 127: 1897S-1901S.

Wolff, G.L. and P. Whittaker. 2005. Dose-response effects of ectopic agouti protein on iron overload and age-associated aspects of the A^{vy}/a obese mouse phenome. Peptides 26: 1697-1711.

Xiang, S. and W. Pan, and A.J. Kastin. 2005. Strategies to create a regenerating environment for the injured spinal cord. Curr. Pharm. Des. 11: 1267-1277.

Zlokovic, B.V. and S. Jovanovic, W. Miao, S. Samara, S. Verma, and C.L. Farrell. 2000. Differential regulation of leptin transport by the choroid plexus and blood-brain barrier and high affinity transport systems for entry into hypothalamus and across the blood-cerebrospinal fluid barrier. Endocrinology 141: 1434-1441.

Section III
Diseases and Conditions

CHAPTER 22

Adipokines and Obesity

Alberto O. Chavez[1] and Devjit Tripathy[1,*]
[1]Division of Diabetes. University of Texas Health Science Center at San Antonio, 7703 Floyd Curl Dr. MS 7886, San Antonio, Texas 78229, USA

ABSTRACT

Obesity is associated with a significant increase in morbidity and mortality. Over the last decade, adipose tissue has emerged as a major player in the regulation of lipid and glucose metabolism. It is now considered a major endocrine organ because of its size and secretion of numerous metabolically active molecules, collectively known as adipokines, which play a key role in regulation of energy homeostasis. In obesity, excessive and abnormal distribution of adipose tissue leads to the activation and recruitment of immune cells followed by release of pro-inflammatory adipokines, which act both locally and in distant tissues (i.e., liver, skeletal muscle, endothelium, pancreas) to cause insulin resistance. Collectively, insert these abnormalities eventually lead to the development of type 2 diabetes mellitus, dyslipidemia, endothelial dysfunction with accelerated atherosclerosis, and overt cardiovascular disease. The number of newly identified adipokines continues to expand, whereas the relevance and specific function of each molecule remains to be fully elucidated. The understanding of intricate adipocyte biology and the underlying molecular basis of obesity and its complications will facilitate the development of novel preventive and more effective therapeutic regimens.

INTRODUCTION

Obesity has emerged as the epidemic of the new millennium and represents a major public health problem. In the United States its prevalence has steadily grown

*Corresponding author

to such levels that more than 32% of adults are obese by current standards. It also is a well-known independent risk factor for metabolic syndrome (MS) and as a result is strongly associated with cardiovascular disease (CVD) and decreased life expectancy (Ogden *et al.* 2006).

For many years adipose tissue (AT) was considered as a site in which excessive fat is stored and free fatty acids (FFA) were identified as its only secretory products. However, it has become evident that in addition to FFA, adipocytes synthesize and secrete a host of proteins that collectively are designated as adipokines. These molecules have local autocrine/paracrine effects, as well as systemic effects. Tumor necrosis factor alpha (TNF-α) was the first pro-inflammatory cytokine shown to be constitutively expressed in the adipocytes of obese insulin-resistant animals and neutralization of TNF-α with anti-TNF-α antibody decreased insulin resistance (Hotamisligil *et al.* 1993). Since then, a number of other adipokines have been identified including interleukin 6, plasminogen activator inhibitor 1, resistin, IL-1, leptin, adiponectin, resistin, and visfatin. All of these molecules exhibit positive or negative effects on insulin sensitivity in a paracrine or endocrine manner. Thus, expression and secretion into plasma of these and other cytokines could provide a link between insulin resistance and chronic inflammation in obesity. Because AT plays a critical role in the regulation of fuel homeostasis, this chapter focuses on the mechanisms by which obesity leads to an imbalance between pro- and anti-inflammatory cytokines (adipokines) causing insulin resistance both locally and systemically and eventually type 2 diabetes (T2DM), endothelial dysfunction and increase in overall CVD (Antuna-Puente *et al.* 2008).

CHRONIC LOW-GRADE INFLAMMATION IN OBESITY

Body Fat Distribution and Adipokines Secretion Profile

Not only the total amount of AT, but the distribution of fat in different parts of the body is an important determinant of the adipokine secretion and fat metabolism. Upper body fat, which includes intra-abdominal fat, consists of omental and mesenteric (visceral) fat depots. Increased visceral fat is associated with greater metabolic complications than total or subcutaneous fat. Visceral fat depot correlates more strongly with insulin resistance, dyslipidemia, and diabetes than total body fat. Although the exact mechanism of the strong association between visceral fat and metabolic abnormalities is not known, one hypothesis is that visceral fat produces more metabolically active substances than subcutaneous fat. Excess FFA release, AT inflammation and consequently more adipokine secretion have been proposed to be the possible mediators of adverse effects of excess visceral fat. Animal studies have shown increased expression of inflammatory cytokines in the visceral fat than subcutaneous AT. Of note, loss of subcutaneous fat by liposuction was not associated with any improvement in metabolic parameters in obese individuals (Klein *et al.* 2004). In contrast, thiazolidinedione treatment increases the total amount of body fat, but decreases the visceral fat and improves insulin sensitivity (Miyazaki *et al.* 2002).

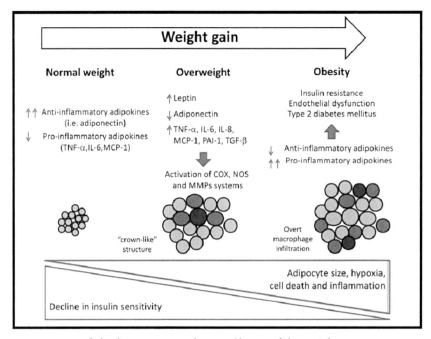

Fig. 1 Patterns of adipokine secretion in the natural history of obesity. Schematic representation of the change in the pattern of secretion of adipokines and insulin sensitivity from the healthy lean state (left) to frank obesity (right). Hypoxia, cell death and adipokines have been shown to trigger macrophage infiltration in adipose tissue and increase in adipokines secretion which results in a vicious circle that activates inflammation that eventually progresses to type 2 diabetes mellitus and cardiovascular disease. COX = cycloxygenase, NOS = nitric oxide synthase, MMP = matrix metalloproteinase, TGF-beta = transforming growth factor-beta. Dark circle = dead adipocyte, red circles = macrophages (M1 stage).

Color image of this figure appears in the color plate section at the end of the book.

Mechanisms of Macrophage Infiltration in Adipose Tissue

An early finding in obesity research was the fact that AT is in an increased state of inflammation. Although this low-grade systemic inflammatory state had been clinically and biochemically described, it was only in recent years that macrophage infiltration was identified as the primary cause of the chronic immune activation of AT (Weisberg *et al.* 2003). Additionally, the other main determinant of this low-grade inflammation is the increased expression of pro-inflammatory adipokines induced by various local and systemic mechanisms. This increase in humoral response by AT is clearly linked to insulin resistance, the development of T2DM and premature CVD.

Despite extensive research, the sequence of events that direct macrophages to infiltrate AT to exert their pro-inflammatory actions has not been completely elucidated, although extensive data has become available to support several hypotheses. Initial reports have suggested that during the natural history of obesity and AT expansion following over nutrition, there is an imbalance between the size of an increasingly hypertrophic adipocyte and the oxygen delivery through the local vasculature. This local hypoxia may trigger the synthesis and release of several inflammatory adipokines and eventually lead to impaired cell function and death, which in turn will attract macrophages to phagocytose cellular debris. The importance of this mechanism has been supported by studies that reveal aggregation of macrophages in crown-like structures around dead adipocytes. These subtypes of macrophages have been reported to have a marked pro-inflammatory phenotype (M1) in contrast to other anti-inflammatory macrophages (M2) within the AT. This switching of macrophage populations and the imbalance between M1 and M2 phenotypes in the obese fat has been proposed to favor a pro-inflammatory state (Lumeng *et al.* 2007).

Pro-inflammatory Adipokines in Obesity

TNF-α was one of the first adipokines described that linked adipokines to insulin resistance. TNF-α is a 26 kDa transmembrane protein, which is released into the circulation as a 17 kDa soluble protein after extracellular cleavage by a metalloproteinase. Although circulating TNF-α concentrations are low, and have no clear correlation with insulin resistance, the tissue expression levels of TNF-α correlate inversely with insulin sensitivity. TNF-α inhibits insulin signaling by blocking insulin receptor tyrosine kinase activity and inducing serine phosphorylation of IRS-1. In obese humans, TNF-α is over-expressed in AT. TNF-α induces AT lipolysis by activation of JNK pathway. Administration of soluble forms of TNF-α receptors blocks these effects (Barnes and Miner 2009).

TNF-α also promotes insulin resistance by enhancing the secretion of other inflammatory adipokines in AT. TNF-α is known to suppress adiponectin production by adipocytes, while inducing the secretion of other adipokines by AT such as IL-6, MCP-1 and PAI-1, primarily through activation of Foxo-1 and therefore causing insulin resistance, in addition to its ability to increase lipolysis and FFA load to the liver with the resultant increase in the rate of triglyceride (TG) synthesis and VLDL secretion. Thus, TNF-α elevation has major implications in lipid metabolism. Moreover, an impaired regulation of the TIMP-3/TACE protelolytic system that regulates TNF-α concentrations and actions has been recently described in obese T2DM subjects and up-regulated in response to an acute elevation of FFA concentrations in healthy subjects (Monroy *et al.* 2009).

IL-6 is a 22-27 kDa protein with various degrees of glycosylation and closely related to TNF-α. Plasma levels of IL-6 correlate positively with fat mass, insulin resistance and plasma FFA levels. In obese animals, central administration of IL-6 leads to increased energy expenditure and decreased body fat. Alternatively, in humans it induces dose-dependent increase in plasma glucose and severity of insulin resistance. IL-6 has been shown to increase plasma FFA concentrations secondary to increased adipose lipolysis of TG. IL-6 induces insulin resistance by down-regulation of IRS-1 and up-regulation of the suppressor of cytokine signaling 3, a negative regulator of insulin signaling. Moreover, IL-6 inhibits adiponectin production by adipocytes that might contribute to IL-6-induced insulin resistance (Hoene and Weigert 2008).

As discussed before, macrophage infiltration is a major determinant of inflammation in AT. MCP-1 plays an important role in recruitment of macrophages into AT. Mice lacking MCP-1 or MCP-1 receptors have reduced macrophage infiltration and are relatively more insulin sensitive. Enlarged adipocytes seen in obese individuals were thought to be the primary source of MCP-1 and other cytokines, but recent studies have shown that bone-marrow–derived macrophages are equally important. Expression of MCP-1 appears to be related to AdipoR1 and AdipoR2 activation. Mice lacking AdipoR1 and AdipoR2 display increased MCP-1 expression and macrophage infiltration. Human studies have shown increased expression of MCP-1 gene expression in AT in obese individuals supporting its role in local inflammation and increased adipokine secretion (Dahlman *et al.* 2005).

PAI-1 is elevated in subjects with metabolic complications of obesity (i.e., metabolic syndrome) and is expressed and secreted by the endothelial cells within the stromal fraction of the AT. Its effect as inhibitor of both serine proteases and plasminogen activators may contribute to the development of a pro-thrombotic state. The best-characterized factors that regulate gene expression of PAI-1 by binding to its promoter are TGF-β and thrombospondin, the latter being also identified as a novel adipokine expressed preferentially in the adipocytes within the AT of obese insulin-resistant subjects with numerous effects on angiogenesis, cell proliferation and wound healing (Varma *et al.* 2008, Alessi *et al.* 2007).

Additionally, elevated levels of leptin have been shown to modulate macrophage diapedesis, increase expression of vascular adhesion molecules and stimulate local over-expression of other adipokines such as monocyte chemotactic protein 1 (MCP-1), which leads to macrophage infiltration. Alternatively, an early infiltration by T lymphocytes by still unclear mechanisms has been observed during initial stages of obesity. Also, early activation of pre-adipocytes with expression of certain antigenic and functional features of macrophages has been observed (Gutierrez *et al.* 2009).

OBESITY AND THE REGULATION OF LIPID AND GLUCOSE METABOLISM BY ADIPOKINES

Obesity and Insulin Resistance

Insulin resistance is the decreased ability of insulin to elicit its physiological effects in target organs and is a hallmark of obesity. It constitutes a well-known risk factor and usually precedes the onset of T2DM. Increased release of adipokines by AT macrophages and adipocytes causes insulin resistance in a number of ways.

In the postprandial state, AT accounts for less than 4-5% of whole body glucose uptake, while skeletal muscle accounts for ~80-90% (DeFronzo 2004). From the quantitative standpoint, therefore, it is clear that resistance to effect of insulin in AT with respect to glucose metabolism cannot explain the systemic (whole body, which primarily reflects muscle) insulin resistance observed in obesity and type 2 diabetes. This raises an important question: how does inflammation in the AT lead to systemic (muscle and liver) insulin resistance in humans? Possible explanations include the following: (1) decreased insulin responsiveness in AT leads to increased lipolysis and elevated plasma FFA concentrations, which are known to cause insulin resistance in muscle and liver, and (2) cytokines elaborated by the AT and resident macrophages are released into the circulation and act at distal sites (i.e., skeletal muscle and liver) to impair insulin signaling.

Table I Key features of metabolic complications of obesity

- Adipose tissue is formed by adipocytes (~50%), resident macrophages (10%) and stromal cells (preadipocytes, endothelium and epithelial cells) that secrete a myriad adipokines and growth factors that tightly regulate energy metabolism.
- Enlarged adipocyte and AT expansion with activated macrophage infiltration initiate an inflammatory cascade and disregulated lipolytic activity mediated by locally secreted adipokines with deleterious effects on lipoprotein production and clearance, as well as insulin secretion and glucose uptake.
- The chronic low-grade inflammation of obesity is characterized by decreased concentrations of adiponectin, increment of circulating levels of leptin, resistin, TNF-α, IL-6, MCP-1, PAI-1 and activation of the RAS in the AT.
- Clinical manifestations of this adipokine imbalance include hypertension, dyslipidemia, impaired glucose tolerance and T2DM, endothelial dysfunction and premature CVD, collectively known as the insulin resistance (metabolic) syndrome.

The table lists the sequence of events that lead to the initial disruption of the tight regulation of energy homeostasis, before the appearance of clinical manifestations of metabolic syndrome and CVD. AT = adipose tissue; TNF-alpha = tumor necrosis factor alpha; IL-6 = interleukin 6; MCP = monocyte chemotactic protein 1; PAI-1 = plasminogen activator inhibitor 1; RAS = renin-angiotensin system; T2DM = type 2 diabetes mellitus; CVD = cardiovascular disease.

Adipokines with Central and Peripheral Effects on Insulin Sensitivity in Obesity

Leptin is a 16 kDa peptide produced primarily by AT and controls energy expenditure and food intake. Leptin communicates the repletion of peripheral energy stores to the brain, thereby suppressing feeding and permitting energy expenditure through a variety of neuroendocrine and autonomic mechanisms. In rodents the lack of (*ob/ob*) or resistance to effect of leptin because of lack of leptin receptors (*db/db*) leads to marked obesity (Friedman and Halaas 1998). Absence of leptin or mutations in leptin receptors have been reported to lead to massive obesity in humans (Farooqi *et al.* 2007). However, these mutations are extremely rare and in fact a common variety of obesity is associated with increased leptin levels. Leptin effects are mediated by its receptors located in the central nervous system, AT and skeletal muscle. Leptin signaling leads to activation of signal transducer and activator of transcription-3 (STAT-3) and various kinases

Table 2   Adipokines secreted by adipose tissue and its effects in energy homeostasis

Pro-inflammatory cytokines associated to impaired insulin resistance and action
  1. Leptin
  2. TNF-alpha
  3. IL-6
  4. Resistin

Acute phase protein and vascular reactivity
  1. PAI-1
  2. C-reactive protein

Enzymatic systems
  1. RAS
  2. TIMP-3/TACE
  3. NOS pathway
  4. COX pathway

Complement-like factors
  1. Adiponectin
  2. Adipsin

Others
  1. RBP4
  2. Lipocalin-2
  3. Visfatin
  4. Apelin
  5. Omentin
  6. Vaspin

This table presents the most widely studied adipokines involved in the complications of obesity by structure and action. Adipokines that are newly identified nature and thus not fully characterized as metabolic regulators or have a controversial role are listed as Others.

including PI-3 kinase, and AMP-activated protein kinase (AMPK). Additionally recent studies have shown that central administration of leptin leads to activation of sympathetic nervous system. Leptin is also an important regulator of the reproductive system and leptin deficiency is associated with hypogonadism.

High levels of circulating leptin levels are seen in the common form of obesity. Failure of high circulating leptin levels in obesity to promote weight loss suggests a state of "leptin resistance", the molecular mechanism of which remains poorly understood. Studies in rodent models of obesity shed some light on this phenomenon. High-fat diet in rodents increase circulating leptin levels and impair leptin transport across the blood-brain barrier. In obese subjects, the ability of leptin to activate AMP kinase is impaired. When leptin was first discovered it was presumed to have enormous therapeutic potential for obesity. However, so far the use of leptin has been limited to rare human syndromes of complete leptin deficiency and to lipodystrophy (Oral *et al.* 2002). Therapy with human leptin completely reversed the hyperphagia, obesity, and hypogonadism associated with congenital leptin deficiency. Similarly, chronic treatment with leptin ameliorated several metabolic abnormalities in patients with congenital lipodystrophy (Oral *et al.* 2002). However, obese individuals are resistant to the satiety and weight-reducing effect of leptin. A recent combination of leptin and amylin has been shown to restore hypothalamic responsiveness to leptin and shown promising results in the diet-induced obese rodent models. Results from ongoing human clinical trials are still awaited.

Adiponectin is an important adipokine that was initially discovered in 1995 as a protein secreted exclusively by adipocytes and related in structure to C1q, but regulated by insulin with clear effects in energy homeostasis (Scherer *et al.* 1995). Since then, the detailed biology of this key protein has been largely studied and today is the best-characterized anti-inflammatory adipokine with beneficial effects on insulin sensitivity. Adiponectin is the most abundant gene product in AT. Circulating adiponectin concentrations are high (500-3000 µg/L) accounting for ~0.01% of total plasma protein. Adiponectin forms a characteristic homotrimer in circulation. Adiponectin circulating in serum exists primarily in three main species: a low-molecular-mass (LMM) trimer of ≈67 kDa, a hexamer of ≈140 kDa, and a high-molecular-mass (HMM) multimer of >300 kDa.

In obese, insulin-resistant and T2DM individuals, the circulating adiponectin levels are consistently low. Adiponectin is induced during adipocyte differentiation and its secretion is stimulated by insulin and thiazolidinediones. Adiponectin exerts its effects primarily by interaction with its receptors AdipoR1 and AdipoR2 in skeletal muscle and activation of AMPK, the cellular fuel gauge which inhibits acetyl-CoA carboxylase and increases fatty acid beta oxidation. Circulating adiponectin levels are decreased in insulin-resistant states such as obesity and T2DM and the adiponectin levels correlate inversely with insulin resistance. Moreover, plasma adiponectin levels have been shown to be decreased in CVD and hypertension. However, it is not clear whether reduced adiponectin levels are a cause or consequence of insulin resistance. Administration of adiponectin to non-obese diabetic mice leads to an insulin-independent decrease in plasma glucose concentrations, circulating adiponectin concentrations are decreased in obesity, and weight loss leads to increase in adiponectin levels. Therefore, strategies to increase adiponectin are a logical approach for treatment of obesity and T2DM (Gable *et al.* 2006).

Resistin is a member of the cysteine-rich secretory protein family FIZZ. It is formed mainly in the AT and initially was shown to inhibit insulin action. Plasma resistin concentrations are increased in genetic and diet-induced obese mice and its expression is markedly down-regulated by treatment with TZD. In humans, resistin is produced primarily by infiltrating macrophages; however, its role in insulin resistance still remains controversial. Some studies have shown that resistin is positively correlated with insulin resistance, while others show no association between plasma resistin concentration and insulin resistance in humans. Recently, resistin was shown to influence pro-inflammatory cytokine release from human adipocytes via activation of nuclear factor-κB and JNK signaling pathways (Barnes and Miner 2009).

Retinol binding protein 4 (RBP4), a fat-derived adipokine, was shown to be associated with insulin resistance and decreased expression of GLUT-4 (Yang *et al.* 2005). In skeletal muscle, RBP4 causes insulin resistance by impairing insulin signaling; in the liver, RBP4 increases gluconeogenesis. Some studies in humans have shown that subjects with obesity, T2DM, hypertension, and polycystic ovarian syndrome have increased serum RBP4 levels and an inverse correlation has been observed between the plasma RBP4 concentration and insulin sensitivity. Moreover, interventions such as weight loss and gastric banding, which lead to improved insulin sensitivity, have been shown to decrease RBP4 levels (Graham *et al.* 2006, Weiping *et al.* 2006). However, these data have not been replicated in all populations (von Eynatten *et al.* 2007) and we have recently shown that although plasma RBP4 levels are elevated in impaired glucose tolerance and T2DM, they are not associated with obesity, insulin resistance or impaired insulin secretion in a Mexican-American population (Chavez *et al.* 2009).

Novel Adipokines in Obesity

Lipocalins are a family of proteins with a unique function of transporting small hydrophobic molecules, such as retinols. These proteins have been associated with many biological processes and the number of identified members of the family continues to grow. Lipocalin-2 has been identified as an adipokine with potential as metabolic regulator and it has been compared in function and relevance to retinol binding protein 4 (Wang *et al.* 2007). Although its role is not clear, it has been shown to be elevated in obese, insulin-resistant, and diabetic subjects, while treatment with TZD decreases its expression. Also, initial studies suggest that is up-regulated by insulin via phosphoinositide-3 kinase and mitogen-activated protein kinase signaling pathways. Therefore, it could be used as a marker for obesity-induced insulin resistance and to evaluate outcomes of therapeutic interventions in CVD (Zhang *et al.* 2008).

Visfatin is a nicotinamide phosphoribosyltransferase (NAMPT) enzyme that catalyzes the first step in the biosynthesis of NAD from nicotinamide. It was originally described as a putative cytokine that proved to enhance the maturation of B cell precursors and later was considered to have a role as an insulin mimetic adipokine eliciting the action through interaction with the insulin receptor (Fukuhara *et al.* 2005). Although it surged as a novel molecule with potential effects in the regulation of insulin sensitivity and energy metabolism, its role as of today is somewhat controversial; the original findings showing interactions with the insulin signaling pathway through binding and modulation of the insulin receptor have been largely questioned and are yet to be confirmed.

Vaspin is a novel protein (Serpina12) member of the serine protease inhibitors known as serpins and has been recently identified as a promising molecule to study the complex interactions between adipokines and glucose and lipid metabolism. Serpins are involved in many biological functions and vaspin has been shown to be up-regulated in obese subjects, whereas it is not detectable in healthy lean individuals. On the other hand, subjects with worsening T2DM have also shown low concentrations of vaspin that normalize after treatment with a TZD (pioglitazone), and treatment with vaspin has been shown to increase glucose uptake by improving insulin sensitivity (Youn *et al.* 2008). Thus, vaspin may play a key role in the pathogenesis of metabolic syndrome.

Lastly, Omentin-1 is a protein recently identified by sequence tag analyses from a human omental AT cDNA library with predominant expression in visceral white adipose tissue. Although its role is not fully understood, initial reports consistently show that its expression is down-regulated in obesity and insulin resistance and that omentin acts as an insulin sensitizer through modulation of Akt signaling and regulation of GLUT-4 expression. Furthermore, *in vitro* experiments revealed that treatment with recombinant omentin-1 enhances insulin-stimulated glucose uptake in human subcutaneous and omental adipocytes (Tan *et al.* 2008).

TARGETING ADIPOSE TISSUE TO TREAT OBESITY AND ITS ASSOCIATED MORBIDITIES

After reviewing the diverse mechanisms by which AT and specifically adipokines regulate energy metabolism, it becomes clear that intervention in these early steps would have a tremendous impact on prevention and treatment of obesity and its devastating complications, either by promoting a healthier environment for the adipocyte/vascular stroma unit (i.e., preventing local hypoxia, inducing macrophage emigration or switching to M2 phenotype) or by blocking the release of pro-inflammatory adipokines. This field warrants extensive additional research, as new drugs appear on the horizon.

CONCLUSION

Obesity is characterized by expansion of AT, inducing hypoxia in the hypertrophied adipocytes and cell death, thus promoting the migration of activated macrophages that turn into an inflammatory phenotype. The combination of these events results in an increased expression and secretion of a myriad pro-inflammatory adipokines by AT and the resident macrophages, and the decrease of adipocyte anti-inflammatory molecules such as adiponectin. This increase in adipocyte-derived humoral factors impairs lipid and glucose metabolism in an autocrine and paracrine manner, as well as the molecular mechanisms of auto-regulation of food intake and energy metabolism. Identification of the nature and function of these adipokines will help to develop new therapy to block the deleterious effects of adipokines on insulin sensitivity, endothelial function and other obesity-related morbidities. However, the role that each newly discovered AT-derived molecule plays in the regulation of energy homeostasis should be analyzed cautiously, since no single adipokine explains completely the metabolic disturbances observed in obesity and the disrupted interactions between multiple immune and metabolic pathways.

Summary Points

- Adipose tissue is a major endocrine organ that secretes a number of molecules collectively known as adipokines.
- Adipokines exert local and systemic effects to regulate energy metabolism.
- Obesity is strongly associated with components of metabolic syndrome and increased cardiovascular morbidity.
- Excess of adipose tissue leads to a heightened state of low-grade systemic chronic inflammation, mainly due to activation of the immune system by adipokines.
- The activated immune cell subtypes cause infiltration of adipose tissue by macrophages, which in turn secrete other adipokines with deleterious effects on vasculature and distal metabolically active tissues.
- The imbalance between pro-inflammatory and anti-inflammatory adipokines causes insulin resistance, increased lipolysis and impairment in glucose and lipid metabolism.

- Leptin is a major adipokine secreted by adipocytes with major effects in the regulation of food intake and energy expenditure and modulates secretion of other inflammatory cytokines by modulation of T-cell immune response.
- Adiponectin is the major anti-inflammatory adipokine, and its concentrations are usually impaired in obesity, insulin resistance and type 2 diabetes mellitus. It increases insulin sensitivity by increasing energy expenditure and lipid oxidation and increasing the expression of PPAR-alpha regulated genes.
- TNF-alpha, IL-6, PAI-1, MCP-1, and resistin are the most extensively characterized pro-inflammatory adipokines secreted by AT and are chiefly responsible for the low-grade inflammation of obesity.

Abbreviations

AMPK	:	AMP-activated protein kinase
AT	:	adipose tissue
CVD	:	cardiovascular disease
FFA	:	free fatty acids
IL-6	:	interleukin 6
MCP-1	:	monocyte chemotactic protein 1
MS	:	metabolic syndrome
PAI-1	:	plasminogen activator inhibitor 1
RBP4	:	retinol binding protein 4
T2DM	:	type 2 diabetes mellitus
TACE	:	TNF-alpha converting enzyme
TG	:	triglycerides
TIMP-3	:	tissue inhibitor of metalloproteinase 3
TNF-α	:	tumor necrosis factor alpha
TZD	:	thiazolidinedione
VLDL	:	very low density lipoprotein

Definition of Terms

Adipokine: Any cytokine secreted by the adipose tissue. Usually, they exert autocrine and paracrine regulatory signals, but also carry out intercellular communication between distant metabolically active tissues.

Cytokine: Any of a number of molecules, usually polypeptides that are secreted by different cells throughout the body, which play a major role in signaling and intercellular communication. These proteins usually act in an autocrine or paracrine manner.

Lipolysis: Breakdown of adipocyte-stored triacylglycerols into free fatty acids as a source of energy.

Obesity: Medical condition characterized by an excessive accumulation of body fat and associated with a significant increase in morbidity and reduced life expectancy.

Systemic inflammation: Chronic condition characterized by chronic elevation of inflammation-related molecules (i.e., interleukin 6, TNF-α) commonly seen in obesity and associated with endothelial dysfunction, insulin resistance and cardiovascular disease.

References

Alessi, M.C. and M. Poggi, and I. Juhan-Vague. 2007. Plasminogen activator inhibitor-1, adipose tissue and insulin resistance. Curr. Opin. Lipidol. 18: 240-245.

Antuna-Puente, B. and B. Feve, S. Fellahi, and J.P. Bastard. 2008. Adipokines: the missing link between insulin resistance and obesity. Diabetes Metab. 34: 2-11.

Barned, K.M. and J.L. Miner. 2009. Role of resistin in insulin sensitivity in rodents and humans. Curr. Protein Pept. Sci. 10: 96-107.

Chavez, A.O. and D.K. Coletta, S. Kamath,D.T. Cromack, A. Monroy, F. Folli, R.A. DeFronzo, and D. Tripathy. 2009. Retinol-binding protein 4 is associated with impaired glucose tolerance but not with whole body or hepatic insulin resistance in Mexican Americans. Am. J. Physiol. Endocrinol. Metab. 296: E758-764.

Dahlman, I. and M. Kaaman, T. Olsson, G.D. Tan, A.S. Bickerton, K. Wahlen, J. Andersson, E.A. Nordstrom, L. Blomqvist, A. Sjogren, M. Forsgren, A. Attersand, and P. Arner. 2005. A unique role of monocyte chemoattractant protein 1 among chemokines in adipose tissue of obese subjects. J. Clin. Endocrinol. Metab. 90: 5834-5840.

DeFronzo, R.A. 2004. Pathogenesis of type 2 diabetes mellitus. Med. Clin. North. Am. 88: 787-835.

Farooqi, I.S. and T. Wangensteen, S. Collins, W. Kimber, G. Matarese, J.M. Keogh, E. Lank, B. Bottomley, J. Lopez-Fernandez, I. Ferraz-Amaro, M.T. Dattani, O. Ercan, A.G. Myhre, L. Retterstol, R. Stanhope, J.A. Edge, S. Mckenzie, N. Lessan, M. Ghodsi, V. De Rosa, F. Perna, S. Fontana, I. Barroso, D.E. Undlien, and S. O'Rahilly. 2007. Clinical and molecular genetic spectrum of congenital deficiency of the leptin receptor. N. Engl. J. Med. 356: 237-247.

Friedman, J.M. and J.L. Halaas. 1998. Leptin and the regulation of body weight in mammals. Nature 395: 763-770.

Fukuhara, A. and M. Matsuda, M. Nishizawa, K. Segawa, M. Tanaka, K. Kishimoto, Y. Matsuki, M. Murakami, T. Ichisaka, H. Murakami, E. Watanabe, T. Takagi, M. Akiyoshi, T. Ohtsubo, S. Kihara, S. Yamashita, M. Makishima, T. Funahashi, S. Yamanaka, R. Hiramatsu, Y. Matsuzawa, and I. Shimomura. 2005. Visfatin: a protein secreted by visceral fat that mimics the effects of insulin. Science 307: 426-430.

Gable, D.R. and S.J. Hurel, and S.E. Humphries. 2006. Adiponectin and its gene variants as risk factors for insulin resistance, the metabolic syndrome and cardiovascular disease. Atherosclerosis 188: 231-244.

Graham, T.E. and Q. Yang, M. Bluher, A. Hammarstedt, T.P. Ciaraldi, R.R. Henry, C.J. Wason, A. Oberbach, P.A. Jansson, U. Smith, and B.B. Kahn. 2006. Retinol-binding protein 4 and insulin resistance in lean, obese, and diabetic subjects. N. Engl. J. Med. 354: 2552-2563.

Gutierrez, D.A. and M.J. Puglisi, and A.H. Hasty. 2009. Impact of increased adipose tissue mass on inflammation, insulin resistance, and dyslipidemia. Curr. Diab. Rep. 9: 26-32.

Hoene, M. and C. Weigert. 2008. The role of interleukin-6 in insulin resistance, body fat distribution and energy balance. Obes. Rev. 9: 20-29.

Hotamisligil, G.S. and N.S. Shargill, and B.M. Spiegelman. 1993. Adipose expression of tumor necrosis factor-alpha: direct role in obesity-linked insulin resistance. Science 259: 87-91.

Klein, S. and L. Fontana, V.L. Young, A.R. Coggan, C. Kilo, B.W. Patterson, and B.S. Mohammed. 2004. Absence of an effect of liposuction on insulin action and risk factors for coronary heart disease. N. Engl. J. Med. 350: 2549-2557.

Lumeng, C.N. and J.L. Bodzin, and A.R. Saltiel. 2007. Obesity induces a phenotypic switch in adipose tissue macrophage polarization. J. Clin. Invest. 117: 175-184.

Miyazaki, Y. and A. Mahankali, M. Matsuda, S. Mahanlaki, J. Hardies, K. Cusi, L.J. Mandarino, and R.A. DeFronzo. 2002. Effect of pioglitazone on abdominal fat distribution and insulin sensitivity in type 2 diabetic patients. J. Clin. Endocrinol. Metab. 87: 2784-2791.

Monroy, A. and S. Kamath, A.O. Chavez, V.E. Centonze, M. Veerasamy, A. Barrentine, J.J. Wewer, D.K. Coletta, C. Jenkinson, R.M. Jhingan, D. Smokler, S. Reyna, N. Musi, R. Khokka, M. Federici, D. Tripathy, R.A. DeFronzo, and F. Folli. 2009. Impaired regulation of the TNF-alpha converting enzyme/tissue inhibitor of metalloproteinase 3 proteolytic system in skeletal muscle of obese type 2 diabetic patients: a new mechanism of insulin resistance in humans. Diabetologia 52: 2169-2181.

Ogden, C.L. and M.D. Carroll, L.R. Curtin, M.A. McDowell, C.J. Tabak, and K.M. Flegal. 2006. Prevalence of overweight and obesity in the United States, 1999-2004. JAMA 295: 1549-1555.

Oral, E.A. and V. Simha, E. Ruiz, A. Andewelt, A. Premkumar, P. Snell, A.J. Wagner, A.M. DePaoli, M.L. Reitman, S.I. Taylor, P. Gorden, and A. Garg. 2002. Leptin-replacement therapy for lipodystrophy. N. Engl. J. Med. 346: 570-578.

Scherer, P.E. and S. Williams, M. Fogliano, G. Baldini, and H.F. Lodish. 1995. A novel serum protein similar to C1q, produced exclusively in adipocytes. J. Biol. Chem. 270: 26746-9.

Tan, B.K. and R. Adya, S. Farhatullah, K.C. Lewandowski, P. O'Hare, H. Lehnert, and H.S. Randeva. 2008. Omentin-1, a novel adipokine, is decreased in overweight insulin-resistant women with polycystic ovary syndrome: *ex vivo* and *in vivo* regulation of omentin-1 by insulin and glucose. Diabetes 57: 801-808.

Varma, V. and A. Yao-Borengasser, A.M. Bodles, N. Rasouli, B. Phanavanh, G. Nolen, E.M. Kern, R. Nagarajan, H.J. Spencer 3rd, M.J. Lee, S.K. Fried, R.E. McGehee Jr., C.A. Peterson, and P.A. Kern. 2008. Thrombospondin-1 is an adipokine associated with obesity, adipose inflammation, and insulin resistance. Diabetes 57: 432-439.

Von Eynatten, M. and P.M. Lepper, D. Liu, K. Lang, M. Baumann, P.P. Nawroth, A. Bierhaus, K.A. Dugi, U. Heemann, B. Allolio, and P.M. Humpert. 2007. Retinol-binding protein 4 is associated with components of the metabolic syndrome, but not with insulin resistance, in men with type 2 diabetes or coronary artery disease. Diabetologia 50: 1930-1937.

Wang, Y. and K.S. Lam, E.W. Kraegen, G. Sweeney, J. Zhang, A.W. Tso, W.S. Chow, N.M. Wat, J.Y. Xu, R.L. Hoo, and A. Xu. 2007. Lipocalin-2 is an inflammatory marker closely associated with obesity, insulin resistance, and hyperglycemia in humans. Clin. Chem. 53: 34-41.

Weiping, L. and C. Qingfeng, M. Shikun, L. Xiurong, Q. Hua, B. Xiaoshu, Z. Suhua, and L. Qifu. 2006. Elevated serum RBP4 is associated with insulin resistance in women with polycystic ovary syndrome. Endocrine 30: 283-288.

Weisberg, S.P. and D. McCann, M. Desai, M. Rosenbaum, R.L. Leibel, and A.W. Ferrante Jr. 2003. Obesity is associated with macrophage accumulation in adipose tissue. J. Clin. Invest. 112: 1796-808.

Yang, Q. and T.E. Graham, N. Mody, F. Preitner, O.D. Peroni, J.M. Zabolotny, K. Kotani, L. Quadro, and B.B. Kahn. 2005. Serum retinol binding protein 4 contributes to insulin resistance in obesity and type 2 diabetes. Nature 436: 356-362.

Youn, B.S. and N. Kloting, J. Kratzsch, N. Lee, J.W. Park, E.S. Song, K. Ruschke, A. Oberbach, M. Fasshauer, M. Stumvoll, and M. Bluher. 2008. Serum vaspin concentrations in human obesity and type 2 diabetes. Diabetes 57: 372-377.

Zhang, J. and Y. Wu, Y. Zhang, D. Leroith, D.A. Bernlohr, and X. Chen. 2008. The role of lipocalin 2 in the regulation of inflammation in adipocytes and macrophages. Mol. Endocrinol. 22: 1416-1426.

CHAPTER 23

Nonalcoholic Fatty Liver Disease and Adipokines

Stergios A. Polyzos,[1] Jannis Kountouras[1,*] and Christos Zavos[1]

[1]Second Medical Clinic, Medical School, Aristotle University of Thessaloniki, Ippokration Hospital, 49 Konstantinoupoleos Street, Thessaloniki 54124, Greece

ABSTRACT

Nonalcoholic fatty liver disease (NAFLD) has become the commonest form of liver disease in developed countries, affecting 20-30% of the general population. The pathogenesis is thought to be a multiple-hit process involving insulin resistance (IR), oxidative stress, apoptosis and adipokines. Adipose tissue has recently emerged as an active endocrine organ producing multiple proteins collectively referred to as adipokines. The release of adipokines plays an essential role in the pathogenesis of IR syndrome, including NAFLD, because they alter insulin sensitivity in insulin-targeted organs, such as the skeletal muscle and the liver. Some adipokines, including adiponectin, visfatin and acylation-stimulating protein, may positively influence insulin sensitivity, whereas others, including tumor necrosis factor α, interleukin 6 and resistin, influence insulin sensitivity negatively. Furthermore, classical cytokines, produced by immune cells infiltrating adipose tissue, are involved in liver inflammation and play an important role in the pathogenesis of NAFLD. The dynamic balance and interactions among various adipokines/cytokines, which improve or worsen IR, lead to the final beneficial or detrimental effect on NAFLD.

This chapter summarizes the cross-talk between adipokines and NAFLD. It focuses on the pathophysiology and most recent experimental and clinical studies by which the relationship between adipokines and NAFLD is explained.

*Corresponding author

Understanding of interplay among various adipokines in NAFLD may result in the dawn of a new era in the management of the disease.

INTRODUCTION

Nonalcoholic fatty liver disease (NAFLD) is one of the commonest causes of chronic liver disease in Western countries, affecting 20-30% of the general population. Its incidence in both adults and children is rising, linked with age and burgeoning epidemics of obesity and diabetes mellitus type 2 (DM-2) (Bellentani and Marino 2009). NAFLD refers to a wide spectrum of liver damage ranging from simple nonalcoholic fatty liver (NAFL) to nonalcoholic steatohepatitis (NASH) and NASH-related cirrhosis with its complications (Table 1).

Table I   Key features of NAFLD

- NAFLD encompasses a wide histological spectrum, including NAFL, NASH and NASH-related cirrhosis.
- IR is a key factor in the pathogenesis of NAFLD.
- As their size increases, the cellular homeostasis and the secretory profile of adipocytes is increasingly dysregulated, thereby aggravating IR and NAFLD.
- High TNF-α and low adiponectin create a metabolic milieu favoring NAFLD.
- Adipokine and NAFLD interplay constitutes a challenging area of research, which may lead to new therapeutical interventions.

IR, insulin resistance; NAFL, simple nonalcoholic fatty liver; NAFLD, nonalcoholic fatty liver disease; NASH, nonalcoholic steatohepatitis; TNF, tumor necrosis factor.

The natural history of NAFLD depends on the histologic subtype. NAFL has a generally benign long-term prognosis. Only a minority of patients with NAFL develop advanced liver disease (progression to cirrhosis in 3%), but this condition is causing increasing alarm because of its marked prevalence. In contrast, NASH is a progressive fibrotic disease and a leading cause of cryptogenic cirrhosis. NASH-related cirrhosis may have a similar prognosis to cirrhosis from other causes, leading to liver failure and hepatocellular carcinoma, and may recur after transplantation. Cirrhosis and liver-related death in NASH patients occur in 20% and 12% respectively over a 10-year period (Argo and Caldwell 2009). In addition to higher liver-related morbidity and mortality, patients with NAFLD appear to have a higher all-cause mortality (Bellentani and Marino 2009).

Although the pathogenesis of NAFLD remains elusive, the prevailing theory is the "multi-hit hypothesis" (Polyzos *et al.* 2009b). The first hepatic insult (hit) is the dysregulation of fatty acid metabolism, which leads to hepatic steatosis (NAFL). Numerous hepatocellular adaptations follow and signaling pathways are altered, which render hepatocytes susceptible to further possible insults ("multiple hits"). Such insults may be environmental or genetic perturbations, pro-inflammatory cytokines or oxidative stress, which lead to liver cell inflammation and necrosis with fibrogenic cascade activation, resulting in fibrosis and cirrhosis in a number of patients. Insulin resistance (IR) is nearly universal in NAFLD and is thought to play an important role in its pathogenesis, so that NAFLD is considered the hepatic component of IR or metabolic syndrome (Table 1). IR plays an important role in the first hit, as well as the progression from NAFL to NASH.

THE ENDOCRINE FUNCTION OF ADIPOSE TISSUE

Adipose tissue has recently emerged as endocrine tissue by producing multiple proteins, collectively referred to as adipokines. Adipokines are considered to play an important role in the pathogenesis of IR syndrome, including NAFLD, because they alter insulin sensitivity in insulin-targeted organs, such as the skeletal muscle and the liver. Furthermore, classical cytokines are involved in inflammation [tumor necrosis factor (TNF) α, interleukin (IL) 1β, IL-6, IL-8, IL-10, transforming growth factor (TGF) β and the acute-phase response (plasminogen activator inhibitor 1, haptoglobin, serum amyloid A). There is a vicious cycle between adipokines and cytokines in obesity: hypertrophic adipocytes release chemokines that induce macrophage accumulation in adipose tissue. Accumulated macrophages produce pro-inflammatory cytokines and nitric oxide, and these inflammatory changes induce dysregulation of adipokine production and secretion, characterized by a decrease in insulin-sensitizing and anti-inflammatory adipokines, and an increase in pro-inflammatory adipokines (Kamada *et al.* 2008); this so-called "low-grade" inflammatory process contributes to NAFLD.

Additional factors strengthening the role of adipokines in IR and NAFLD, as well as the interaction between adipocytes and inflammatory cells, are the suppressor of cytokine signaling (SOCS) proteins. SOCS proteins play an essential role in the pathogenesis of IR states, including NAFLD, by concordantly attenuating cytokine and insulin signaling. Over-expression of SOCS1 and SOCS3 in the liver causes IR and an increase in sterol regulatory element binding protein (SREBP) 1c, the key regulator enzyme of free fatty acid (FFA) synthesis in liver.

Understanding the interplay among various adipokines in NAFLD may result in new treatment approaches signaling the dawn of a new era in the management of NAFLD. Adipokine levels in IR states and NAFLD are summarized in Table 2. The effect of the most extensively studied adipokines in liver steatosis, inflammation and fibrosis is summarized in Table 3.

ADIPONECTIN AND NAFLD

Contrary to other adipokines, adiponectin is paradoxically decreased with increasing fatty mass. Apart from mature adipocytes, liver cells may also produce adiponectin when challenged with fibrotic stimuli (Polyzos *et al.* 2010b). Adiponectin plays a major role in IR and NAFLD (Fig. 1). It acts mainly through two transmembrane receptors (AdipoR1, AdipoR2). An additional cell membrane receptor for adiponectin is T-cadherin, but it lacks an intracellular domain and

Table 2 Serum levels of adipokines in IR and NAFLD

Adipokine	Insulin resistance	NAFLD
Acylation-stimulating protein	↓	↓
Adipsin	↓	na
Adiponectin	↓	↓
Adrenomedullin	↓	na
Apelin	↑	↑
Chemerin	↑	na
Interleukin 1	↑	↑
Interleukin 6	↑	↑
Interleukin 8	↑	↑
Interleukin 18	↑	↑
Monocyte chemoattractant protein 1	↑	↑
Leptin	↓	↔
Omentin	↓	na
Resistin	↑	↑
Retinol binding protein 4	↑	↔
Tumor necrosis factor α	↑	↑
Transforming growth factor	↑	↑
Vaspin	↑	na
Visfatin	↓	↓

Serum levels of adipokines in both insulin resistance states and NAFLD, as derived from experimental and/or clinical studies. ↔, controversial; ↓, decreased; ↑, increased.

Table 3 The effect of adipokines in liver histology

Adipokine	Steatosis	Inflammation	Fibrosis
Adiponectin	↓	↓	↔
Leptin	↓	↔	↑
Resistin	↑	↑	↑
TNF-α	↑	↑	↑

The effect of the most commonly studied adipokines in parameters of liver histology, including liver steatosis, inflammation and fibrosis, as derived from experimental and/or clinical studies. ↔, controversial; ↓, decrease; ↑, increase; TNF, tumor necrosis factor.

its effect on cellular signaling is unknown. It seems that adiponectin achieves its function in the liver via activating 5-AMP-activated protein kinase (AMPK) and peroxisome proliferator-activated receptor (PPAR) α pathways. As a consequence, glucose uptake and FFA oxidation are increased in the skeletal muscle and the liver, and gluconeogenesis, FFA influx and *de novo* lipogenesis are decreased in the liver. Furthermore, adiponectin has antifibrotic action in the liver by down-regulating the expression of aldehyde oxidase 1, TGF-β and connective tissue growth factor and diminishing the nuclear translocation of Smad2, which mediates the signaling of TGF-β. Adiponectin also has anti-inflammatory action in the liver by suppressing TNF-α and other pro-inflammatory cytokines and inducing anti-inflammatory cytokines, such as IL-10, thereby suppressing nuclear factor-kappaB (NFκB) signaling pathway. When NFκB signaling is suppressed, reactive oxygen metabolite (ROM) production, oxidative stress and cell injury are decreased (Polyzos *et al.* 2010b).

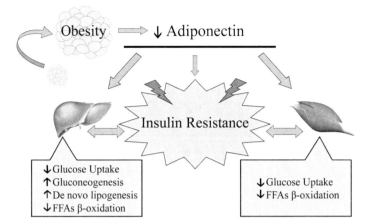

Fig. 1 The effect of obesity on adiponectin, insulin resistance and fatty liver. Adiponectin is decreased with increasing fatty mass. It leads to IR and metabolic consequences in insulin-dependent tissues, including the liver and the skeletal muscles. Hepatic glucose uptake and FFAs oxidation are decreased, and gluconeogenesis and *de novo* lipogenesis are increased, resulting in intrahepatic lipid accumulation and fatty liver. Similarly, skeletal muscle glucose uptake and FFA oxidation are decreased. The final result is hyperglycemia and decreased FFA clearance, which both further deteriorate IR and set up a vicious cycle among obesity, IR and fatty liver. FFA, free fatty acids; IR, insulin resistance.

Reprinted from Polyzos *et al.* Diabetes Obes. Metab. 2010, 12(5): 365-383, with permission.

In the IR setting, there is increasing evidence for the hepatoprotective role of adiponectin in NAFLD. By acting on different liver cells, adiponectin has pleiotropic actions. Specifically, adiponectin has insulin-sensitizing, antifibrogenic and anti-inflammatory properties, by acting on hepatocytes, hepatic stellate cells (HSCs) and hepatic macrophages (Kupffer cells), respectively. The effect of adiponectin on sinusoidal endothelial cells has not been investigated yet (Polyzos *et al.* 2010b). It is noteworthy that adiponectin has also an anti-apoptotic effect on hepatocytes (Jung *et al.* 2009). Importantly, adiponectin is required for normal progress of liver regeneration. Adiponectin-null mice exhibit impaired liver regeneration and increased hepatic steatosis. Increased expression of SOCS3 and subsequently reduced activation of signal transducer and activator of transcription (STAT)-3 in adiponectin-null mice might contribute to the alteration of the liver regeneration capability and hepatic lipid metabolism after partial hepatectomy.

Serum adiponectin does not seem to follow a linear pattern within NAFLD spectrum. It is similar between body mass index–matched controls and NAFL (or higher in controls than NAFL), higher in NAFL than NASH, and lower in NASH than NASH-related cirrhosis, thereby following a parabolic rather than a linear distribution by increasing NAFLD severity (Polyzos *et al.* 2009a). There are two possible mechanisms proposed for the paradoxical increase in adiponectin in cirrhosis: the decreased hepatic clearance of adiponectin and/or a compensatory increase toward the overwhelming production of pro-inflammatory cytokines in cirrhosis. Apart from serum adiponectin, liver expression of adiponectin, liver AdipoR1 and AdipoR2 and adiponectin gene single nucleotide polymorphisms has recently been studied in NAFLD (Polyzos *et al.* 2010b). However, data are currently controversial and premature for solid conclusions.

The hepato-protective effect of adiponectin and its insulin-sensitizing action make exogenous adiponectin administration an attractive therapeutic potential in NAFLD. However, this will be difficult to put into clinical practice, because of the large polypeptidic structure of adiponectin and the need for post-translational modification, since the bacterially produced adiponectin is functionally inactive (Polyzos *et al.* 2010a). Apart from the direct administration of adiponectin, therapeutic strategies have focused on the indirect up-regulation of adiponectin through the administration of various therapeutic agents and/or lifestyle modifications, including diet and exercise (Fig. 2) (Polyzos *et al.* 2010b). Although

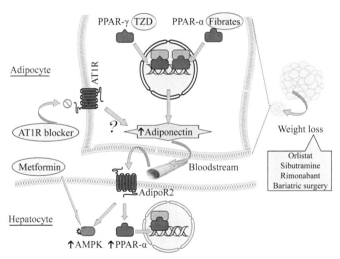

Fig. 2 Therapeutic strategies up-regulating adiponectin or interplaying with adiponectin signaling in nonalcoholic fatty liver disease.

Orlistat, sibutramine, rimonabant and bariatric surgery are believed to up-regulate adiponectin in adipocyte mainly by decreasing fatty mass. TZDs (currently pioglitazone and rosiglitazone) are PPAR-γ agonists that possibly up-regulate adiponectin gene transcription. Fibrates are PPAR pan-agonists (-α, -γ, -δ), mainly PPAR-α, that may also have an impact on adiponectin gene transcription. Metformin acts via AMPK activation, as adiponectin does. AT1R blockers block AT1R and up-regulate adiponectin, but the intracellular pathway remains to be elucidated. The TZDs-adiponectin sequence is of special interest, because PPAR-γ and PPAR-α are successively activated in adipocyte and hepatocyte respectively. AMPK, 5-AMP-activated protein kinase; AT1R, angiotensin II type 1 receptor; PPAR, peroxisome proliferator-activated receptor; TZDs, thiazolidinediones. Reprinted from Polyzos *et al.*, Diabetes Obes. Metab. 2010, 12(5): 365-383, with permission.

there is currently no proven pharmacotherapy for the treatment of NASH, the use of recombinant adiponectin or treatments up-regulating adiponectin appears to be promising in the management of NAFLD.

TNF-α AND NAFLD

Kupffer cells are the main source of TNF-α in the liver. It acts by binding two receptors of cell membrane, TNF-receptors 1 and 2. TNF-α acts by activating pro-inflammatory pathways, such as NFκB and c-jun N-terminal kinase (Polyzos *et al.* 2009b). The lipogenic action of TNF-α is partly achieved by the up-regulation of SREBP-1, thereby increasing *de novo* lipogenesis. There is a vicious cycle between hypertrophic adipocytes in obesity and macrophages/Kuppfer cells, critically involved in the pathogenesis of NAFLD: adipocytes in obesity release large quantities of FFAs through macrophage-induced lipolysis. FFAs serve as a ligand for toll-like receptor 4 in macrophages/Kuppfer cells, which further increases the production of TNF-α through the NFκB pathway, thereby increasing macrophage-induced lipolysis in adipocytes (Kamada *et al.* 2008).

It seems that TNF-α plays an important role in the pathogenesis of NAFLD, including hepatic inflammation and fibrogenesis. Experimental administration of TNF-α induced IR, whereas neutralization of TNF-α improved insulin sensitivity. Mice lacking TNF-α or TNF-receptors have improved insulin sensitivity. Experimental administration of infliximab, a monoclonal antibody against TNF-α, reduces necrosis, inflammation, steatosis and fibrosis, in parallel with the expression of TNF-α, IL-6, IL-1β and SOCS-3 in the liver (Polyzos *et al.* 2009b). Patients with NASH have increased liver expression of TNF-α and TNF-receptors, as well as increased expression of TNF-α in adipose tissue, associated with increased liver fibrosis. Single nucleotide polymorphisms of the TNF-α gene and the TNF-α promoter also correlate with the progression of NAFLD. Administration of pentoxifylline, a TNF-α inhibitor, improves aminotransferase levels and IR in NASH patients (Kamada *et al.* 2008). TNF-α, as well as IL-6, suppresses adiponectin transcription and secretion, thereby enhancing IR. On the other hand, adiponectin inhibits the expression and secretion of TNF-α, whereas it induces the production of the anti-inflammatory cytokines IL-10 and IL-1 receptor antagonist (IL-1Ra) in human immune cells. This represents an example of the complicated interplay among the different beneficial and detrimental adipokines reflected in IR and NAFLD (Polyzos *et al.* 2009c).

LEPTIN AND NAFLD

Leptin acts by binding to its receptor (ObR), which belongs to the cytokine receptor class I family and has various isoforms. The leptin signaling is transmitted by the janus kinase/STAT pathway. Secretion of leptin is proportional to the fatty mass and provides anti-obesity signals, regulating food intake, sympathetic tone, and energy expenditure in conditions of energy excess, through hypothalamic pathways (Marra and Bertolani 2009). However, these favorable effects are observed in lean individuals, but not in obese patients, because they develop leptin resistance. Leptin also has insulin-sensitizing action, mainly through AMPK activation, as adiponectin does. As a consequence, leptin stimulates glucose uptake and FFA oxidation in the skeletal muscles and the liver and decreases *de novo* lipogenesis in the liver (inhibition of SREBP-1 expression). Leptin seems to have also a crucial role in the central nervous system regulation of glucose metabolism, which may be associated with the pathogenesis of IR in obesity, DM-2 and NAFLD (Al-Dokhi 2009).

The close relationship of leptin with adipose tissue and fat stores of the body suggests its involvement in the etiology and pathogenesis of NAFLD. High leptin levels and leptin resistance in obesity might contribute to the development of NAFLD, having a negative impact on insulin signaling. Experimental studies showed marked steatosis and severe NASH development in a leptin-deficient model. Leptin is also considered to be an essential mediator of liver fibrosis, by increasing the expression of procollagen-I and TGF-β1, and activating HSCs. Activated HSCs acquire the ability to secrete leptin, which further promotes their own activation and liver fibrosis, thereby establishing a vicious cycle. Furthermore, leptin injections increased serum TNF-α, further amplifying inflammation and fibrosis. Although polymorphism of human ObR gene may contribute to the onset of NAFLD by regulating lipid metabolism and affecting insulin sensitivity, human studies examining the association between NAFLD and serum leptin provided conflicting results, and no association was found between leptin and the severity of liver fibrosis (Tsochatzis *et al.* 2009). Studies on a larger scale with homogenous population and carefully matched healthy controls are needed to elucidate the role of leptin and leptin resistance in NAFLD.

RESISTIN AND NAFLD

Experimental studies suggested that resistin acts by inhibiting AMPK activation and inducing SOCS-3 production, thereby aggravating IR (Al-Dokhi 2009). Interestingly, resistin is down-regulated by thiazolidinediones (TZDs), further supporting the role of resistin in IR (Kamada *et al.* 2008). The liver is the major target of resistin action, where it increases hepatic IR. In the skeletal muscle, resistin decreases the uptake and oxidation of FFAs. Furthermore, resistin stimulates TNF-α and IL-6 production by macrophages (Tsochatzis *et al.* 2009). Consecutively, TNF-α and IL-6 stimulate resistin by monocytes (Kaser *et al.* 2003), in an increasingly inflammatory milieu. Finally, resistin may play a role in HSC activation, thus participating in liver fibrogenesis. However, data from human studies remained controversial until the most recent clinical studies. Serum resistin was increased in NAFLD compared with healthy individuals and independently associated with necroinflammation in nondiabetic NAFLD patients (Cengiz *et al.* 2009). On the contrary, no difference in serum resistin was found among NAFL and NASH patients and obese controls in another study with severely obese individuals subjected to bariatric surgery (Argentou *et al.* 2009). This discrepancy may reflect mainly methodological differences, i.e., study populations.

RETINOL BINDING PROTEIN (RBP) 4 AND NAFLD

Elevated serum RBP-4 impairs post-receptor insulin signaling in the skeletal muscle and the liver, thereby leading to IR. Treatment with TZDs normalizes serum RBP-4 levels and reverses IR (Yang *et al.* 2005). In healthy individuals, serum RBP-4 positively correlates with IR and liver fat. Accordingly, serum RBP-4 was indicated as a robust marker of IR in nondiabetic individuals (Marra and Bertolani 2009). However, data from clinical studies with NAFLD are still controversial. Some authors suggested that increased serum RBP-4 is independently associated with NAFLD (Wu *et al.* 2008), whereas others found no difference between NAFLD and healthy controls (Cengiz *et al.* 2009) or lower serum RBP-4 in patients with NAFLD than in controls (Schina *et al.* 2009). However, in the last study, RBP-4 liver tissue expression was increased and correlated with NAFLD histology. Other studies also reported an inverse relationship between RBP-4 levels and degree of liver damage. Therefore, RBP4 might be a novel non-invasive marker of NAFLD severity.

VISFATIN AND NAFLD

Visfatin, an adipokine with insulin-mimetic and antagonistic effects, acts by binding the insulin receptor, but at a different site than insulin. It is increased in obesity and this is regarded as a compensatory mechanism to preserve insulin sensitivity. Visfatin increases glucose uptake by the skeletal muscle and adipose tissue and inhibits glucose release by the liver, thereby decreasing IR (Fukuhara *et al.* 2005). Although serum visfatin is positively correlated with obesity and DM-2, there are limited data for NAFLD. In one study, visfatin was lower in NASH than in NAFL patients, but not independently associated with presence of NASH (Jarrar *et al.* 2008). In another study, visfatin was a predictor for portal inflammation, but not of lobular inflammation or steatosis grade (Aller *et al.* 2009).

INTERLEUKINS AND NAFLD

A diversity of interleukins, including IL-1, IL-6, IL-8 and IL-18, mainly secreted by inflammatory cells infiltrating adipose tissue, may play a role in the pathogenesis of NAFLD.

IL-6 has been proposed to be one of the mediators that link obesity-related chronic inflammation with IR and NAFLD. It was initially supported that IL-6 has hepatoprotective action in steatotic liver, but a paradoxical effect of short- and long-term IL-6 exposure seems to exist, since long-term IL-6 exposure may result in liver injury and apoptosis (Tsochatzis *et al.* 2009). Liver IL-6 expression was increased in NASH compared with NAFL or controls and positively correlated to the degree of inflammation and fibrosis (Wieckowska *et al.* 2008). IL-6 genetic polymorphisms are involved in inflammation and IR and associated with NASH. IL-6 neutralizing antibodies lead to improved IR and increased suppression of hepatic glucose release *in vitro* (Klover *et al.* 2005).

IL-8 is produced by a host of cells including monocytes, macrophages, Kupffer cells and hepatocytes, can activate neutrophils and plays a key role in ROM production. ROMs cause tissue damage and are largely implicated in the pathogenesis of NAFLD (Polyzos *et al.* 2009b). IL-8 was higher in NAFLD patients than in both obese and non-obese controls, but no difference between NAFL and NASH was found (Jarrar *et al.* 2008). Interestingly, liver expression of IL-8 and IL-6 was significantly higher in NAFLD than in controls and was associated with activated complement (C)3 liver deposition. Activation of the C system is associated with disease severity and is a newly proposed pathogenetic mechanism in NAFLD (Rensen *et al.* 2009).

IL-1α is also a key mediator of inflammation. IL-1 increases IR by reducing insulin receptor substrate 1 expression activating NFκB pathway. Basal IL-1α was similar between NAFL, NASH and obese controls subjected to bariatric surgery, but was stimulated *in vitro* more in NASH than in NAFL and control, which might play a role in inflammatory cell infiltration in the liver of patients with NAFLD. Augmented IL-1α and TNF-α production occurs in obese NAFLD patients, which might play a pathophysiological role upon inflammatory leukocyte infiltration of the liver (Poniachik *et al.* 2006). IL-1Ra is an IL-1 receptor antagonist and negatively regulates IL-1 signaling. Adiponectin partly exerts its anti-inflammatory effect by inducing IL-1Ra expression (Polyzos *et al.* 2009b).

IL-18 has pleiotropic pro-inflammatory actions and induces the synthesis and release of other pro-inflammatory cytokines, chemokines and nitric oxide. Importantly, Kupffer cells have the potential to induce liver injury by IL-18 production. Serum IL-18 was higher in patients with NAFLD than in controls (Vecchiet *et al.* 2005). Enhanced expression of IL-18 was also reported in patients with chronic hepatitis C and NAFLD. Moreover, IL-18 was strongly expressed in the liver and decreased after rosiglitazone treatment in a rat model with NAFLD (Wang *et al.* 2008).

OTHER ADIPOKINES AND NAFLD

Serum levels of other adipokines in IR states and NAFLD are summarized in Table 2. For most of them data are limited and, most importantly, it is unknown whether each of them has a causative effect on IR and NAFLD or is only a biomarker.

CONCLUSION

The list of adipokines is continuously enriched and they constitute a challenging area of research, which will cast light on the pathophysiology of IR and NAFLD, thereby leading to new therapeutical interventions. Deeper knowledge of the pathophysiology of the endocrine function of adipose tissue could result in individual adipokine profile analysis, and thus individualization of therapeutic strategy in NAFLD.

Summary Points

- The importance of NAFLD lies in its marked prevalence worldwide, the link between NASH and cryptogenic cirrhosis and related complications, and higher all-cause mortality of affected patients.
- IR is a key factor in the pathogenesis of NAFLD, so that NAFLD is considered the hepatic component of IR syndrome.
- The dynamic interplay among various adipokines, which improve or worsen IR, leads to the final net result in NAFLD, which may be beneficial or detrimental, respectively.
- Adiponectin has hepato-protective and insulin-sensitizing action, which makes exogenous adiponectin administration an attractive therapeutic potential in NAFLD.
- TNF-α and other pro-inflammatory cytokines are associated with increased hepatic inflammation and fibrogenesis in NAFLD.

Abbreviations

AT1R	:	angiotensin II type 1
C	:	complement
DM-2	:	diabetes mellitus type 2
FFAs	:	free fatty acids
HSCs	:	hepatic stellate cells
IL	:	interleukin
IL-1Ra	:	IL-1 receptor antagonist
IR	:	insulin resistance
NAFL	:	simple nonalcoholic fatty liver
NAFLD	:	nonalcoholic fatty liver disease
NASH	:	nonalcoholic steatohepatitis
NFκB	:	nuclear factor-kappaB
PPAR	:	peroxisome proliferator-activated receptor

RBP : retinol-binding protein

ROMs : reactive oxygen metabolites

SOCS : suppressor of cytokine signaling

SREBP : sterol regulatory element binding protein

STAT : signal transducer and activator of transcription

TGF : transforming growth factor

TNF : tumor necrosis factor

TZDs : thiazolidinediones

Definition of Terms

Insulin receptor substrates: Proteins playing a key role in transmitting intracellular signaling from the insulin receptor to both metabolic and mitogenic insulin pathways.

Nuclear factor-kappaB: Intracellular protein complex stimulated by various factors (i.e., cytokines, bacterial and viral antigens, free radicals) and affecting cellular response through gene transcription.

Sterol regulatory element binding proteins: Transcription factors that bind to the sterol regulatory region of DNA, thereby up-regulating the synthesis of enzymes implicated in sterol biosynthesis and *de novo* lipogenesis.

Suppressors of cytokine signaling: Intracellular proteins that negatively regulate intracellular adipokines/cytokines signaling.

Thiazolidinediones: Medications that act by binding to PPAR-γ, thereby affecting various gene transcription. Currently used in the treatment of DM-2, TZDs have shown promising results in the treatment of NAFLD.

Acknowledgements

We thank the journal *Diabetes, Obesity and Metabolism* for their kind permission to reproduce figures 1 and 3 from the article "The role of adiponectin in the pathogenesis and treatment of nonalcoholic fatty liver disease", by S.A. Polyzos, J. Kountouras, C. Zavos, E. Tsiaous, 2010, 12(5): 365-383.

References

Al-Dokhi, L.M. 2009. Adipokines and etiopathology of metabolic disorders. Saudi. Med. J. 30: 1123-1132.

Aller, R. and D.A. de Luis, O. Izaola, M.G. Sagrado, R. Conde, M.C. Velasco, T. Alvarez, D. Pacheco, and J.M. Gonzalez. 2009. Influence of visfatin on histopathological changes of non-alcoholic fatty liver disease. Dig. Dis. Sci. 54: 1772-1777.

Argentou, M. and D.G. Tiniakos, M. Karanikolas, M. Melachrinou, M.G. Makri, C. Kittas, and F. Kalfarentzos. 2009. Adipokine serum levels are related to liver histology in severely obese patients undergoing bariatric surgery. Obes. Surg. 19: 1313-1323.

Argo, C.K. and S.H. Caldwell. 2009. Epidemiology and natural history of non-alcoholic steatohepatitis. Clin. Liver Dis. 13: 511-531.

Bellentani, S. and M. Marino. 2009. Epidemiology and natural history of non-alcoholic fatty liver disease (NAFLD). Ann. Hepatol. 8 Suppl 1: 4-8.

Cengiz, C. and Y. Ardicoglu, S. Bulut, and S. Boyacioglu. 2010. Serum retinol-binding protein 4 in patients with nonalcoholic fatty liver disease: does it have a significant impact on pathogenesis? Eur. J. Gastroenterol. Hepatol. 22: 813-819.

Fukuhara, A. and M. Matsuda, M. Nishizawa, K. Segawa, M. Tanaka, K. Kishimoto, Y. Matsuki, M. Murakami, T. Ichisaka, H. Murakami, E. Watanabe, T. Takagi, M. Akiyoshi, T. Ohtsubo, S. Kihara, S. Yamashita, M. Makishima, T. Funahashi, S. Yamanaka, R. Hiramatsu, Y. Matsuzawa, and I. Shimomura. 2005. Visfatin: a protein secreted by visceral fat that mimics the effects of insulin. Science 307: 426-430.

Jarrar, M.H. and A. Baranova, R. Collantes, B. Ranard, M. Stepanova, C. Bennett, Y. Fang, H. Elariny, Z. Goodman, V. Chandhoke, and Z.M. Younossi. 2008. Adipokines and cytokines in non-alcoholic fatty liver disease. Aliment. Pharmacol. Ther. 27: 412-421.

Jung, T.W. and Y.J. Lee, M.W. Lee, S.M. Kim, and T.W. Jung. 2009. Full-length adiponectin protects hepatocytes from palmitate-induced apoptosis via inhibition of c-Jun NH terminal kinase. FEBS J. 276: 2278-2284.

Kamada, Y. and T. Takehara, and N. Hayashi. 2008. Adipocytokines and liver disease. J. Gastroenterol. 43: 811-822.

Kaser, S. and A. Kaser, A. Sandhofer, C.F. Ebenbichler, H. Tilg, and J.R. Patsch. 2003. Resistin messenger-RNA expression is increased by proinflammatory cytokines *in vitro*. Biochem. Biophys. Res. Commun. 309: 286-290.

Klover, P.J. and A.H. Clementi, and R.A. Mooney. 2005. Interleukin-6 depletion selectively improves hepatic insulin action in obesity. Endocrinology 146: 3417-3427.

Marra, F. and C. Bertolani. 2009. Adipokines in liver diseases. Hepatology 50: 957-969.

Polyzos, S.A. and J. Kountouras, and C. Zavos. 2009a. Non-linear distribution of adiponectin in patients with nonalcoholic fatty liver disease limits its use in linear regression analysis. J. Clin. Gastroenterol. (in press).

Polyzos, S.A. and J. Kountouras, and C. Zavos. 2009b. Nonalcoholic fatty liver disease: the pathogenetic roles of insulin resistance and adipocytokines. Curr. Mol. Med. 72: 299-314.

Polyzos, S.A. and J. Kountouras, and C. Zavos. 2009c. The multi-hit process and the antagonistic roles of tumor necrosis factor-alpha and adiponectin in nonalcoholic fatty liver disease. Hippokratia 13: 127-127.

Polyzos, S.A. and J. Kountouras, and C. Zavos. 2010a. Adiponectin as a potential therapeutic agent for nonalcoholic steatohepatitis. Hepatol. Res. 40: 446-447.

Polyzos, S.A. and J. Kountouras, C. Zavos, and E. Tsiaousi. 2010b. The role of adiponectin in the pathogenesis and treatment of nonalcoholic fatty liver disease. Diabetes Obes. Metab. 12: 365-383.

Poniachik, J. and A. Csendes, J.C. Diaz, J. Rojas, P. Burdiles, F. Maluenda, G. Smok, R. Rodrigo, and L.A. Videla. 2006. Increased production of IL-1alpha and TNF-alpha in lipopolysaccharide-stimulated blood from obese patients with non-alcoholic fatty liver disease. Cytokine 33: 252-257.

Rensen, S.S. and Y. Slaats, A. Driessen, C.J. Peutz-Kootstra, J. Nijhuis, R. Steffensen, J.W. Greve, and W.A. Buurman. 2009. Activation of the complement system in human nonalcoholic fatty liver disease. Hepatology 50: 1809-1817.

Schina, M. and J. Koskinas, D. Tiniakos, E. Hadziyannis, S. Savvas, B. Karamanos, E. Manesis, and A. Archimandritis. 2009. Circulating and liver tissue levels of retinol-binding protein-4 in non-alcoholic fatty liver disease. Hepatol. Res. 39: 972-978.

Tsochatzis, E.A. and G.V. Papatheodoridis, and A.J. Archimandritis. 2009. Adipokines in nonalcoholic steatohepatitis: from pathogenesis to implications in diagnosis and therapy. Mediators Inflamm. 2009: 831670.

Vecchiet, J. and K. Falasca, P. Cacciatore, P. Zingariello, M. Dalessandro, M. Marinopiccoli, E. D'Amico, C. Palazzi, C. Petrarca, P. Conti, E. Pizzigallo, and M.T. Guagnano. 2005. Association between plasma interleukin-18 levels and liver injury in chronic hepatitis C virus infection and non-alcoholic fatty liver disease. Ann. Clin. Lab. Sci. 35: 415-422.

Wang, H.N. and Y.R. Wang, G.Q. Liu, Z. Liu, P.X. Wu, X.L. Wei, and T.P. Hong. 2008. Inhibition of hepatic interleukin-18 production by rosiglitazone in a rat model of nonalcoholic fatty liver disease. World J. Gastroenterol. 14: 7240-7246.

Wieckowska, A. and B.G. Papouchado, Z. Li, R. Lopez, N.N. Zein, and A.E. Feldstein. 2008. Increased hepatic and circulating interleukin-6 levels in human nonalcoholic steatohepatitis. Am. J. Gastroenterol. 103: 1372-1379.

Wu, H. and W. Jia, Y. Bao, J. Lu, J. Zhu, R. Wang, Y. Chen, and K. Xiang. 2008. Serum retinol binding protein 4 and nonalcoholic fatty liver disease in patients with type 2 diabetes mellitus. Diabetes Res. Clin. Pract. 79: 185-190.

Yang, Q. and T.E. Graham, N. Mody, F. Preitner, O.D. Peroni, J.M. Zabolotny, K. Kotani, L. Quadro, and B.B. Kahn. 2005. Serum retinol binding protein 4 contributes to insulin resistance in obesity and type 2 diabetes. Nature 436: 356-362.

CHAPTER 24

Endocrine Disruptors, Adipokines, and the Metabolic Syndrome

Eric R. Hugo[1] and Nira Ben-Jonathan[1,*]
[1]Department of Cancer and Cell Biology, University of Cincinnati, 3125 Eden Avenue, Cincinnati, OH 45267, USA

ABSTRACT

Increasing evidence indicates that adipose tissue is responsive to endocrine-disrupting compounds (EDCs) from the environment, which include industrial chemicals, phytoestrogens, and pharmaceutical estrogens. Many of these compounds interact with a variety of hormone receptors and cause perturbations in their signaling pathways. EDCs affect adipocyte proliferation and differentiation and alter the activity of enzymes involved in lipid and glucose metabolism in these cells. In addition, EDCs perturb the normal expression and release of several adipokines. Among EDCs, bisphenol A (BPA), a constituent of polycarbonate plastics and epoxy resins, has been studied most extensively. BPA at environmentally relevant concentrations suppresses the release of adiponectin, an insulin-sensitizing adipokine, and stimulates the release of IL-6 and TNFα, two inflammatory cytokines. Dysregulation of these adipokines/cytokines underlies the development of the metabolic syndrome, which predisposes individuals to obesity-related diabetes and cardiovascular disease. Hence, these data suggest that EDCs contribute to this debilitating condition. An extended examination of the overall effect of a variety of EDCs on adipose tissue homeostasis is warranted.

*Corresponding author

INTRODUCTION

Over the past two decades the incidence of obesity has reached pandemic levels. In addition to its toll on human morbidity and mortality, excess adiposity and its associated diseases have a serious economic impact. Whereas food over-consumption and sedentary lifestyles have been commonly implicated as the main causes for the vast increase in obesity, it has become increasingly clear that other factors are involved. A major factor that is often overlooked is the ubiquitous exposure of humans to numerous endocrine-disrupting compounds (EDCs).

Selected Features of Human Adipose Tissue

Until recently, the primary function of adipose tissue was thought to be storage of energy in the form of lipids. With the discovery of leptin in 1994, a previously unappreciated function of adipose tissue as an endocrine organ was recognized. Since then, over 100 adipose tissue–derived regulatory proteins have been identified. Some of these are unique to adipose tissue, while others are produced by many sites.

Adipose tissue is broadly divided into two categories: brown adipose tissue (BAT) and white adipose tissue (WAT). BAT, found more prominently in infants, is a specialized tissue involved in thermogenesis, whereas WAT is the main site of energy storage. In humans, WAT is distributed into visceral and subcutaneous depots, which differ in composition and function (Wajchenberg *et al.* 2002). Visceral fat comprises 6% of total body fat in women but 20% in men, reflecting the greater propensity of men to accumulate excess abdominal fat. The secretory activity in WAT differs by depot, with visceral tissue contributing more pro-inflammatory proteins than subcutaneous tissue.

Table I   Key aspects of adipose tissue response to endocrine-disrupting compounds

- Adipose tissue depots are significant components of the endocrine system and play a major role in metabolic homeostasis.
- Adipose tissue responds to nutrients and energy demands by secreting a variety of cytokines, adipokines, and hormones.
- Nutrient sensing in fat is largely accomplished through the activity of nuclear hormone receptors, including several receptors for steroid hormones.
- Human activity has generated thousands of synthetic molecules that are present in the environment. Many of these interact with hormone receptors and mimic or antagonize the activity of the endogenous ligands.
- EDCs can interfere with the normal activity of nutrient-sensing receptors in adipose tissue, leading to dysregulation of adipocyte differentiation, lipid accumulation and adipokine production.
- Perturbed endocrine functions of adipose tissue can significantly contribute to the metabolic syndrome, with increased risk of developing diabetes and cardiovascular disease.

This table lists salient factors implicating environmental contaminants as contributing factors to the obesity epidemic and its associated diseases. EDCs: endocrine-disrupting compounds.

Adipose tissue is composed of several cell types that are loosely embedded in a collagen matrix (Ailhaud *et al.* 1992). The major cell type is the very large, terminally differentiated adipocyte that carries out most adipose tissue functions, including glucose metabolism, maintenance of free fatty acid (FFA) levels, and adipokine secretion. Adipose tissue also contains proliferation-competent preadipocytes that are committed to the adipocyte lineage, endothelial and mast cells, fibroblasts, and infiltrating macrophages (Rink *et al.* 1996). Stromal macrophages contribute to the overall metabolic activity of adipose tissue by releasing pro-inflammatory cytokines such as TNFα and IL-6, which act in a paracrine manner on mature adipocytes. Obesity results from increased adipocyte size (hypertrophy) and number (hyperplasia). In humans, increases in adipocyte number occur first during gestation and infancy and again during adolescence. Increases in fat mass later in life are mainly due to hypertrophy. In extreme obesity, however, total adipocyte number can increase as much as three-fold.

The Metabolic Syndrome and Its Associated Diseases

The metabolic syndrome is a group of conditions that put the individual at high risk for heart disease and type 2 diabetes (T2D). Characteristics of this condition include hypertension, hyperglycemia, elevated triglyceride and FFA levels, reduced HDL, and excessive abdominal fat (Maury and Brichard 2010). Excess visceral fat contributes more significantly to this syndrome than subcutaneous fat. For example, abdominal fat is more lipolytically active by virtue of its higher sensitivity to β-adrenergic agents and lower sensitivity to α2-adrenergic agents and insulin. Elevated FFAs entering hepatic portal circulation inhibit insulin clearance and result in hyperinsulinemia. Chronic elevation of FFAs also reduces pancreatic β-cell function and impairs glucose metabolism and insulin sensitivity in both the liver and skeletal muscles (Arner 2001).

If left untreated, the metabolic syndrome can progress to T2D and cardiovascular disease, both of which are chronic disorders that require long-term management. Unmanaged T2D can have devastating consequences to the patient, including blindness, renal failure, and amputation of the extremities. About 8% of the US population has diabetes, increasing to over 20% after age 60 (Unger and Parkin 2009). Cardiovascular disease is another sequela of the metabolic syndrome. Coronary artery disease, arteriosclerosis, and heart valve disease are all aspects of this condition. Elevated FFA, hypertriglyceridemia, and increased pro-inflammatory cytokines all contribute to the development of this disease. In addition, C-reactive protein, secreted by the liver in response to adipokines, affects the development of the disease.

CHARACTERISTICS OF SELECTED ADIPOKINES

Adipose tissue secretes many polypeptides, collectively known as adipose-derived cytokines or adipokines, that act locally as autocrine or paracrine factors and distally as hormones. They exert profound effects on the metabolism, feeding behavior, inflammation, and vascular functions. Many aspects of the metabolic syndrome can be explained by disturbances in adipokine homeostasis. Only those adipokines known to be affected by endocrine disruptors will be discussed here.

Adiponectin

Adiponectin is a 224 amino acid protein selectively secreted by mature adipocytes (Maury and Brichard 2010). Its circulating levels constitute as much as 0.01% of total serum proteins. Despite its origin in mature adipocytes, excess adiposity paradoxically leads to lower serum adiponectin levels. Adiponectin circulates as trimers, hexamers, and higher molecular weight complexes (HMW). Although the precise functions of these isoforms are unclear, the ratio of HMW to other forms changes in various diseases and following treatment with some antidiabetic drugs. The two known receptors for adiponectin, AdipoR1 and AdipoR2, are expressed in many tissues, including cardiac and skeletal muscle, liver, vascular endothelium, and the hypothalamus (Kadowaki *et al.* 2006).

Adiponectin levels correlate inversely with obesity, glucose, FFA, and insulin levels. Increased serum adiponectin is positively associated with insulin sensitivity and HDL levels. Adiponectin stimulates β-oxidation and glucose metabolism in muscle while reducing glucose output and increasing insulin responsiveness in liver. Adiponectin production is regulated at two levels: transcription and assembly of the multimers (Kadowaki *et al.* 2006). Transcription is controlled in part by activation of PPARγ, and it is through this receptor that the antidiabetic thiazolidinediones increase adiponectin synthesis. Because assembly requires the correct formation of interchain disulfide bonds, a role for protein disulfide isomerase has been implicated. Unlike many protein hormones, adiponectin is not stored in secretory granules, suggesting constitutive secretion, with most regulatory events occurring transcriptionally or during assembly.

Leptin

Leptin is a 16 kDa protein encoded by the *ob* gene. Adipocytes are the major source of circulating leptin, although low levels of synthesis are found in other tissues (Maury and Brichard 2010). Deficiency in leptin or the leptin receptor leads to a rapid onset of extreme obesity. Leptin acts on the central nervous system to limit food intake and increase energy expenditure, and other neuroendocrine functions. Peripherally, leptin affects fertility, pancreatic islet function, and glucose metabolism in muscle. In humans, leptin levels increase in extreme obesity, indicating the development of leptin resistance.

Inflammatory Cytokines

TNFα is the most potent and extensively studied pro-inflammatory cytokine. It is initially expressed as a 212 amino acid transmembrane protein, which is proteolytically cleaved to yield a 51 kDa homotrimer (Fain 2006). TNFα is expressed by both adipose macrophages and adipocytes. It has been debated whether adipocyte TNFα is cleaved, or acts in an autocrine/paracrine manner while still attached to the membrane. TNFα signals through two receptors TNFR1 and TNFR2, both of which are expressed in adipose tissue. Increased TNFα production correlates strongly with adiposity and insulin resistance (Ryden and Arner 2007). TNFα down-regulates the glucose transporter, interferes with insulin receptor phosphorylation and signaling, and inhibits transcription factors that affect insulin sensitivity, thus playing a significant role in insulin signaling. It also inhibits adiponectin synthesis and stimulates lipolysis, leading to increased FFAs, which further increase insulin resistance. In addition, TNFα increases the production of pro-inflammatory cytokines such as IL-1 and IL-6.

IL-6 is intimately involved in hematopoiesis, inflammation, and the immune response. Within adipose tissue, the cytokine is produced by preadipocytes, by macrophages, and to a lesser extent by mature adipocytes (Heinrich *et al.* 2003). Locally, IL-6 increases lipolysis, suppresses adiponectin release, antagonizes insulin-stimulated glucose uptake, and inhibits lipoprotein lipase activity. Serum IL-6 levels are elevated in obesity and contribute to the chronic inflammatory state that accompanies the metabolic syndrome. IL-6 signals through the IL-6Rα receptor and the gp130 signal transduction protein. Although gp130 expression is ubiquitous, IL-6R expression is limited to a subset of tissues.

Other Adipokines

Monocyte chemoattractant protein 1 (MCP-1) is a 13 kDa protein secreted by a variety of tissues. In adipose tissue, it is the primary promoter of inflammatory macrophage infiltration. MCP-1 secretion increases with obesity, leading to increased macrophage numbers and a concomitant increase in macrophage-derived pro-inflammatory cytokines (Kim *et al.* 2006). Vascular endothelial growth factor (VEGF), which stimulates new blood vessel growth necessary for fat mass enlargement, is made by several cell types within adipose tissue and its production increases in obesity (Gealekman *et al.* 2008).

Receptors Involved in Adipose Tissue Functions

Adipose tissue expresses multiple receptors that are involved in nutrient sensing, metabolic homeostasis, and gender-specific fat deposition. Many of these belong to the nuclear hormone receptor (NRs) family, which share features such as a lipophilic ligand binding domain, a sequence-specific DNA binding domain, and protein domains enabling hetero- or homo-dimerization and interactions with co-activators and co-repressors (Grun and Blumberg 2006). About half of the known human NRs are expressed in adipose tissue, but some are still classified as "orphan receptors" without a known natural ligand.

Peroxisome proliferator-activated receptor gamma (PPARγ) plays key roles in nutrient sensing and preadipocyte differentiation (Grun and Blumberg 2006). Its natural ligands include long chain fatty acids and prostaglandins, and it is also targeted by several anti-diabetic drugs. Activation of PPARγ is essential for adipogenesis and directly affects expression of many adipokines. PPARγ forms heterodimers with retinoid X receptors (RXR) before transcription of adipocyte-specific genes can occur. PPARγ has been implicated as the target of several EDCs, with organotin compounds interfering with the normal function of the receptor by improperly activating the dimers (Grun and Blumberg 2006).

Adipose tissue responds to sex steroids in a gender- and depot-dependent manner. This response is mediated by the androgen receptor (AR), as well as by five structurally related nuclear receptors, of which ERα is the best recognized and most highly expressed (Rodriguez-Cuenca *et al.* 2006). As illustrated in Fig. 1, we found that in addition to ERα, human adipose tissue expresses significant amounts of ERβ, members of the estrogen-related receptors (ERR), as well as the

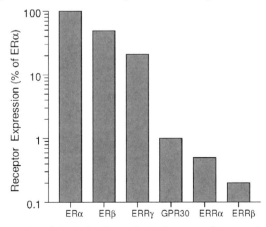

Fig. 1 Relative expression of classical and non-classical receptors for estrogens in human visceral adipose tissue, as determined by quantitative RT-PCR. ERα/β, estrogen receptor α or β; GPR30, G protein-coupled receptor 30; ERR, estrogen-related receptor. The figure was redrawn from Hugo *et al.* (2008) and Ben-Jonathan *et al.* (2009).

estrogen-responsive G-protein coupled receptor 30, or GPR30 (Hugo *et al.* 2008). The activity of most NRs is modulated by the presence or absence of nuclear co-activators and co-repressor, whose expression varies by cell and tissue type.

Other receptors, involved directly or indirectly in adipokine production, include those for leptin, adiponectin, IL-6, MCP-1, angiotensin, and TNFα. Additionally, toll-like receptors are expressed by adipocytes, stromovascular cells, and infiltrating macrophages, enabling adipose tissue to respond directly to FFAs. Expression and secretion of the adipokines is also influenced to a lesser degree by receptors for growth hormone, thyroid hormone, prolactin and glucocorticoids (Maury and Brichard 2010).

ENDOCRINE DISRUPTORS

EDCs that disrupt the normal function of adipose tissue are divided into two major classes: estrogenic and non-estrogenic. Both classes of compounds are widely found in the environment and are derived from several sources, including industrial releases, phytoestrogens, and pharmaceutical estrogens. Given the critical role of energy metabolism in reproduction, it is not surprising that adipose tissue is sensitive to estrogens through a variety of receptors. Estradiol (E2) plays two significant roles in adipose tissue function: metabolic regulation and fat distribution. Although xenoestrogens have been reported to bind to both ERα and ERβ, they have much lower binding affinities to the classical ERs than E2. Figure 2 illustrates the structural heterogeneity of representative EDCs and the multitude of membrane, cytoplasmic and nuclear receptors that mediate their actions.

Organotins

Organotins are used as antifouling agents and fungicides in a variety of crop and industrial applications (Grun and Blumberg 2006). It is from these diverse sources that organotins enter the food chain. Epidemiological studies have shown a measurable burden in the population. Tributyltin chloride (TBT) was reported to promote adipogenesis in 3T3-L1 adipocytes (Inadera and Shimomura 2005), with more recent studies demonstrating that this is done by activating both PPARγ and RXRs, leading to heterodimer formation. PPARγ/RXR activation is thought to be critical for the initiation of the genetic program leading to adipogenesis.

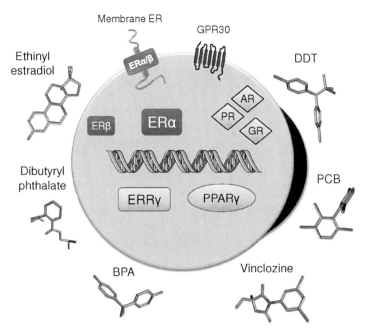

Fig. 2 A model of the adipocyte illustrating the various receptors, located in the membrane, cytoplasm or nucleus, which have been reported to mediate the action of endocrine-disrupting compounds (EDC). The structure of selected EDCs is shown outside of the adipocyte. BPA, bisphenol A; DDT, dichlorodiphenyltrichloroethane; ER, estrogen receptor; ERR, estrogen-related receptor; GPR30, G protein-coupled receptor 30; GR, glucocorticoid receptor; PCB, polychlorinated biphenyl; PPAR, peroxisome proliferator-activated receptor; PR, progesterone receptor.

Color image of this figure appears in the color plate section at the end of the book.

Phthalates

Phthalates are small molecular weight compounds used as softening agents in many plastics. Since they are not covalently bound to the polymers, they can leach out into products packaged in phthalate-containing plastics. Phthalates can enter the body either through ingestion or transdermally through cosmetic preparations. Their metabolites are found in most tested individuals, indicating a widespread exposure. Although not all the molecular targets of phthalates are known, they were reported to activate the metabolic sensors PPARs, RXR, and PXR (Masuyama *et al.* 2000), thereby influencing both adipogenesis and adipokine expression.

Alkylphenols

Alkylphenols, of which octyl- and nonylphenol have been best studied, are prevalent in the environment because of their widespread use as surfactants. In adipose tissue, alkylphenols are thought to act primarily through ERs, although in other tissues they activate additional NRs (CAR and PXR), thus additional signaling pathways cannot be ruled out (Masuyama *et al.* 2000). Alkylphenols promote adipogenesis and increase adipokine levels via increased fat mass.

Phytoestrogens

Phytoestrogens such as the soy compounds genistein and coumestrol act as natural selective estrogen response modifiers. Depending upon their target receptors, they exhibit either estrogenic or antiestrogenic properties (Cederroth *et al.* 2009). Originally, these dietary estrogens were considered to have beneficial health effects, but most recently they have been examined for their endocrine-disrupting properties.

Pharmaceutical Estrogens

Ethinyl estradiol (EE2), a major component in oral contraceptives, has been detected in the environment at increasingly higher levels through waste treatment facilities (Falconer 2006). This estrogen is widely prescribed as hormone replacement therapy to post-menopausal women and for the treatment of polycystic ovary syndrome. Although therapeutic levels of EE2 were shown to affect several adipokines, its activity at environmentally relevant levels has not been examined and warrants exploration given its widespread presence in the environment.

BISPHENOL A: A UBIQUITOUS ENDOCRINE DISRUPTOR

Bisphenol A is the best-characterized EDC having a direct effect on adipose tissue. BPA is a small (228 Da) molecule used as a monomer in the production of polycarbonate plastics. It is found in numerous consumer products, including food and water containers and baby bottles. BPA is also a major component of epoxy lining of food and beverage cans, medical tubing, and dental fillings (Ben-Jonathan and Steinmetz 1998, Welshons *et al.* 2006). Small amounts of BPA are released from the polymers to food or water especially upon heating. BPA has been measured in serum, breast milk, and urine at levels ranging from 0.3 to 5 ng/ml (approximately 1-20 nM) in studies conducted worldwide (Welshons *et al.* 2006). A recent study reported that >90% of individuals sampled (n = 2517) had measurable levels of BPA in their urine (Calafat *et al.* 2008).

Effects of BPA on Multiple Biological Systems

BPA was first detected upon noticing increased proliferation of breast cancer cells cultured in medium made from water autoclaved in polycarbonate bottles (Krishnan *et al.* 1993). Shortly thereafter, BPA was detected in canned food and dental resin. Concurrent with these observations, our laboratory was studying interactions between prolactin (PRL) and E2. Upon observing a strong structural similarity between BPA and the potent estrogen diethylstilbestrol, we set out to examine whether BPA altered PRL production. This study (Steinmetz *et al.* 1997) was the first to show estrogen-like properties of BPA within the neuroendocrine axis *in vitro* and *in vivo*. Subsequently, we reported on the effects of BPA on the reproductive tract in female rats (Steinmetz *et al.* 1998) and the induction of prolonged hyperprolactinemia following treatment of neonatal rats with BPA (Khurana *et al.* 2000). We also found that some rat strains (i.e., Fischer 344) were exquisitely sensitive to the estrogen-mimetic effects of BPA (Long *et al.* 2000).

BPA has been intensely debated with regard to its adverse affects on fetal development, reproductive fecundity and carcinogenesis (Vandenberg *et al.* 2009). Much of the controversy is due to differences in data interpretation and experimental paradigms. Because BPA production is a multibillion dollar industry, the chemical and food industries have a vested interest in classifying BPA as safe. Much of the debate centers on the micromolar doses of BPA used in many studies, which are three orders of magnitude higher than environmentally relevant levels. Another issue is the often observed non-monotonic response to BPA in many model systems. Unlike classical toxins where "the dose makes the poison", BPA responses often show a U-shaped or inverted U-shaped curves. Thus, extrapolation from action, or lack of action, of BPA at high doses to its presumed bioactivity at low doses is unwarranted. Nonetheless, BPA at micromolar concentrations was reported to accelerate adipogenesis and increase lipid accumulation in 3T3-L1 adipocytes (Masuno *et al.* 2005).

BPA and Adipokine Release

Using breast, subcutaneous and visceral adipose tissue explants as well as isolated mature adipocytes, we recently reported that BPA at 1 and 10 nM concentrations inhibits adiponectin release (Hugo *et al.* 2008, Ben-Jonathan *et al.* 2009). Notably, both BPA and E2 showed a clear U-shaped dose-response curve (Fig. 3). We subsequently found that BPA stimulated the release of two inflammatory cytokines, IL-6 and TNFα. Unexpectedly, ICI 182780, a potent inhibitor of ERα and ERβ, mimicked the effects of both BPA and E2 on the release of all three adipokines. The mechanism by which BPA exerts its actions remains unclear. While BPA binds to both ERα and ERβ, its binding affinity is ≈10,000-fold lower than that of E2, suggesting that it should mimic or compete with estrogens only at the μM range. Yet, numerous reports using a variety of cells have shown BPA actions at nM concentrations are often similar to, or stronger than, E2.

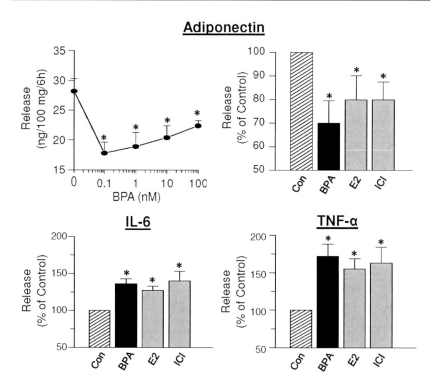

Fig. 3 Comparison of the effects of BPA (BPA), estradiol (E2), and ICI 182,780 (ICI) on adipokine release from human adipose tissue explants. Upper left panel: Dose-dependent effects of BPA on adiponectin release from a representative patient (means ± SEM of 5 replicates). Other panels: Values are means ± SEM; $N = 4$ patients. Subcutaneous adipose tissue explants, obtained from patients undergoing abdominoplasty, were incubated for 6 h with the various compounds. Conditioned media were analyzed for adiponectin, interleukin 6 (IL-6) and tumor necrosis factor α (TNFα) by their respective ELISAs. *significant ($P < 0.05$) difference from controls (Con). The figure was redrawn from Ben-Jonathan et al. (2009).

As we recently reviewed (Ben-Jonathan et al. 2009), several mechanisms have been proposed to reconcile the above incongruity. One, differences in binding to serum proteins, transport and metabolism, as compared to estrogens, account for the low-dose effect of BPA in animal studies, especially during fetal development. Two, BPA binds differently within the ligand binding domain of ERs and recruits a different set of co-regulators. Three, BPA elicits rapid responses via non-genomic mechanisms by activating membrane-anchored ERs, unidentified non-classical membrane ERs, or GPR30. Four, BPA binds to ERRγ, which has a relatively high binding affinity to BPA. Finally, a unique receptor, perhaps one of the "orphan receptors" may mediate the actions of BPA. The numerous actions attributed to BPA in adipose tissue are illustrated in Fig. 4.

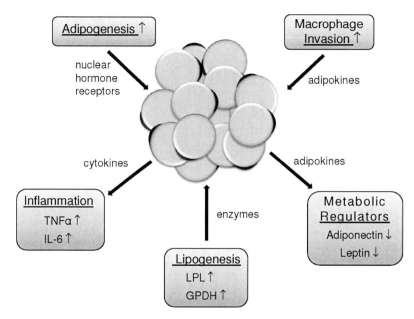

Fig. 4 A model depicting the multiple effects ascribed to bisphenol A on adipose tissue. BPA increases inflammatory cytokine production while reducing levels of the protective adipokine adiponectin. BPA stimulates adipogenesis by promoting preadipocyte differentiation and stimulates lipid production by increasing lipogenesis. Increased TNFα production stimulates macrophage infiltration, which further contributes to the inflammatory response.

Summary Points

- Adipose tissue is an endocrine organ that produces and responds to a large number of hormones, adipokines, and cytokines.
- The incidence of obesity, the metabolic syndrome, type 2 diabetes and cardiovascular disease has greatly increased over the past two decades, reaching epidemic proportions in many countries worldwide. Changes in lifestyle and diet do not fully account for this phenomenon.
- Throughout the last 50 years, thousands of man-made compounds have been released into the environment, many of which alter the normal endocrine function of adipose tissue in many species including humans.
- Bisphenol A (BPA), a highly prevalent compound, alters the release of adiponectin, IL-6, and TNFα in human adipose tissue. Such alterations can contribute to the development of the metabolic syndrome and its sequelae.

Abbreviations

AR	:	androgen receptor
BAT	:	brown adipose tissue
BPA	:	bisphenol A
E2	:	17-β estradiol
EDC	:	endocrine-disrupting compound
ER	:	estrogen receptor
ERR	:	estrogen related receptor
FFA	:	free fatty acid
IL-6	:	interleukin 6
MCP-1	:	monocyte chemoattractant protein 1
NR	:	nuclear hormone receptor
PPARγ	:	peroxisome proliferator-activated receptor gamma
RXR	:	retinoid X receptor
T2D	:	type 2 diabetes
TBT	:	tributyltin
TNFα	:	tumor necrosis factor alpha
WAT	:	white adipose tissue

Definition of Terms

Bisphenol A: A monomer of polycarbonate plastics, some epoxy resins, and polyvinyl chloride used in numerous household goods.

Endocrine-disrupting compound (EDC): Low molecular weight compounds present in the environment that either mimic or antagonize normal hormonal signaling.

Metabolic syndrome: An acquired disorder defined by the presence of any three of the following characteristics: abdominal obesity, hypertension, hypertriglyceridemia, hyperglycemia, and low HDL.

Nuclear hormone receptors: A family of receptor proteins with common structural features that are involved in nutrient sensing, metabolic homeostasis, and gender-specific adipose deposition. This protein family has the greatest number of identified targets of EDCs.

References

Ailhaud, G. and P. Grimaldi, and R. Negrel. 1992. Cellular and molecular aspects of adipose tissue development. Annu. Rev. Nutr. 12: 207-233.

Arner, P. 2001. Regional differences in protein production by human adipose tissue. Biochem. Soc. Trans. 29: 72-75.

Ben-Jonathan, N. and E.R. Hugo, and T.D. Brandebourg. 2009. Effects of bisphenol A on adipokine release from human adipose tissue: Implications for the metabolic syndrome. Mol. Cell Endocrinol 304: 49-54.

Ben-Jonathan, N. and R. Steinmetz. 1998. Xenoestrogens: The emerging story of bisphenol A. Trend. Endocrinol. Metab. 9: 124-128.

Calafat, A.M. and X. Ye, L.Y. Wong, J.A. Reidy, and L.L. Needham. 2008. Exposure of the U.S. population to bisphenol A and 4-tertiary-octylphenol: 2003-2004. Environ. Health Perspect. 116: 39-44.

Cederroth, C.R. and J. Auger, C. Zimmermann, F. Eustache, and S. Nef. 2009. Soy, phyto-oestrogens and male reproductive function: a review. Int. J. Androl (in press).

Fain, J.N. 2006. Release of interleukins and other inflammatory cytokines by human adipose tissue is enhanced in obesity and primarily due to the nonfat cells. Vitam. Horm. 74: 443-477.

Falconer, I.R. 2006. Are endocrine disrupting compounds a health risk in drinking water? Int. J. Environ. Res. Public Health 3: 180-184.

Gealekman, O. and A. Burkart, M. Chouinard, S.M. Nicoloro, J. Straubhaar, and S. Corvera. 2008. Enhanced angiogenesis in obesity and in response to PPARgamma activators through adipocyte VEGF and ANGPTL4 production. Am. J. Physiol Endocrinol. Metab. 295: E1056-E1064.

Grun, F. and B. Blumberg. 2006. Environmental obesogens: organotins and endocrine disruption via nuclear receptor signaling. Endocrinology 147: S50-S55.

Heinrich, P. C. and I. Behrmann, S. Haan, H. M. Hermanns, G. Muller-Newen, and F. Schaper. 2003. Principles of interleukin (IL)-6-type cytokine signalling and its regulation. Biochem. J. 374: 1-20.

Hugo, E.R. and T.D. Brandebourg, J.G. Woo, J. Loftus, J.W. Alexander, and N. Ben-Jonathan. 2008. Bisphenol A at environmentally relevant doses inhibits adiponectin release from human adipose tissue explants and adipocytes. Environ. Health Perspect. 116: 1642-1647.

Inadera, H. and A. Shimomura. 2005. Environmental chemical tributyltin augments adipocyte differentiation. Toxicol. Lett. 159: 226-234.

Kadowaki, T. and T. Yamauchi, N. Kubota, K. Hara, K. Ueki, and K. Tobe. 2006. Adiponectin and adiponectin receptors in insulin resistance, diabetes, and the metabolic syndrome. J. Clin. Invest 116: 1784-1792.

Khurana, S. and S. Ranmal, and N. Ben Jonathan. 2000. Exposure of newborn male and female rats to environmental estrogens: delayed and sustained hyperprolactinemia and alterations in estrogen receptor expression. Endocrinology 141: 4512-4517.

Kim, C.S. and H.S. Park, T. Kawada, J.H. Kim, D. Lim, N.E. Hubbard, B.S. Kwon, K.L. Erickson, and R. Yu. 2006. Circulating levels of MCP-1 and IL-8 are elevated in human obese subjects and associated with obesity-related parameters. Int. J. Obes. (Lond) 30: 1347-1355.

Krishnan, A.V. and P. Stathis, S.F. Permuth, L. Tokes, and D. Feldman. 1993. Bisphenol-A: an estrogenic substance is released from polycarbonate flasks during autoclaving. Endocrinology 132: 2279-2286.

Long, X. and R. Steinmetz, N. Ben Jonathan, A. Caperell-Grant, P.C. Young, K.P. Nephew, and R.M. Bigsby. 2000. Strain differences in vaginal responses to the xenoestrogen bisphenol A. Environ. Health Perspect. 108: 243-247.

Masuno, H. and J. Iwanami, T. Kidani, K. Sakayama, and K. Honda. 2005. Bisphenol a accelerates terminal differentiation of 3T3-L1 cells into adipocytes through the phosphatidylinositol 3-kinase pathway. Toxicol. Sci. 84: 319-327.

Masuyama, H. and Y. Hiramatsu, M. Kunitomi, T. Kudo, and P.N. MacDonald. 2000. Endocrine disrupting chemicals, phthalic acid and nonylphenol, activate Pregnane X receptor-mediated transcription. Mol. Endocrinol. 14: 421-428.

Maury, E. and S.M. Brichard. 2010. Adipokine dysregulation, adipose tissue inflammation and metabolic syndrome. Mol. Cell. Endocrinol. 314: 1-16.

Rink, J.D. and E.R. Simpson, J.J. Barnard, and S.E. Bulun. 1996. Cellular characterization of adipose tissue from various body sites of women. J. Clin. Endocrinol. Metab. 81: 2443-2447.

Rodriguez-Cuenca, S. and M. Gianotti, P. Roca, and A.M. Proenza. 2006. Sex steroid receptor expression in different adipose depots is modified during midpregnancy. Mol. Cell Endocrinol. 249: 58-63.

Ryden, M. and P. Arner. 2007. Tumour necrosis factor-alpha in human adipose tissue from signalling mechanisms to clinical implications. J. Intern. Med. 262: 431-438.

Steinmetz, R. and N.G. Brown, D.L. Allen, R.M. Bigsby, and N. Ben Jonathan. 1997. The environmental estrogen bisphenol A stimulates prolactin release *in vitro* and *in vivo*. Endocrinology 138: 1780-1786.

Steinmetz, R. and N.A. Mitchner, A. Grant, D.L. Allen, R.M. Bigsby, and N. Ben Jonathan. 1998. The xenoestrogen bisphenol A induces growth, differentiation, and c-fos gene expression in the female reproductive tract. Endocrinology 139: 2741-2747.

Unger, J. and C.G. Parkin. 2009. Appropriate, timely, and rational treatment of type 2 diabetes mellitus: Meeting the challenges of primary care. Insulin 4: 144-157.

Vandenberg, L.N. and M.V. Maffini, C. Sonnenschein, B.S. Rubin, and A.M. Soto. 2009. Bisphenol-A and the great divide: a review of controversies in the field of endocrine disruption. Endocr. Rev. 30: 75-95.

Wajchenberg, B.L. and D. Giannella-Neto, M.E. da Silva, and R.F. Santos. 2002. Depot-specific hormonal characteristics of subcutaneous and visceral adipose tissue and their relation to the metabolic syndrome. Horm. Metab Res. 34: 616-621.

Welshons, W.V. and S.C. Nagel, and F.S. Vom Saal. 2006. Large effects from small exposures. III. Endocrine mechanisms mediating effects of bisphenol A at levels of human exposure. Endocrinology 147: S56-S69.

CHAPTER 25

Adipokines and the Heart

Gianluca Iacobellis

Endocrinology, Department of Medicine, McMaster University, Hamilton, ON, Canada

ABSTRACT

Adipose tissue plays an important role in the development of cardiovascular diseases. Abdominal and cardiac visceral fat depots, such as the epicardial adipose tissue, may indirectly or locally modulate myocardium and coronary arteries. Adipose tissue, including the epicardial fat, is a metabolically active organ that generates a variety of adipokines, bioactive molecules that may significantly influence cardiac function and morphology. The physiological and pathophysiological role of the adipokines is still not completely known. In fact, while some adipokines may cause cardiovascular damage, others actually display clear cardioprotective properties. In particular, epicardial adipocytes can secrete both pro- and anti-inflammatory adipokines and therefore a double role, harmful and protective, has been attributed to this visceral fat depot. Adipokines can be locally secreted through paracrine or vasocrine mechanisms. Co-occurrence of cardiovascular and metabolic abnormalities and increased or abnormal fat mass can produce increased, decreased or abnormal secretion of adipokines.

EFFECT OF ISOLATED EXCESS ADIPOSITY ON THE HEART

The relationship of adiposity and the heart is controversial and not completely known. Isolated excess adiposity, in the absence of the common cardio-metabolic obesity-related co-morbidities, is related to adaptive functional and morphological modification of heart (Iacobellis *et al.* 2002, 2006, Iacobellis and Sharma 2007), as summarized in Table 1. Increased local cardiac fat depots may also contribute to the development of left ventricle (LV) changes (Iacobellis *et al.* 2004, 2007, Iacobellis and Willens 2009, Iacobellis 2009).

Table 1 Morphological and functional cardiac changes in isolated adiposity

Normal LV mass
Appropriate eccentric LV hypertrophy (mostly in subjects with BMI > 50 kg/m^2)
Normal RV diameter
Increased RV mass
Left atrium enlargement
Aortic root enlargement
Normal global and segmental LV contractility
Hyperdynamic systole
Normal or lower end and circumferential systolic stress
Increased pre-load
Impaired LV and RV diastolic filling
Impaired LV and RV diastolic relaxation

This table summarizes the most common morphological and functional cardiac changes that can be observed in subjects with isolated and metabolically healthy obesity. LV, left ventricle; BMI, body mass index; RV, right ventricle.

EPICARDIAL ADIPOSE TISSUE AND ADIPOKINES

A considerable amount of evidence suggests the existence of paracrine interactions between epicardial adipose tissue, the true visceral fat of the heart, and the myocardium (Iacobellis and Sharma 2006, Iacobellis *et al.* 2005a) (Fig. 1). No muscle fascia divides epicardial fat and myocardium, and therefore the two tissues share the same microcirculation (Fig. 2). As a result of the close anatomical relationship between epicardial adipose tissue and the adjacent myocardium, epicardial fat could locally modulate cardiac morphology and function (Table 2). Epicardial fat is metabolically very active (Iacobellis *et al.* 2008). It is a source of several pro-inflammatory and pro-atherogenic cytokines, as well as tumor necrosis factor alpha (TNF-α), monocyte chemoattractant protein 1 (MPC-1), interlukin 6 (IL-6), nerve growth factor (NGF), resistin, visfatin, omentin, leptin, plasminogen activator inhibitor 1 (PAI-1), and angiotensinogen. However, epicardial fat also produces anti-inflammatory, anti-atherogenic adipokines, such as adiponectin and adrenomedullin. A double, detrimental and protective, cardiac effect has been attributed to the epicardial fat (Iacobellis and Barbaro 2008). Adipokines secreted from epicardial adipocytes could affect the heart and coronary arteries through paracrine and/or vasocrine mechanisms (Fig. 3). It is reasonable to hypothesize that inflammatory signals from the epicardial fat could act reciprocally because of atherogenic inflammation in the underlying plaques. The presence of inflammatory cells in epicardial adipose tissue could also reflect the response to plaque rupture. It is also plausible that the paracrine release of cytokines from periadventitial epicardial fat could traverse the coronary wall by diffusion from outside to inside and interact with cells in each of its layers. Inflammatory adipokines might be released from epicardial tissue directly into vasa vasorum and be transported downstream into the arterial wall, according to a "vasocrine signaling" mechanism.

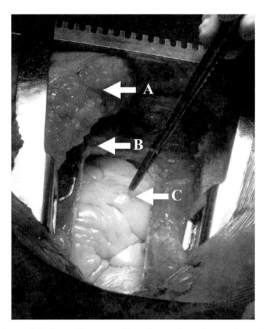

Fig. 1 Adipose tissue of the heart. Large cardiac adipose tissue in a patient with coronary artery disease during a triple coronary artery bypass. The adipose tissue abnormally covers the entire heart in this patient. (A) Adipose tissue deposited in the chest under the skin (subcutaneous). (B) Pericardial adipose tissue. (C) Epicardial adipose tissue pulled up by forceps. (This figure is unpublished in this current version and comes from the author's own collection. Patient's informed consent had been obtained.)

Color image of this figure appears in the color plate section at the end of the book.

ADIPOKINES AND THE CARDIOVASCULAR SYSTEM

There is now compelling evidence that adipose tissue is a metabolically active organ that generates a variety of bioactive molecules collectively called "adipokines" (Gualillo *et al.* 2007), which may significantly affect or modulate cardiac function and morphology (Frühbeck 2004). The physiological and pathophysiological role of the adipokines is still not completely known. In fact, while some adipokines may cause cardiovascular damage, others actually display clear cardioprotective properties. Adipokines can modulate the cardiovascular system and participate either directly or indirectly in the regulation of several cardiovascular processes.

However, several adipokines have both cardioprotective and harmful properties, as summarized in Table 3, suggesting a dichotomous and still unclear role.

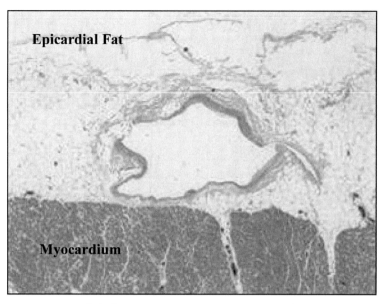

Fig. 2 Human epicardial adipose tissue. Microscopic appearance of epicardial adipose tissue (with a large lipid vacuole inside) and underlying myocardium in a patient with coronary artery disease. Note that no muscle fascia separates the epicardial fat and myocardium. (This figure is unpublished in this modified version and comes from the author's own collection. Patient's informed consent had been obtained.)

Table 2 Characteristics and properties of epicardial adipose tissue

True visceral fat of the heart
Anatomic and functional proximity to the myocardium
Located in the atrioventricular and interventricular grooves
Small adipocytes
High free fatty acids release
Physiological role as fatty acids buffer and local energy source
Metabolically very active and source of several adipokines
Pathological role as source of pro-atherogenic and pro-inflammatory adipokines
Protective role as source of anti-atherogenic and anti-inflammatory adipokines
Local secretion of adipokines into the coronary circulation
Involved in cardiac hypertrophy and diastolic dysfunction
Can be measured with echocardiography, computed tomography and magnetic resonance imaging
Marker of visceral adiposity
Correlate of coronary artery disease, atherosclerosis and metabolic syndrome

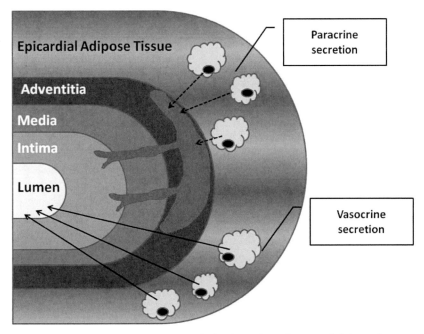

Fig. 3 Paracrine or vasocrine secretion of adipokines from the epicardial fat. A diagram of two possible mechanisms by which adipokines might reach the coronary artery lumen from the epicardial fat. Adipokines from periadventitial epicardial fat could traverse the coronary wall by diffusion from outside to inside by a paracrine mechanism. Adipokines might be released from epicardial tissue directly into vasa vasorum and be transported downstream into the arterial wall by a vasocrine mechanism. (This artwork is unpublished and comes from the author's own collection.)

Color image of this figure appears in the color plate section at the end of the book.

ADIPOKINES WITH CARDIOPROTECTIVE PROPERTIES

Adiponectin and the Heart

Adiponectin is mainly expressed by adipocytes and can be found in three oligomeric forms (Scherer *et al.* 1995). Adiponectin shows cardioprotective, anti-inflammatory and anti-atherogenic properties. Adiponectin constitutes a cardioprotective factor, since this adipokine modulates myocardial remodeling after ischemic injury through activated protein kinase and cyclooxigenase-2-dependent mechanisms, attenuates cardiac hypertrophy and interstitial fibrosis and protects against cardiomyocyte and capillary loss after myocardial infarction. Adiponectin-deficient mice exhibit reduced endothelium-dependent vasodilation on an atherogenic diet, increased leukocyte-endothelium adhesiveness, increased

Table 3 Adipokines and their cardiovascular effects

Adiponectin	Cardioprotective, anti-inflammatory, anti-atherogenic
Angiotensin II	Vasoconstrictor, increases blood pressure and participates in vascular remodeling and myocardial hypertrophy, involved in adipogenesis
Apelin	Possible double role: blood pressure control and cardiac contractility stimulation
Cardiotrophin 1	Possible double role: myocardial survival and proliferation, myocardial hypertrophy
CRP	Pro-inflammatory
Endothelin 1	Vasoconstrictor, induces adipose tissue lipolysis
Ghrelin	Possible cardioprotective effects
IL-6	Pro-inflammatory
Leptin	Possible double role: contributes to the regulation of blood pressure, heart rate, myocardial hypertrophy, regulates normal cardiovascular function
Omentin	Insulin-mimetic
Osteopontin	Possible double role: myocardial survival and proliferation, myocardial remodeling and hypertrophy, pro-inflammatory
PAI-1	Potent inhibitor of fibrinolysis, pro-atherogenic
RBP4	Insulin resistance
Resistin	Pro-inflammatory, insulin-resistance, possible role in congestive heart failure
TNF-α	Pro-inflammatory, insulin resistance
Visfatin	Insulin-mimetic

CRP, C-reactive protein; IL-6, interleukin; PAI-1, plasminogen activator inhibitor 1; RBP4, retinol-binding protein 4; TNF- α, tumour necrosis factor alpha.

neointimal hyperplasia after acute vascular injury and increased blood pressure values. The beneficial effects of adiponectin against endothelial dysfunction are mediated by its ability to increase nitric oxide (NO) bioavailability.

The local production of adiponectin by cardiomyocytes suggests its autocrine-paracrine effect in the heart. The production source of adiponectin in the coronary circulation has been also suggested to be the epicardial adipose tissue (Iacobellis *et al.* 2005b). Local secretion of harmful, inflammation-producing cytokines from the epicardial fat could be predominant and therefore down-regulate adiponectin production in severe and unstable coronary artery disease. Secretion of adiponectin, an anti-inflammatory cytokine, from the epicardial fat in the coronary circulation could be up-regulated and therefore higher than pro-inflammatory adipokine production only in response to certain stimuli that may occur with improved hemodynamic conditions (Iacobellis *et al.* 2009a). High plasma adiponectin concentrations are associated with lower risk of acute coronary syndrome, myocardial infarction and coronary heart disease. Low plasma adiponectin concentration is associated with impaired endothelial-dependent vasodilation, concentric hypertrophy and diastolic dysfunction. Hypoadiponectinemia has been closely linked to impaired endothelial-dependent vasodilation in both normal subjects and patients with hypertension and type 2 diabetes.

Adrenomedullin and the Heart

Adrenomedullin is a potent vasodilatator and antioxidative peptide. Adrenomedullin is also produced by subcutaneous and visceral adipose tissue, including the epicardial fat.

Although it seems to have predominantly a cardioprotective role, several cardiac physiological and pathophysiological properties have been attributed to adrenomedullin (Cheung *et al.* 2004). Ischemic and hypoxic conditions seem to stimulate the production and secretion of adrenomedullin. The increase in circulating adrenomedullin levels and adipose tissue expression might possibly be induced by cerebral or cardiac ischemia. However, the relationship of adrenomedullin with coronary artery disease is partly unclear. It has been suggested that adrenomedullin synthesis from the epicardial fat could have a protective effect on the coronary arteries (Iacobellis *et al.* 2009b). Chronic and stable coronary artery disease is associated with low intracoronaric and systemic adrenomedullin levels. Coronary artery disease may also down-regulate epicardial fat adrenomedullin gene and protein expression. In fact, intracoronary adrenomedullin concentrations are reported to be significantly lower in subjects with coronary artery disease than in individuals without coronary artery disease. It has been suggested that epicardial fat adrenomedullin synthesis increases in response to chronic hypoxemia. Recent studies also showed that adrenomedullin is able to up-regulate M2 muscarinic receptors in cardiomyocytes derived from the murine P19 cell line. Adrenomedullin promotes angiogenesis and may therefore participate in the angiogenesis under certain conditions.

Circulating adrenomedullin levels have been suggested as promising independent predictors of future cardiovascular events in patients with multiple cardiovascular risk factors. However, its use in clinical practice will need further investigations.

Apelin and the Heart

Apelin was identified as the endogenous ligand of the orphan G protein-coupled receptor, the human apelin J receptor. Similarities have been found between the structure and anatomical distribution of apelin and its receptor and that of angiotensin II and the AT1 receptor, providing clues about the physiological functions of this molecule. The cardiovascular system appears to be a primary target of apelin, since apelin J, the apelin receptor, is expressed in the heart and the media layer of human coronary arteries, aorta and saphenous vein grafts. The intravenous administration of apelin in rats has been shown to decrease mean arterial pressure through NO-dependent mechanisms. Apelin-12 has been shown to dilate peripheral veins more efficiently than Ca^{2+} antagonists. This hypotensive effect is accompanied by a slight increase in heart rate, which results from the baroreceptor reflex-mediated stimulation of the sympathetic nervous system. In this sense, it has been reported that apelin increases myocardial contractility in isolated perfused rat hearts. Cardiac effects of apelin are further reflected in a study reporting that circulating apelin levels, atrial apelin, and atrial and ventricular apelin receptor expression are dramatically decreased in patients with heart failure. It has been recently reported that in ischemic myocardium of isolated rat heart apelin expression is up-regulated but returns to baseline after reperfusion. During this period of reduced apelin expression, administration of exogenous apelin-13 attenuated the ischemic/reperfusion injury, reducing the infarct size. The evidence that apelin acts as a cardioprotective factor and that the sensitivity to apelin might be altered in disease states makes the apelin–apelin receptor system a promising target for the development of drugs to attenuate the ischemic/reperfusion injury in patients with heart failure.

Ghrelin and the Heart

Ghrelin is mainly produced by the stomach and intestine, but it is also synthesized by the adipose tissue and cardiomyocytes. Ghrelin mainly acts on the pituitary and hypothalamus to stimulate growth hormone (GH) release, food intake and weight gain. GH and its mediator, insulin-like growth factor 1, are anabolic hormones that are essential for myocardial development and performance. Ghrelin has cardiovascular effects through GH-dependent and -independent mechanisms. Ghrelin is synthesized in cardiomyocytes and operates as an endogenous cardioprotective factor (Soeki *et al.* 2008). The presence of ghrelin receptors in cardiac ventricles seems to provide evidence for direct cardioprotective effects of ghrelin. Ghrelin protects cardiomyocytes and endothelial cells against apoptosis through the activation of an intracellular survival pathway. Clinically, intravenous administration of ghrelin in healthy individuals and patients with chronic heart failure improves the cardiac function by increasing the cardiac index and stroke volume index. The beneficial hemodynamic effects of ghrelin on patients with congestive heart failure seem to be attributable to both an inotropism of GH and a fall in cardiac overload. Moreover, ghrelin administration after myocardial infarction has been shown to attenuate LV enlargement and myocardial fibrosis in rodents. Together, these results suggest that ghrelin has therapeutic potential in the treatment of severe heart failure. However, the potential of ghrelin in the treatment of severe cardiac heart failure is currently under investigation.

ADIPOKINES WITH DICHOTOMOUS CARDIAC EFFECTS

Cardiotrophin 1 and the Heart

Cardiotrophin 1 (CT-1) is a member of the IL-6 family of cytokines that is involved in cardiac growth and dysfunction (Calabrò *et al.* 2009). CT-1 is secreted by the adipose tissue and progressively increases, along with differentiation time from preadipocyte to mature adipocyte in 3T3-L1 cells. CT-1 is also secreted by cardiomyocytes. Cardiomyocytes CT-1 may play an autocrine role during cardiac chamber growth and morphogenesis by promoting the survival and proliferation of immature myocytes, most likely via gp130-dependent signaling pathways. CT-1 seems to have a potential double effect on the heart. In fact, while CT-1 provides myocardial protection by promoting cell survival and proliferation, it also induces myocyte hypertrophy and collagen synthesis, thereby participating in the progression of ventricular remodeling. Interestingly, while activation of the p42/p44 mitogen-activated protein kinase pathway is necessary for the survival-promoting effects of CT-1 in cardiac cells, it is not required for its hypertrophic effect. CT-1 plasma levels are associated with glucose levels and elevated in patients with hypertension and coronary artery diseases, and they are also correlated with the severity of valve diseases and heart failure.

Leptin and the Heart

Leptin, the *OB* gene product, is mainly produced by adipose tissue and participates in the control of body weight, regulation of food intake and energy expenditure. Leptin secretion is proportional to the amount of adipose tissue, and its plasma concentrations are markedly increased in obese individuals. Hyperleptinemia is suggested to play an important role in the pathogenesis of obesity-associated hypertension. Intracerebroventricular and intravenous administration of leptin increases the mean arterial pressure and heart rate as well as the sympathetic outflow to kidneys, adipose tissue, skeletal vasculature and adrenal medulla in rodents. Leptin has also been shown to contribute to blood pressure homeostasis inducing a depressor response attributable to the vasodilation of conduit and resistance vessels. In the aorta and coronary arteries, leptin induces vasodilation via NO, whereas in mesenteric arteries leptin causes relaxation through endothelium-derived hyperpolarizing factor. Leptin is also synthesized in cardiomyocytes and released to the coronary effluent, suggesting the hypothesis that cardiac leptin may exert physiological effects to the myocardium. In fact, leptin exerts several cardiac actions, such as decreasing the contractility of ventricular myocytes via NO, promoting the hypertrophy of rat cardiomyocytes via activation of the mitogen-activated protein kinase system or reducing the myocardial injury after ischemia-reperfusion (Smith *et al.* 2006). The angiogenic and wound-healing properties of leptin may considerably improve the infarcted tissue repair.

Osteopontin and the Heart

Osteopontin is a phosphoprotein identified originally in osteoblasts and osteoclasts, but also secreted by adipocytes. Circulating osteopontin levels are increased in obesity. Osteopontin influences the cardiovascular function, playing a role in atherosclerosis, LV hypertrophy and cardiac fibrosis (Zahradka 2008). Furthermore, myocardial osteopontin expression overlaps with the development of heart failure. A double cardiac role has been attributed to osteopontin, as well. In fact, osteopontin operates in the post-infarcted heart by coordinating the intracellular signals required to integrate myofibroblast proliferation, migration and extracellular matrix deposition with the recruitment of macrophages and initiation of collateral vessel formation, thus ensuring that the mechanical properties of the heart are not further compromised.

ADIPOKINES WITH INSULIN MIMETIC ACTIONS

Visfatin and the Heart

Visfatin is a novel adipokine that is produced mainly in visceral adipose tissue. Visfatin is considered a marker of visceral abdominal fat. However, epicardial fat also expresses visfatin in humans (Fain *et al.* 2008). Nevertheless, no significant difference in visfatin mRNA expression among epicardial, subcutaneous and omental fat has been described. Recently, it has been reported that visfatin up-regulates matrix metalloproteinases and vascular endothelial growth factor (VEGF), key molecules of the angiogenic process, in human micro- and macro-vascular endothelial cells, revealing a novel insight into the potential role of visfatin in cardiovascular disease.

Omentin and the Heart

Omentin is an adipokine expressed in omental and epicardial adipose tissue (Fain *et al.* 2008) in humans and may regulate insulin action. Omentin mRNA seems to be predominantly expressed in epicardial and omental human fat. Its cardiovascular role is still unknown.

ADIPOKINES WITH PRO-INFLAMMATORY AND PRO-ATHEROGENIC PROPERTIES

Tumor Necrosis Factor alpha and the Heart

TNF-α is secreted by several adipose tissue depots. Epicardial adipose tissue is an important source of pro-inflammatory factors, including TNF-α (Mazurek *et al.* 2003). In patients with significant coronary artery disease, epicardial adipose tissue exhibited significantly higher levels of TNF-α than subcutaneous fat. TNF-α is associated with obesity, inflammation and insulin resistance.

Interleukin 6 and the Heart

IL-6 is produced by human adipose tissue. IL-6 reportedly has negative inotropic and hypertrophic effects on the heart. Sustained volume overload is capable of inducing persistent up-regulation of some cardiac cytokines, such as IL-6. Increased IL-6 production by adipocytes may contribute to the obesity-associated insulin-resistance. Inhibition of IL-6 signaling is suggested as a potential therapy for hypertension and cardiac hypertrophy.

Monocyte Chemoattractant Protein 1 and the Heart

Adipocytes, particularly visceral adipocytes, are able to produce large amounts of MCP-1, a chemokine directly involved in ventricular remodeling. Elevated MCP-1 levels are correlated with established risk factors for atherosclerosis, increased risk of death or myocardial infarction. Human epicardial fat secretes MCP-1. In patients with significant coronary artery disease epicardial adipose tissue exhibited significantly higher levels of MCP-1 than subcutaneous fat.

C-reactive Protein and the Heart

Adipose tissue can be a source of C-reactive protein synthesis that inversely correlates with adiponectin, an endogenous adipocyte-derived anti-inflammatory protein. Adipose C-reactive protein may be a marker of a chronic inflammatory state that can trigger acute coronary syndrome.

Resistin and the Heart

Resistin is an adipokine expressed by macrophages and adipocytes. In humans, resistin is strongly expressed in macrophages and, to a lesser extent, by adipocytes.

Although it has been frequently associated with inflammation and insulin resistance, its physiological and pathophysiological role is unclear (Gómez-Ambrosi and Frühbeck 2001). High plasma levels of resistin correlate with pro-atherogenic inflammatory markers, unstable angina, congestive heart failure and coronary atherosclerosis. Macrophages infiltrating human atherosclerotic aneurysms have been shown to secrete resistin. In turn, resistin increases pro-inflammatory markers, up-regulates the expression of adhesion molecules, promotes the release of endothelin 1 and stimulates the synthesis of MCP-1 in human endothelial cells, which contributes to macrophage recruitment and further increased production of pro-inflammatory cytokines. Resistin increases pro-inflammatory markers, up-regulates the expression of adhesion molecules, promotes the release of endothelin 1 and stimulates the synthesis of MCP-1 in human endothelial cells. Resistin also induces endothelial dysfunction. Moreover, resistin induces endothelial dysfunction in isolated coronary artery rings and worsens cardiac ischemia-reperfusion injury in isolated perfused rat hearts. Recently, very elevated serum resistin concentrations were independently associated with risk for incident heart failure in older persons, independently of markers of inflammation and insulin resistance, and adiposity measures (Kritchevsky *et al.* 2009). This finding that links the elevated resistin levels to a higher risk of congestive heart failure may actually suggest some hypotheses. Among the different mechanisms that could explain the association between resistin and congestive heart failure the possibility that resistin concentrations could reflect cardiac load is not excluded.

Nerve Growth Factor and the Heart

NGF is a neurotrophic peptide also produced by human adipose tissue including the epicardial fat. NGF is elevated in subjects with obesity and may contribute to coronary inflammation and atherosclerosis.

Plasminogen Activator Inhibitor 1 and the Heart

PAI-1 is a major regulator of the fibrinolytic system. PAI-1 is mainly synthesized by hepatocytes and endothelial cells, but it is also secreted by adipocytes. The increased gene expression and secretion of PAI-1 by adipose tissue contribute to its elevated plasma levels in obesity. Increased triglyceride accumulation and infiltration within the adipocyte is considered one of the triggers for increased expression and production of inflammatory cytokines and PAI-1.

ADIPOKINES ASSOCIATED WITH HYPERTENSION AND INSULIN RESISTANCE

Renin-Angiotensin System and the Heart

Adipose tissue expresses renin-angiotensin system (RAS) and angiotensinogen. Angiotensinogen produced in the adipose tissue and angiotensinogen-derived peptides, as well as angiotensin II, may influence adipogenesis and blood pressure (Engeli *et al.* 2003). Increased angiotensinogen production by adipose tissue supports a role of adipose RAS system in hypertensive obese patients. Visceral adipose tissue expresses higher levels of angiotensinogen. Adipose RAS seems to be differently regulated in subcutaneous adipose tissue from obese patients with hypertension with respect to normotensive obese patients and controls. Epicardial fat expresses angiotensinogen. Increased angiotensinogen production in epicardial adipose tissue has been reported to contribute to the development of postoperative insulin resistance.

Endothelin 1 and the Heart

Endothelin 1 (ET-1) is a potent vasoconstrictor peptide. ET-1 is also produced by human adipocytes. ET-1 has profound effects on adipose tissue metabolism. Elevated circulating or adipose-derived ET-1 could promote systemic insulin resistance. ET-1 locally secreted by cardiomyocytes may contribute to cardiac hypertrophy as an autocrine/paracrine factor.

Retinol-binding Protein 4 and the Heart

Retinol-binding protein 4 (RBP4) is a protein secreted by adipocytes. Elevated RBP4 levels have been suggested as predictors of diabetes, insulin resistance and associated cardiovascular risk factors. Elevated RBP4 has been correlated with inflammatory markers in diabetic patients with dilated cardiomyopathy.

Summary Points

- Isolated and uncomplicated excess adiposity is associated with adaptive morphological and functional cardiac changes.
- Adipose tissue secretes bioactive molecules called adipokines.
- Adipokines can, locally or indirectly, either cause heart damage or exert clear cardioprotective actions.
- Increased epicardial fat, the cardiac visceral fat, can locally and mechanically affect heart morphology and function.
- Epicardial fat, the visceral fat depot of the heart, secretes both pro- and anti-inflammatory adipokines that through paracrine or vasocrine pathways can affect and modulate the myocardium and coronary arteries.

Abbreviations

ACE	:	angiotensin-converting enzyme
CT-1	:	cardiotrophin 1
ET-1	:	endothelin 1
GH	:	growth hormone
LV	:	left ventricle
MPC-1	:	monocyte chemoattractant protein 1
IL-6	:	interlukin 6
NGF	:	nerve growth factor
PAI-1	:	plasminogen activator inhibitor 1
RAS	:	renin-angiotensin system

RBP4	:	retinol-binding protein 4
TNF-α	:	tumor necrosis factor alpha
VEGF	:	vascular endothelial growth factor

Definition of Terms

Epicardial adipose tissue: The fat depot that surrounds the heart. It secretes both harmful and protective adipokines that reach the heart and the coronary arteries.

Pro-inflammatory adipokines: Bioactive molecules secreted by the adipose cells. They can cause or contribute to local inflammatory and atherogenic processes.

Anti-inflammatory adipokines: Bioactive molecules secreted by the adipose cells. They can reduce or contribute to reduce inflammatory and atherogenic processes. These adipokines may have protective effects to the internal organs.

Atherosclerosis: A disease of the large and medium-sized arteries characterized by a gradual build-up of fatty and inflammatory plaques within the arterial wall that may result in a significant reduction of the vessel lumen, impairing blood flow and perfusion to the distal tissues including the heart and the brain. Major clinical complications of atherosclerosis are coronary artery diseases and stroke.

Cardiac hypertrophy: A compensatory mechanism of the heart to a chronic and long-term increased volume and pressure overload. Heart walls and chambers can be thicker and larger and have a different geometric pattern. This mechanism can become maladaptive when cardiac hypertrophy is associated with reduced or impaired heart performance.

References

Calabrò, P. and G. Limongelli, L. Riegler, V. Maddaloni, R. Palmieri, E. Golia, T. Roselli, D. Masarone, G. Pacileo, P. Golino, and R. Calabrò. 2009. Novel insights into the role of cardiotrophin-1 in cardiovascular diseases. J. Mol. Cell. Cardiol. 46: 142-148.

Cheung, B.M. and C.Y. Li, and L.Y. Wong. 2004. Adrenomedullin: its role in the cardiovascular system. Sem. Vasc. Med. 4: 129-134.

Engeli, S. and P. Schling, K. Gorzelniak, M. Boschmann, J. Janke, G. Ailhaud, M. Teboul, F. Massiera, and A.M. Sharma. 2003. The adipose-tissue renin-angiotensin-aldosterone system: role in the metabolic syndrome? Int. J. Biochem. Cell. Biol. 35: 807-825.

Fain, J.N. and H.S. Sacks, B. Buehrer, S.W. Bahouth, E. Garrett, and R.Y. Wolf. 2008. Identification of omentin mRNA in human epicardial adipose tissue: comparison to omentin in subcutaneous, internal mammary artery periadventitial and visceral abdominal depots. Int. J. Obes. (Lond). 32: 810-815.

Frühbeck, G. 2004. The adipose tissue as a source of vasoactive factors. Curr. Med. Chem. Cardiovasc. Hematol. Agents 2: 197-208.

Gómez-Ambrosi, J. and G. Frühbeck. 2001. Do resistin and resistin-like molecules also link obesity to inflammatory diseases? Ann. Intern. Med. 135: 306-307.

Gualillo, O. and J.R. González-Juanatey, and F. Lago. 2007. The emerging role of adipokines as mediators of cardiovascular function: physiologic and clinical perspectives. Trends Cardiovasc. Med. 17: 275-283.

Iacobellis, G. and M.C. Ribaudo, G. Leto, A. Zappaterreno, E. Vecci, and U. Di Mario. 2002 Influence of excess fat on cardiac morphology and function: study in uncomplicated obesity. Obes. Res. 10: 767-773.

Iacobellis, G. and M.C. Ribaudo, A. Zappaterreno, C.V. Iannucci, and F. Leonetti. 2004. Relation between epicardial adipose tissue and left ventricular mass. Am. J. Cardiol. 94: 1084-1087.

Iacobellis, G. and D. Corradi, and A.M. Sharma. 2005a. Epicardial adipose tissue: anatomic, biomolecular and clinical relationships with the heart. Nat. Clin. Pract. Cardiovasc. Med. 2: 536-543.

Iacobellis, G. and D. Pistilli, M. Gucciardo, F. Leonetti, F. Miraldi, G.L. Brancaccio and C.R.T. Di Gioia. 2005b. Adiponectin expression in human epicardial adipose tissue *in vivo* is lower in patients with coronary artery disease. Cytokine 29: 251-255.

Iacobellis, G. and A.M. Sharma. 2006. Adiposity of the heart. Ann. Intern. Med. 145: 554-555.

Iacobellis, G. and C.M. Pond, and A.M. Sharma. 2006. Different "weight" of cardiac and general adiposity in predicting left ventricle morphology. Obesity (Silver Spring) 14: 1679-1684.

Iacobellis, G. and A.M. Sharma. 2007. Obesity and the heart: redefinition of the relationship. Obes. Rev. 8: 35-39.

Iacobellis, G. and F. Leonetti, N. Singh, and A.M. Sharma. 2007. Relationship of epicardial adipose tissue with atrial dimensions and diastolic function in morbidly obese subjects. Int. J. Cardiol. 115: 272-273.

Iacobellis, G. and G. Barbaro. 2008. The double role of epicardial adipose tissue as pro- and anti-inflammatory organ. Horm. Metab. Res. 40: 442-445.

Iacobellis, G. and Y.J. Gao, and A.M. Sharma. 2008. Do cardiac and perivascular adipose tissue play a role in atherosclerosis? Curr. Diab. Rep. 8: 20-24.

Iacobellis, G. 2009. Relation of epicardial fat thickness to right ventricular cavity size in obese subjects. Am. J. Cardiol. 104: 1601-1602.

Iacobellis, G. and H. Willens. 2009. Echocardiographic epicardial fat: a review of research and clinical applications. J. Am. Soc. Echocardiogr. 23: 1311-1319.

Iacobellis, G. and C.R. di Gioia, D. Cotesta, L. Petramala, C. Travaglini, V. De Santis, and C. Letizia. 2009a. Epicardial adipose tissue adiponectin expression is related to intracoronary adiponectin levels. Horm. Metab. Res. 41: 227-231.

Iacobellis, G. and C.R. Gioia, M. Di Vito, L. Petramala, D. Cotesta, V. De Santis, and C. Letizia. 2009b. Epicardial adipose tissue and intracoronary adrenomedullin levels in coronary artery disease. Horm. Metab. Res. 41: 855-860.

Kritchevsky, S.B. and R.S. Vasan, P.W. Wilson, and T.B. Harris. Health ABC Study. 2009. Serum resistin concentrations and risk of new onset heart failure in older persons: the health, aging, and body composition (Health ABC) study. Arterioscler. Thromb. Vasc. Biol. 29: 1144-1149.

Mazurek, T. and L. Zhang, A. Zalewski, J.D. Mannion, J.T. Diehl, and H.L. Arafat. 2003. Human epicardial adipose tissue is a source of inflammatory mediators. Circulation. 108: 2460-2466.

Scherer, P.E. and S. Williams, M. Fogliano, G. Baldini, and H.F. Lodish. 1995. A novel serum protein similar to C1q, produced exclusively in adipocytes. J. Biol. Chem. 270: 26746-26749.

Smith, C.C. and M.M. Mocanu, S.M. Davidson, A.M. Wynne, J.C. Simpkin, and D.M. Yellon. 2006. Leptin, the obesity-associated hormone, exhibits direct cardioprotective effects. Br. J. Pharmacol. 149: 5-13.

Soeki, T. and I. Kishimoto, D.O. Schwenke, T. Tokudome, T. Horio, M. Yoshida, H. Hosoda, and K. Kangawa. 2008. Ghrelin suppresses cardiac sympathetic activity and prevents early LV remodeling in rats with myocardial infarction. Am. J. Physiol. Heart Circ. Physiol. 294: H426-H432.

Zahradka, P. 2008. Novel role for osteopontin in cardiac fibrosis. Circ. Res. 102: 270-272.

CHAPTER 26

Adipokines as Mediators of Rheumatic Disease

Deborah M. Levy[1] and Earl D. Silverman[1,*]

[1]Division of Rheumatology, Hospital for Sick Children, University of Toronto, 555 University Ave, Toronto, ON M5G 1X8, Canada

ABSTRACT

The rheumatic diseases are a diverse group of diseases characterized by a pro-inflammatory state and the development of autoimmunity. Adipokines can act as cytokines to alter the immune system and are likely important in the pathogenesis and/or clinical syndromes of rheumatic diseases. In addition they are important factors leading to the metabolic syndrome and atherosclerosis (both commonly seen in patients with rheumatic diseases).

Leptin can act as a cytokine and is produced by inflammatory cells following stimulation by multiple cytokines. Leptin can activate monocytes/macrophages, neutrophils, basophils, eosinophils, natural killer cells, and dendritic cells and is generally pro-inflammatory. Leptin-deficient mice are protected against autoimmune diseases and cardiovascular disease, which suggests an important role for leptin in human rheumatic diseases. However, the concentration of circulating leptin levels has not been consistently correlated with disease activity in patients with rheumatic diseases. Adiponectin is produced by cells important in rheumatic diseases (skeletal muscle, osteoblasts, osteoclasts, chondrocytes, synovial fibroblasts and endothelial cells); it can alter monocyte, macrophage, dendritic, T, B and endothelial cell function. Adiponectin plays a counter-regulatory role in inflammation and may be an effector mechanism of the anti-inflammatory effects of glucocorticoids. Murine studies have suggested that the therapeutic administration of adiponectin may decrease disease severity and damage in systemic lupus erythematosus (SLE). Adiponectin likely plays an important role in rheumatic diseases.

---

*Corresponding author

Resistin is found in the synovial fluid of patients with rheumatoid arthritis (RA) and causes production of pro-inflammatory cytokines. Bone marrow cells and macrophages are capable of producing resistin in response to inflammation and its production is increased by TNF-α and glucocorticoids. In mice, resistin causes synovitis and pannus formation in the joint. However, as there are few studies of resistin in rheumatic diseases its role is not yet defined. Elevated circulating visfatin levels are found in RA and SLE and correlate with damage in RA. Visfatin can stimulate the production of pro-inflammatory cytokines, up-regulate co-stimulatory molecules, and activate antigen-presenting cells. However, to date there are few studies of visfatin in rheumatic diseases.

This chapter outlines the role of adipokines in rheumatic diseases.

INTRODUCTION

This chapter addresses proven, proposed, and potential roles of adipokines in human rheumatic diseases. Animal models of disease, although important in delineating potential pathophysiological roles of adipokines in disease, are discussed only briefly and in the context of improving our understanding of human disease, as animal models are addressed in separate chapters.

Rheumatic diseases are common as up to 1% of adults will develop rheumatoid arthritis (RA), and systemic lupus erythematosus (SLE) occurs in approximately 1 in 1000 females. The rheumatic diseases are pro-inflammatory states associated with significant morbidity and, in particular, premature atherosclerosis that has become increasingly recognized as a significant cause of mortality in patients with rheumatic diseases. Cardiovascular mortality is not generally a result of direct cardiac involvement but rather is usually due to ischemic heart disease secondary to coronary atherosclerosis. Compared to non-diseased controls, RA patients have a more than two-fold risk for myocardial infarction (MI) and SLE patients are five to six times as likely to experience a coronary event as the general population with the relative risk of death due to coronary heart disease as high as 17 and death due to stroke 7.9. This increased burden of cardiovascular disease in patients with RA and SLE is likely the result of a combination of factors including the chronic inflammation associated with these diseases, and the higher prevalence of the metabolic syndrome (obesity, dyslipidemia, hypertension and insulin resistance) seen in these patients and therefore adipokines are likely important.

OVERVIEW OF ROLE OF ADIPOKINES IN RHEUMATIC DISEASES

Although adipokines were originally discovered as a result of their effect on appetite and obesity, investigators have subsequently demonstrated their role in the metabolic syndrome and atherosclerosis, and they can act as cytokines. This chapter focuses on the role of adipokines, specifically adiponectin, leptin, resistin and visfatin, in rheumatic diseases, their potentials as biomarkers of disease activity and their potential role in the immune dysregulation and cardiovascular disease seen in these patients. Figure 1 and Table 1 give overviews of the rheumatic diseases associated with the four individual adipokines.

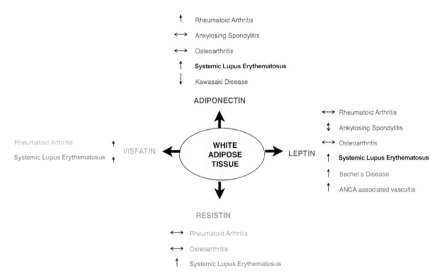

Fig. 1 Levels of adiponectin, leptin, resistin and visfatin in rheumatic diseases. The upward arrow indicates increased levels as compared to controls while the downward arrow indicates decreased levels as compared to controls.

Color image of this figure appears in the color plate section at the end of the book.

Adiponectin

Adiponectin is a member of the complement factor C1q and tumour necrosis factor alpha (TNF-α) superfamily of molecules and may be considered an important molecule linking metabolism and the immune system. Low adiponectin levels are associated with obesity, type-2 diabetes and atherosclerosis, while increased adiponectin levels are found in inflammatory diseases. Increased levels are thought to play a counter-regulatory role. Circulating adiponectin levels are altered by glucocorticoids, drugs that are frequently used in rheumatic diseases, and may be an effector mechanism of the anti-inflammatory effects of glucocorticoids. Adiponectin is produced not only by adipose tissue but also by other cells important in rheumatic diseases including skeletal muscle, osteoblasts, osteoclasts, chondrocytes, synovial fibroblasts and endothelial cells. Although adiponectin induces both pro-inflammatory and anti-inflammatory molecules,

Table 1 Summary of adipokines in rheumatic diseases

CYTOKINE	Rheumatoid arthritis (RA)	Ankylosing spondylitis (AS)	Systemic lupus Erythematosus (SLE)	Osteoarthritis (OA)	Other diseases*
ADIPONECTIN	Increased levels with increased disease duration and damage Treatment with anti-TNF-α agents: no effect on levels	Levels not different from controls	Increased levels as compared to controls Increased levels seen in active renal disease Decreased levels associated with increased BMI and CRP	Levels not different from controls Increased levels associated with increased disease activity Increased serum as compared to synovial fluid	Decreased levels in active KD KD: decreased levels associated with increased IL-6 levels
LEPTIN	Increased serum as compared to synovial fluid levels Increased levels associated with increased disease severity Not changed after anti-TNF-α treatment	Both increased and decreased serum levels reported as compared controls Increased levels associated with increased disease activity	Increased levels as compared to controls Increased levels associated with increased CRP and ESR Increased levels associated increased lipid levels	Levels not different from controls Increased levels associated with disease severity Increased in synovial fluid as compared to serum levels Increased levels associated with increased BMI	Increased levels associated with increased disease activity in BD Decreased levels associated with increased disease activity in AAV
RESISTIN	Not different from controls Increased synovial fluid as compared to serum levels Increased levels associated with increased disease activity Levels decreased after anti-TNF-α therapy	No data	Increased levels associated with increased ESR Increased levels associated with increased steroid dose Increased levels associated with decreased BMD and decreased HDL	Not different from controls Increased synovial fluid as compared to serum levels Increased levels associated with increased disease severity	No data

VISFATIN	No data	Increased levels associated with increased CRP	No data	Increased levels as compared to controls	No data
		No change after anti-TN-α treatment			

*Other diseases: Behcet's disease (BD), Kawasaki disease (KD), ANCA-associated vasculitis (AAV). TNF-α, tumour necrosis factor alpha; BMI, body mass index; CRP, C-reactive protein; IL-6, interleukin 6; ESR, erythrocyte sedimentation rate; BMD, bone mineral density; HDL, high density lipoprotein.

adiponectin appears to have a predominantly anti-inflammatory function (Tilg and Moschen 2006). Adiponectin alters macrophage TNF-α production and increases macrophage interleukin 6 (IL-6), interleukin 8 (IL-8) and chemokine production. Adiponectin induces TNF-α to alter adherence of monocytes to endothelial cells through decreased expression of surface markers including vascular adhesion molecule 1 (VCAM-1) and intracellular adhesion molecule 1 (ICAM-1). Figure 2 reviews the importance of adiponectin in altering monocyte, macrophage, dendritic, T, B and endothelial cell function. The alteration of the function of these cells and the effects of adiponectin in the pathogenesis of rheumatic diseases are individually reviewed in the following sections.

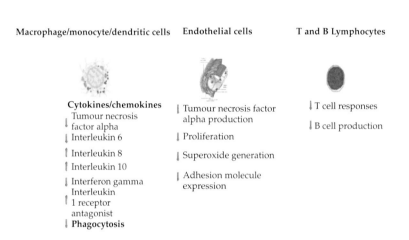

Macrophage/monocyte/dendritic cells   Endothelial cells   T and B Lymphocytes

Cytokines/chemokines
↓ Tumour necrosis factor alpha
↓ Interleukin 6
↑ Interleukin 8
↑ Interleukin 10
↓ Interferon gamma
↑ Interleukin 1 receptor antagonist
↓ **Phagocytosis**

↓ Tumour necrosis factor alpha production
↓ Proliferation
↓ Superoxide generation
↓ Adhesion molecule expression

↓ T cell responses
↓ B cell production

Fig. 2   Known immunological effects of adiponectin on macrophages/monocytes/dendritic cells, endothelial cells and T and B cells as the effects pertain to rheumatic diseases. The upward arrow indicates enhancement and the downward arrow indicates down-regulation. Although macrophages, monocytes and dendritic cells each serve different function, the effects of adiponectin on both cytokine/chemokine production and phagocytosis are similar in all three cells and therefore they are put under one column for simplicity. The overall effect of adiponectin on these cells is anti-inflammatory. Similarly, adiponectin decreases endothelial cell proliferation, superoxide generation, and the production of the pro-inflammatory cytokine tumour necrosis factor alpha with an overall anti-inflammatory effect. It also decreases the up-regulation of adhesion molecules and thus will not all allow inflammatory cells to migrate through the vessel wall. The last part of the figure demonstrates that although the mechanism by which adiponectin affects T and B cells is not well studied, the overall effect is to decrease the response of these cells. However, each of these effects alter individual rheumatic diseases as outlined in the text.

Color image of this figure appears in the color plate section at the end of the book.

Leptin

Leptin can affect the immune system as a cytokine. This may not be surprising as its structure resembles the cytokine IL-6 and the leptin receptor is a member of the class I cytokine receptor superfamily. Leptin can be produced by inflammatory cells following stimulation by cytokines such as interleukin 1 (IL-1), IL-6 and lipopolysaccharide and can act in an autocrine fashion to increase IL-6 and TNF-α production (Faggioni *et al.* 2001). Leptin stimulation of macrophages results in release of both pro-inflammatory cytokines IL-6 and TNF-α and the anti-inflammatory cytokine IL-1 receptor antagonist (IL-1Ra). Leptin appears to be increased in many, but not all, acute phase responses. Leptin levels are gender- and age-dependent and are greater in women than in men even after adjustment for body mass index (BMI). The gender differences and puberty-related changes may be important in the development or frequency of autoimmune diseases such as RA or SLE, diseases that are most common in females of child-bearing age.

Leptin has multiple effects on the immune system as it can activate monocytes/ macrophages, neutrophils, basophils, eosinophils, natural killer cells, and dendritic cells. The effect of leptin on lymphocytes is generally pro-inflammatory as it can (1) activate Th1 and inhibit Th2 cells, (2) cause proliferation of naïve and regulatory T cells, and (3) decrease memory cells while increasing B cell proliferation. Leptin deficiency results in an increased risk of infection and is important in the immune deficiency associated with malnutrition. Mice that lack either leptin receptors or leptin have significant impairment of both innate and adaptive immunity. Leptin can also regulate bone remodeling, by targeting osteoblast and osteoclast differentiation. Figure 3 summarizes the importance of leptin in altering important immune function and cytokine production. The importance of these alterations as they pertain to individual rheumatic diseases is reviewed separately.

Resistin

As resistin was more recently discovered than adiponectin and leptin there have been fewer studies on the role of resistin in rheumatic diseases and control of the immune system. Although resistin is most frequently produced by adipose tissue, bone marrow cells and macrophages are capable of producing resistin and in particular in response to inflammation. Unlike leptin or adiponectin, the structure of resistin is not related to cytokines or cytokine receptors. Resistin up-regulates the production of pro-inflammatory cytokines IL-6, TNF-α and interleukin 12 (IL-12) by peripheral blood mononuclear cells (PBMC). Conversely, macrophage resistin production is up-regulated by TNF-α. Resistin production is also significantly increased by glucocorticoids. Resistin has been shown to be present in human atherosclerotic lesions.

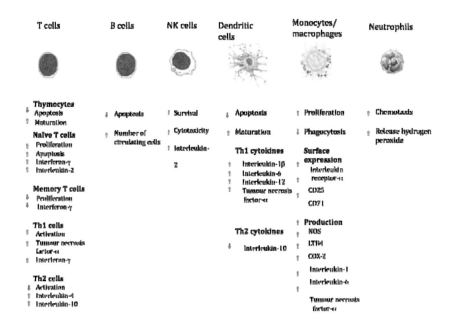

T cells	B cells	NK cells	Dendritic cells	Monocytes/ macrophages	Neutrophils
Thymocytes ↓ Apoptosis ↑ Maturation	↓ Apoptosis	↑ Survival	↓ Apoptosis	↑ Proliferation	↑ Chemotaxis
Naïve T cells ↑ Proliferation ↑ Apoptosis ↑ Interferon-γ ↑ Interleukin-2	↓ Number of circulating cells	↑ Cytotoxicity ↑ Interleukin-2	↑ Maturation **Th1 cytokines** ↑ Interleukin-1β ↑ Interleukin-6 ↑ Interleukin-12 ↑ Tumour necrosis factor-α	↓ Phagocytosis **Surface expression** ↑ Interleukin receptor-α ↑ CD25 ↑ CD71	↑ Release hydrogen peroxide
Memory T cells ↓ Proliferation ↓ Interferon-γ					
Th1 cells ↑ Activation ↑ Tumour necrosis factor-α ↑ Interferon-γ			**Th2 cytokines** ↓ Interleukin-10	↑ **Production** ↑ NOS ↑ LTB4 ↑ COX-2 ↑ Interleukin-1 ↑ Interleukin-6 ↑ Tumour necrosis factor-α	
Th2 cells ↓ Activation ↑ Interleukin-4 ↑ Interleukin-10					

Fig. 3 Immunological effects of leptin in rheumatic diseases. The details for each disease are outlined in the text. The upward arrow indicates enhancement and the downward arrow indicates down-regulation. As can be seen the immunological effects of leptin have been well studied for many cell lines. Leptin decreases apoptosis of thymocytes, naïve T cells and B cells and increases the number of these cells. It promotes the production of the pro-inflammatory cytokines interleukin 2, tumour necrosis factor alpha, and interferon-gamma leading to the activation Th1 cells (pro-inflammatory cells) while also decreasing the activation of Th2 cells although it increases interleukin 10 and interleukin 4 for an overall net pro-inflammatory response from T cells. Many of these cytokines have important roles in the pathogenesis of rheumatic diseases. It also activates natural killer cells and increases their cytotoxicity and survival, which further activates the immune system. The effect of leptin on the antigen-presenting cells, dendritic cells and monocytes/macrophages is pro-inflammatory as it increases the maturation of these cells to become more effective and decreases apoptosis. Furthermore, it results in the production of multiple pro-inflammatory cytokines by these cells as shown in the illustration as well as enhancing the production of the pro-inflammatory molecules/enzymes nitric oxide (NOS), leutriene B4 (LBT4) and cylcooxygenase enzyme 2 (COX-2). The overall effect again is pro-inflammatory. Lastly, it activates neutrophils, enhances phagocytosis, and releases hydrogen peroxide to allow this non-specific effector to perform its killing function better (similar to the effect on natural killer cells).

Color image of this figure appears in the color plate section at the end of the book.

Visfatin

Visfatin was initially discovered as a cytokine that promoted differentiation of B cell precursors and subsequently was shown to be an adipokine. Visfatin was then demonstrated to (1) stimulate the production of pro-inflammatory cytokines, (2) simulate the anti-inflammatory cytokines, (3) up-regulate co-stimulatory molecules, and (4) activate antigen-presenting cells. Endothelial cells stimulated by visfatin have enhanced (1) proliferation, (2) migration, (3) IL-6, IL-8 and monocyte chemotactic protein 1 (MCP-1) production, (4) adhesion molecule expression, and (5) endothelial nitric oxide synthase (eNOS). Neutrophils can produce visfatin following stimulation with pro-inflammatory stimuli.

RHEUMATOID ARTHRITIS

Rheumatoid arthritis (RA) is a chronic inflammatory arthritis affecting both small and large joints (Table 2). Pro-inflammatory cytokines including TNF-α, IL-1 and IL-6 play important roles in the pathogenesis of disease leading to the chronic inflammation. Patients with RA have elevated circulating levels of adiponectin, leptin and visfatin and resistin (Otero *et al.* 2006, Senolt *et al.* 2006, Rho *et al.* 2009). Adiponectin, leptin and resistin have also been detected in the synovial fluid of patients with other forms of arthritis (Schaffler *et al.* 2003).

Adiponectin

This section briefly highlights the importance of adiponectin in RA; please see Fig. 2 of the current chapter and refer to Chapter 34, 'Adiponectin Enhances Inflammation in Rheumatoid Synovial Fibroblasts and Chondrocytes', for more details.

Adiponectin may have two different roles in RA depending on whether it is produced in the circulation or within the joint. Elevated circulating adiponectin levels have been shown in most studies of RA patients and are highest in patients with long-standing RA, associated with more severe RA and with damage (Otero *et al.* 2006, Senolt *et al.* 2006, Ebina *et al.* 2009, Giles *et al.* 2009, Laurberg *et al.* 2009, Rho *et al.* 2009, Targonska-Stepniak *et al.* 2010). Adiponectin is present in the joints of patients with RA as a result of either diffusion of circulating adiponectin or secondary to local synthesis. Synovial fluid (SF) adiponectin levels are elevated in RA patients but the role within the joint is not clear (Senolt *et al.* 2006, Tan *et al.* 2009). In addition to synovial proliferation and cartilage degradation, RA is characterized by juxta-articular bony erosion and therefore any effect of adiponectin on bone cells may be important in RA. Adiponectin alters osteoclasts and may counter the bone resorptive effects of TNF-α in RA patients.

Table 2 Classification criteria for inflammatory and non-inflammatory arthritis

RHEUMATOID ARTHRITIS (RA)[a] RA is diagnosed if ≥ 4 criteria are satisfied. Criteria 1 through 4 must have been present for at least 6 wk.	1. Morning stiffness (≥ 1 h) 2. Arthritis of ≥ 3 joints 3. At least one swollen joint of hand area (wrist, metacarpophalangeal or proximal interphalangeal joints) 4. Symmetric involvement (involvement of same joint areas on both sides of the body) 5. Rheumatoid nodules (subcutaneous nodules over bony prominences or extensor surfaces) 6. Serum rheumatoid factor positivity 7. Radiographic changes (typical of RA)
ANKYLOSING SPONDYLITIS (AS)[b] **Definite AS:** radiographic criterion in addition to at least one clinical criterion **Probable AS:** all clinical criterion alone or radiological without clinical criteria	A. <u>Clinical criteria</u> 1. Low back pain: present for more than 3 mo, improved by exercise but not relieved by rest 2. Limitation of lumbar spine motion in sagittal and frontal planes 3. Limitation of chest expansion relative to normal values for age and sex B. <u>Radiological criterion</u> 1. Sacroiliitis on radiographs
OSTEOARTHRITIS (OA)	**OA of the HAND**[c] Hand pain, aching, or stiffness and 3 or 4 of the following features: 1. Hard tissue enlargement of 2 or more of 10 selected joints 2. Hard tissue enlargement of 2 or more DIP joints 3. Fewer than 3 swollen MCP joints 4. Deformity of at least 1 of 10 selected joints **OA of the HIP**[d] Hip pain and at least 2 of the following 3 features: 1. ESR < 20 mm/h 2. Radiographic femoral or acetabular osteophytes 3. Radiographic joint space narrowing (superior, axial, and/or medial) **OA of the KNEE**[e] Knee pain and at least 3 of the following 6 features: 1. Age > 50 yr 2. Stiffness < 30 min 3. Crepitus 4. Bony tenderness 5. Bony enlargement 6. No palpable warmth

[a] Arnett, F.C., Edworthy, S.M., Bloch, D.A., McShane, D.J., Fries, J.F., Cooper, N.S. et al. 1988. The American Rheumatism Association 1987 revised criteria for the classification of rheumatoid arthritis. Arthritis Rheum. 31: 315-324.
[b] Modified New York Criteria: Van der Linden, S., Valkenburg, H.A., Cats, A. 1984. Evaluation of diagnostic criteria for ankylosing spondylitis: a proposal for modification of the New York criteria. Arthritis Rheum. 27: 361-368.
[c] Altman, R., Alarcón, G., Appelrouth, D., Bloch, D., Borenstein, D., Brandt, K., et al. 1990. The American College of Rheumatology criteria for the classification and reporting of osteoarthritis of the hand. Arthritis Rheum. 33: 1601-1610.
[d] Altman, R., Alarcón, G., Appelrouth, D., Bloch, D., Borenstein, D., Brandt, K., et al. 1991. The American College of Rheumatology criteria for the classification and reporting of osteoarthritis of the hip. Arthritis Rheum. 34: 505-514.
[e] Altman, R., Asch, E., Bloch, D., Bole, G., Borenstein, D., Brandt, K., et al. 1986. The American College of Rheumatology criteria for the classification and reporting of osteoarthritis of the knee. Arthritis Rheum. 29: 1039-1049.

TNF-α is important in the pathogenesis of RA as it is a pro-inflammatory cytokine that plays a key role in joint inflammation and matrix destruction, leading to cartilage destruction and bony erosions. Although adiponectin and TNF-α structurally resemble each other, one antagonizes the action and production of the other. TNF-α blocking agents are playing an increasingly important role in the treatment of severe RA. Most studies have shown that anti-TNF-α treatment resulted in increased adiponectin levels as early as 2 wk and were still found after 1 yr of therapy. In contrast, other studies showed no significant change in adiponectin concentrations early and after up to 26 wk of therapy. Other therapies used with anti-TNF-α therapy and initial adiponectin levels may be important factors in determining the effect of anti-TNF-α therapy. Overall it is likely that adiponectin acts as a counter-regulatory cytokine in RA but currently there is no evidence to suggest that it is a useful biomarker to monitor disease activity or response to therapy.

Leptin

Leptin has multiple immunomodulatory effects that affect both the innate and adaptive immune systems and leptin-deficient mice are resistant to antigen-induced arthritis. Therefore, it is likely important in RA.

Most, but not all, studies of RA patients have demonstrated elevated circulating leptin levels (Anders *et al.* 1999, Salazar-Paramo *et al.* 2001, Bokarewa *et al.* 2003, Gunaydin *et al.* 2006, Otero *et al.* 2006, Wislowska *et al.* 2007). The inconsistency of these studies may be a result of the fact that leptin levels are correlated with fat mass (and BMI) in patients with RA and the controls were not always matched for BMI (Anders *et al.* 1999). Leptin levels tended to be associated with disease activity and to be higher in patients with more severe disease (Targonska-Stepniak *et al.* 2008, Rho *et al.* 2009, Seven *et al.* 2009). Leptin levels were not correlated with inflammatory parameters (Anders *et al.* 1999, Popa *et al.* 2005, Gunaydin *et al.* 2006, Otero *et al.* 2006). Synovial fluid (SF) leptin levels were found to be elevated but were lower than serum levels (Bokarewa *et al.* 2003, Hizmetli *et al.* 2007). SF leptin levels were higher in patients with more severe disease (Bokarewa *et al.* 2003). Although leptin has been shown to alter bone homeostasis, and osteoporosis is a recognized complication of RA, leptin levels have not been correlated with bone mineral density (BMD) (Aguilar-Chavez *et al.* 2009).

Anti-TNF-α therapy of patients with RA did not significantly alter circulating leptin levels. These findings may not be surprising as leptin levels were not correlated with disease activity at the initiation of anti-TNF-α therapy and leptin levels have not been shown to correlate with TNF-α levels or disease activity.

Overall, although a murine model of arthritis suggested that leptin is important in the persistent inflammation and elevated leptin levels are found in patients with RA, the role of leptin is not clear. Furthermore, leptin levels in SF were elevated but lower than systemic levels, suggesting that leptin may not be important at the site of inflammation. Lastly, leptin has not been shown to be a useful biomarker to monitor response to therapy as it has not been shown to change after treatment with anti-TNF-α agents.

Resistin

A role for resistin in inflammatory arthritis was first demonstrated in mice where intra-articular injection resulted in synovitis, synovial lining proliferation and pannus formation (all characteristic of inflammatory arthritis including RA) (Bokarewa *et al.* 2005). Although circulating resistin levels in RA patients were elevated as compared to patients with osteoarthritis, not all studies have found elevated levels as compared to healthy controls (Bokarewa *et al.* 2005, Migita *et al.* 2006, Otero *et al.* 2006, Forsblad d'Elia *et al.* 2008). Most studies demonstrated a significant correlation of serum resistin levels with measures of disease activity and inflammation (Bokarewa *et al.* 2005, Migita *et al.* 2006, Otero *et al.* 2006, Senolt *et al.* 2007, Forsblad d'Elia *et al.* 2008, Gonzalez-Gay *et al.* 2008). Resistin levels were correlated with levels of the anti-inflammatory molecule IL-1Ra at baseline and changes correlated with changes in IL-Ra levels (Forsblad d'Elia *et al.* 2008).

Resistin has been demonstrated in plasma cells, B cells, macrophages, synovial fibroblasts but not in synovial T cells of RA patients (Bokarewa *et al.* 2005). Resistin stimulation of synovial cells results in TNF-α, IL-6, IL-1 and resistin production (Bokarewa *et al.* 2005). In RA, SF resistin levels were higher than levels found in paired blood samples and in controls, and correlated with markers of inflammation (Schaffler *et al.* 2003, Bokarewa *et al.* 2005, Senolt *et al.* 2007). Increased serum resistin levels also correlated with TNF-α levels. Following anti-TNF-α therapy, resistin levels significantly decreased, suggesting that systemic inflammation, possibly TNF-α, can lead to elevated resistin levels or resistin is produced to counteract TNF-α. This latter suggestion is supported by the demonstration that resistin levels were directly correlated with prednisone use and prednisone dose (Forsblad d'Elia *et al.* 2008).

In RA, resistin was associated with inflammation, elevated in SF, present in synovial tissue and changed with anti-TNF-α therapy. Therefore, although the role of resistin in pathogenesis of RA is not clear, it may be a biomarker useful for predicting outcome or response to therapy.

Visfatin

Of the four adipokines discussed in this chapter, the role of visfatin in RA has the fewest studies. Elevated circulating visfatin levels were found in RA patients and levels correlated with C-reactive protein (CRP) (Otero *et al.* 2006, Matsui *et al.* 2008). Visfatin gene expression and protein were elevated in PBMC, synovial tissue and peripheral blood neutrophils as compared to controls (Brentano *et al.* 2007, Matsui *et al.* 2008). Human synovial cells were induced to produce visfatin after toll-like receptor, IL-1, and TNF-α stimulation (Brentano *et al.* 2007). Serum visfatin levels correlated with measures of disease activity in one of two studies while SF levels correlated with measures of inflammation and clinical disease activity in the only study that examined this (Brentano *et al.* 2007, Gonzalez-Gay *et al.* 2010). Visfatin levels were directly correlated with radiological damage (Rho *et al.* 2009). Although elevated visfatin levels were found prior to anti-TNF-α therapy, visfatin levels were not altered following anti-TNF-α therapy.

Although studies are limited, visfatin likely acts as a pro-inflammatory mediator of joint inflammation in RA and further study may clarify its role as a potential therapeutic target.

SYSTEMIC LUPUS ERYTHEMATOSUS

Systemic lupus erythematosus (SLE) is a chronic autoimmune disease with protean manifestations (Table 3). The disease is characterized by the production of multiple autoantibodies with organ system involvement. Patients with SLE are at increased risk for premature atherosclerosis due to multiple contributing factors including chronic inflammation, insulin resistance, altered lipid profile and an increased prevalence of the metabolic syndrome. Since adipokines play a role in atherosclerosis and the metabolic syndrome, adipokines may play important roles not only in SLE per se but also in the metabolic syndrome and early atherosclerosis found in SLE patients.

Table 3   Revised criteria (1997) for classification of systemic lupus erythematosus*

1. Malar rash, sparing nasolabial folds
2. Discoid rash, may be scarring
3. Photosensitivity
4. Oral or nasal ulcers, painless, observed by physician
5. Arthritis, involving ≥ 2 joints, nonerosive
6. Serositis, pleuritis or pericarditis
7. Renal disorder, persistent proteinuria (> 0.5 g/day) or casts
8. Neurological disorder, seizures or psychosis
9. Hematological disorder, hemolytic anemia or leucopenia or lymphopenia or thrombocytopenia
10. Immunological disorder, anti-dsDNA antibodies or anti-Smith antibodies or antiphospholipid antibodies
11. Antinuclear antibodies

Systemic lupus erythematosus is diagnosed if ≥ 4 of 11 criteria are present, serially or simultaneously, during any interval of observation.
*Hochberg, M.C. 1997. Updating the American College of Rheumatology revised criteria for the classification of systemic lupus erythematosus [letter]. Arthritis Rheum. 40: 1725.

Adiponectin

Most studies in adult SLE patients demonstrated that SLE patients had higher circulating adiponectin concentrations compared to healthy controls (Sada *et al.* 2006, Chung *et al.* 2009, De Sanctis *et al.* 2009, Vadacca *et al.* 2009). It is likely that the studies that showed normal adiponectin in SLE patients had fewer patients with active disease or kidney involvement, as disease activity, particularly active renal disease, has been shown to be the major factor associated with elevated adiponectin (Rovin *et al.* 2005). Adiponectin levels were significantly inversely correlated with Homeostasis Model Assessment of Insulin resistance (HOMA-R) (a measure of insulin resistance), BMI and CRP levels in adult SLE patients (Sada *et al.* 2006, Chung *et al.* 2009, De Sanctis *et al.* 2009, Vadacca *et al.* 2009). In contrast, in pediatric patients with SLE, adiponectin levels did not differ from controls and did not correlate with measures of disease activity or insulin resistance (Al *et al.* 2009). In a murine model of SLE, adiponectin-deficient mice developed more severe disease than wild-type and the addition of adiponectin decreased severity.

An important cause of morbidity in lupus is involvement of the kidney (lupus nephritis). Plasma adiponectin levels were found to be increased in patients with lupus nephritis compared to both healthy controls and patients with non-renal SLE. Urinary adiponectin levels were greater in patients with active lupus nephritis compared with active extra-renal lupus or quiescent renal lupus, and decreased rapidly with treatment (Rovin *et al.* 2005).

Overall adiponectin likely plays a counter-regulatory, anti-inflammatory role in SLE and is produced in response to the chronic inflammatory state of SLE. Murine studies have suggested that there may be a role for the therapeutic administration of adiponectin to decrease disease severity and damage in SLE. Adiponectin holds promise as a biomarker in SLE patients.

Leptin

Studies of adults with SLE have demonstrated higher serum leptin levels than healthy controls even after adjustment for BMI, and leptin levels significantly correlated with inflammatory markers but not disease activity (Garcia-Gonzalez *et al.* 2002, Chung *et al.* 2009). Leptin levels were also correlated with blood pressure, total cholesterol, low density lipoprotein (LDL) cholesterol, triglycerides, insulin resistance, and the metabolic syndrome (Chung *et al.* 2009, Vadacca *et al.* 2009). Pediatric SLE patients had elevated leptin levels compared to healthy controls (Al *et al.* 2009). Serial leptin levels did not change significantly over time in pediatric SLE and were not correlated with SLE disease activity (Al *et al.* 2009).

Leptin levels appear to be associated with components of the metabolic syndrome and, therefore, may be associated with cardiovascular risk. Studies are needed to determine whether leptin can be used to identify patients at high risk for cardiovascular disease.

Resistin and Visfatin

Circulating resistin levels have been shown to be similar between SLE patients and controls but were correlated with ESR, complement C3 levels, serum creatinine and corticosteroid dose, and inversely correlated with BMD and high density lipoprotein (HDL) cholesterol (Chung *et al.* 2009, De Sanctis *et al.* 2009, Vadacca *et al.* 2009).

Serum visfatin levels were significantly higher in SLE patients than in controls (Chung *et al.* 2009, De Sanctis *et al.* 2009, Vadacca *et al.* 2009). Visfatin levels were not correlated with any cardiovascular risk factor, measure of insulin resistance or measure of disease activity (Chung *et al.* 2009).

CARDIOVASCULAR DISEASE

Adiponectin

In healthy populations, adiponectin levels have been associated with the metabolic syndrome, the risk of carotid plaque and the risk of and progression of coronary artery disease. Adiponectin has important effects on endothelial function, in the response of endothelium to inflammation, and polymorphisms of the adiponectin gene have been associated with early-onset coronary artery disease. In both adult and pediatric SLE, adiponectin levels were associated with insulin resistance and lipid levels (Sada *et al.* 2006, Al *et al.* 2009). In RA, low adiponectin levels were associated with abnormal lipid profiles (Gonzalez-Gay *et al.* 2008, Rho *et al.* 2010). The association of adiponectin levels and early atherosclerosis in patients with rheumatic diseases requires further exploration.

Leptin

Leptin levels have been associated with insulin resistance, homocysteine levels, coronary artery calcification, carotid intima media thickness, and carotid artery endothelial function. Leptin levels in SLE patients have been correlated with the cardiovascular risk factors, elevated homocysteine levels and insulin resistance (Al *et al.* 2009, Vadacca *et al.* 2009). In RA, elevated leptin levels were associated with insulin resistance, total cholesterol and LDL-cholesterol. Overall leptin may be important in the development of cardiovascular disease in RA and SLE (Rho *et al.* 2010, Targonska-Stepniak *et al.* 2010).

Resistin and Visfatin

Resistin is present in macrophages in arteries with atheroma but not in normal arteries. Resistin was shown to negatively alter endothelial and vascular smooth muscle function. Elevated resistin levels were associated with the metabolic syndrome and an increased number of Framingham cardiovascular risk factors. Visfatin can induce endothelial cells to secrete pro-inflammatory mediators, and plasma visfatin levels have been associated with abnormal endothelial function, the metabolic syndrome, early atherosclerosis, and coronary artery disease. In patients with RA and SLE, resistin and visfatin were not associated with insulin resistance or the metabolic syndrome (Vadacca *et al.* 2009, Gonzalez-Gay *et al.* 2010, Rho *et al.* 2010).

ANKYLOSING SPONDYLITIS

Ankylosing spondylitis (AS) is an inflammatory arthritis that affects predominantly the spine and large joints leading to bony spinal fusion and disability (Table 2). Unlike other autoimmune diseases, it has a male predominance.

In cross-sectional studies, leptin levels were found to be lower in AS patients than in controls, while a study of newly diagnosed patients found higher levels. The different results are likely related to the degree of inflammation, as newly diagnosed patients generally have more active disease than treated patients in cross-sectional studies. Leptin levels were correlated with disease activity in the study of newly diagnosed patients only. Elevated leptin gene expression was found in PBMC of AS patients and leptin stimulation resulted in significantly greater increased IL-6 and TNF-α production. To date it is not clear whether leptin is produced in response to inflammation in AS patients and therefore may be a biomarker in AS or whether leptin is important in the pathogenesis of AS.

Other studies of adipokines in AS are limited, although one study did not find a difference in adiponectin levels between patients and controls.

OSTEOARTHRITIS

Osteoarthritis (OA) is a degenerative joint disease that is characterized by the progressive loss of articular cartilage, sclerosis of subchondral bone, and the formation of osteophytes (Table 2). Although systemic inflammation is not a prominent feature of OA, there may be significant local inflammation.

Adiponectin

Synovial tissue from patients with OA expressed type-1 adiponectin receptors, while adjacent bone rarely expressed either receptor. SF adiponectin levels were higher than in controls, lower than in RA and lower than paired serum concentrations (Chen *et al.* 2006, Presle *et al.* 2006, Senolt *et al.* 2006, 2007, Gegout *et al.* 2008). SF adiponectin was inversely correlated with disease severity but not CRP (Chen *et al.* 2006, Filkova *et al.* 2009). Synovial tissue, synovial fibroblasts and articular adipose tissue from patients with OA expressed messenger ribonucleic acid (mRNA) for adiponectin, while the synovial fibroblasts also expressed mRNA for adiponectin receptors. OA synovial fibroblasts expressed increased mRNA for pro-inflammatory cytokines IL-6 and MMP-1 upon exposure to adiponectin (Ehling *et al.* 2006).

The effect of adiponectin on chondrocytes is less clear. An initial study suggested that adiponectin had an overall anti-inflammatory effect on cultured human chondrocytes, while another study showed a pro-inflammatory effect. The only major difference between the studies was the number of passages of the human chondrocytes; therefore, the effect of adiponectin on human chondrocytes and in OA it is not clear.

Leptin

Animal studies have suggested that leptin is pro-inflammatory and contributes to cartilage damage. Its proposed mechanism of cartilage destruction is its ability to increase nitric oxide synthase, prostaglandin E2 (PGE2), IL-6, IL-8, and MMPs. Leptin is present in OA SF, although levels were lower than in RA SF. Unlike adiponectin, SF levels were higher than blood levels in OA.

Immunostaining of cartilage from OA patients showed significantly more leptin than in normal cartilage and the degree of staining correlated with OA severity (Dumond *et al.* 2003, Gualillo 2007, Simopoulou *et al.* 2007, Gegout *et al.* 2008). Osteophytes and synovium from OA patients could be stimulated to produce leptin and leptin was demonstrated in osteophytes from OA patients (Presle *et al.* 2006). Leptin altered cartilage and matrix metabolism by producing transforming growth factor beta (TGF-β1), insulin-like growth factor 1 (IGF-1) and proteoglycans. It decreased chondrocyte proliferation but increased IL-1β levels and MMP-9 and MMP-13 protein expression and induced synovial fibroblasts to produce IL-8 (Figenschau *et al.* 2001, Dumond *et al.* 2003, Iliopoulos *et al.* 2007, Simopoulou *et al.* 2007, Tong *et al.* 2008). Overall leptin has multiple effects on multiple cells present in and around the joint and, therefore, is likely important in the pathogenesis of osteoarthritis.

Resistin

Serum concentrations of resistin were greater than SF concentrations in OA, and serum concentrations were greater in patients with more severe disease (Chen *et al.* 2006, Presle *et al.* 2006, Senolt *et al.* 2006, 2007, Gegout *et al.* 2008, Filkova *et al.* 2009). Synovial lining cells from OA patients expressed resistin levels similar to those found in synovial lining cells from RA patients. Resistin was expressed in the plasma cells, B cells and fibroblasts but not in T cells within the synovium (Senolt *et al.* 2007). These preliminary studies suggest that resistin is present at the site of inflammation in OA.

Visfatin

Chondrocytes from OA patients intrinsically produce visfatin that was further increased following stimulation by IL-1β. Visfatin was catabolic on human chondrocytes as observed by increasing PGE2 release and MMP-3 and MMP-13 production (Gosset *et al.* 2008). We could not find any studies examining serum or synovial fluid visfatin levels in OA.

ADIPOKINES IN OTHER RHEUMATIC DISEASES

This section briefly outlines the findings in other autoimmune diseases.

Behcet's Disease

Behcet's disease (BD) is a vasculitis characterized by recurrent painful oral and genital ulcers, rash, uveitis, and arthritis and may be accompanied by abdominal pain, blood vessel abnormalities and central nervous system disease. Currently the only published studies of adipokines in BD have been related to leptin.

Multiple studies have demonstrated elevated leptin levels in BD that correlated with disease activity. Patients without ocular involvement were shown to have significantly elevated leptin levels as compared to patients with ocular involvement and healthy controls. Although leptin levels are elevated in active BD, it is not clear whether leptin has a role in the pathogenesis of BD, or whether it can be used as a biomarker.

Systemic Vasculitis

The systemic vasculitides are a group of rheumatic diseases that are characterized by multi-organ system involvement and inflammation of blood vessels (Table 4). They range from chronic, life-threatening illnesses such as antineutrophil cytoplasmic antibody (ANCA)–associated vasculitis (AAV) [Wegner's granulomatosis (WG), Churg-Strauss Syndrome (CSS) and polyarteritis nodosa (PAN)] to self-limited diseases including Kawasaki Disease (KD).

Table 4 Nomenclature for the classification of systemic vasculitis

LARGE VESSEL VASCULITIS	1. Giant cell (temporal) arteritis: arteritis of the aorta and its main branches, affects persons > 50 yr predominantly 2. Takayasu arteritis: inflammation of the aorta and its main branches, predominantly persons <50 years old
MEDIUM VESSEL VASCULITIS	1. Polyarteritis nodosa: necrotizing inflammation of medium and small vessels 2. Kawasaki disease: arteritis of predominantly coronary blood vessels, can affect other medium and small vessels, occurs in childhood
SMALL VESSEL VASCULITIS	1. Wegener's granulomatosis: granulomatous inflammation of respiratory tract, and necrotizing vasculitis of small and medium vessels, commonly with necrotizing glomerulonephritis 2. Churg-Strauss Syndrome: associated with asthma and eosinophilia, granulomatous inflammation often involves the respiratory tract 3. Microscopic polyangiitis: necrotizing vasculitis, often with necrotizing glomerulonephritis, no immune deposits, may have pulmonary capillaritis 4. Henoch-Schonlein Purpura: involves skin, gut and glomeruli, associated with arthritis, predominantly in childhood, IgA dominant immune deposits observed 5. Essential cryoglobulinemic vasculitis: often involves skin and glomeruli 6. Cutaneous leukocytoclastic vasculitis: no systemic vasculitis or glomerulonephritis

Jennette, J.C., Falk, R.J., Andrassy, K., Bacon, P.A., Churg, J., Gross, W.L., *et al.* 1994. Nomenclature of systemic vasculitides: proposal of an international consensus conference. Arthritis Rheum. 37: 187-192.

At diagnosis, patients with active, untreated AAV had decreased leptin levels that increased and normalized with treatment. Leptin levels were inversely correlated with disease activity in patients with or without treatment. Genetic polymorphisms in the leptin receptor have been associated with an increased susceptibility to WG but decreased susceptibility to CSS. Adiponectin levels were significantly lower in patients with acute KD than in patients with other febrile illnesses. These low levels were inversely correlated with IL-6 levels and returned to normal in the convalescent phase.

Overall it is not clear what the role of adipokines is in any of the systemic vasculitides but the only genetic study of leptin receptor in AAV suggests that they may be important in the etiopathogenesis of these diseases.

SUMMARY

Rheumatic diseases are a diverse group of illnesses characterized by systemic involvement and the failure to recognize self and non-self. They are common and associated with significant morbidity and mortality. As the immunoregulatory role of adipokines has been increasingly recognized, there has been increasing interest in the role of adipokines in these diseases, especially RA and SLE.

Leptin is recognized as a pro-inflammatory cytokine with elevated levels leading to the promotion of inflammation. However, the concentration of circulating leptin levels has not been consistently correlated with disease activity. Since leptin-deficient mice are protected against autoimmune diseases and atherosclerosis, despite the inconsistency of the findings regarding serum leptin levels, it has been suggested that antagonizing leptin activity may be of benefit in autoimmune diseases and in particular in the cardiovascular morbidity associated with these diseases. The recent evidence that genetic polymorphisms in leptin receptor influence susceptibility to ANCA-associated vasculitis further suggests the importance of leptin in autoimmune diseases.

Adiponectin, a member of the complement factor C1q and TNF-α superfamily of molecules, may be considered an important molecule linking metabolism and the immune system. Increased levels likely play a counter-regulatory role in inflammation and may be an effector mechanism of the anti-inflammatory effects of glucocorticoids. Cells important in rheumatic diseases such as skeletal muscle, osteoblasts, osteoclasts, chondrocytes, synovial fibroblasts and endothelial cells produce adiponectin. Adiponectin levels are elevated in SF and correlate with disease severity, disease duration but not measures of disease activity in RA. It is likely that adiponectin acts as a counter-regulatory cytokine in RA. Similarly, adiponectin likely plays a counter-regulatory, anti-inflammatory role in SLE and is likely produced in response to the chronic inflammatory state of SLE to decrease inflammation. Murine studies have suggested that the therapeutic administration of adiponectin may decrease disease severity and damage in SLE. Adiponectin holds promise as a biomarker in SLE patients. Adiponectin is also likely to be important in osteoarthritis as both serum and synovial fluid levels are elevated and synovial tissue from OA patients expresses adiponectin receptors.

Many bone marrow–derived cells are capable of producing resistin in response to inflammation, and resistin up-regulates the production of pro-inflammatory cytokines. Resistin production is increased by TNF-α and glucocorticoids. It causes synovitis and pannus formation in mice. Resistin is found in synovial fluid from patients with RA and these cells respond to resistin stimulation by producing inflammatory cytokines. Elevated serum and SF resistin levels are associated with inflammation in RA and are altered with anti-TNF-α therapy. In contrast, resistin levels are not elevated in SLE. The true role of resistin in rheumatic diseases is not yet defined.

To date there are few studies of visfatin in rheumatic diseases. Elevated circulating visfatin levels are found in RA. Human synovial cells produce visfatin after multiple inflammatory stimuli. Serum visfatin levels correlate with damage in RA and elevated SF fluid levels are found. Visfatin likely is pro-inflammatory in RA and may be a mediator of joint inflammation. Elevated serum visfatin levels were found in SLE.

Summary Points

- The rheumatic diseases, a diverse group of illnesses that include rheumatoid arthritis (RA) and systemic lupus erythematosus (SLE), are characterized by systemic inflammation and are associated with significant premature atherosclerosis.
- The adipokines adiponectin, leptin, resistin and visfatin all have multiple effects on the immune system and may up-regulate the production of pro-inflammatory cytokines including IL-6, TNF-α and IL-12 by multiple cells of the immune system; their role in rheumatic diseases likely depends on the location of production and action.
- Adiponectin likely acts as a counter-regulatory cytokine in RA, with potential interactions between adiponectin and pro-inflammatory cytokines, but at present it is not a useful biomarker to monitor disease activity or response to therapy.
- In contrast to RA, adiponectin holds promise as a biomarker in SLE patients and preliminary study demonstrates a potential anti-inflammatory role in decreasing SLE severity and disease damage.
- Leptin has multiple immunomodulatory effects that affect both the innate and adaptive immune systems; however, the role of leptin is not clear in RA, and to date it has not been a useful biomarker to monitor response to therapy.
- As leptin levels are associated with components of the metabolic syndrome it is likely that leptin is important in the cardiovascular disease of SLE.
- Leptin likely has a role in the inflammation of ankylosing spondylitis (AS), although it is unclear whether leptin is produced in response to inflammation in AS (and as such may be a biomarker) or is important in the pathogenesis of AS.
- Emerging data suggests that resistin may have a role as a biomarker to predict outcome or response to therapy in inflammatory and non-inflammatory arthritis.
- Although studies are limited, visfatin likely acts as a pro-inflammatory mediator of joint inflammation in RA and further study may clarify its role as a potential therapeutic target.
- Studies of adipokines in other rheumatic diseases are limited, but preliminary data suggests that leptin may play a role in the pathogenesis of Behcet's disease and the systemic vasculitides.

Abbreviations

AAV	:	ANCA-associated vasculitis
ANCA	:	antineutrophil cytoplasmic antibody
AS	:	ankylosing spondylitis
BMD	:	bone mineral density
BMI	:	body mass index
CRP	:	C-reactive protein
CSS	:	Churg Strauss syndrome
eNOS	:	endothelial nitric oxide synthase
ESR	:	erythrocyte sedimentation rate
HDL	:	high density lipoprotein
HOMA-R	:	homeostatis model assessment of insulin resistance
ICAM	:	intracellular adhesion molecule
IGF	:	insulin growth factor
IL	:	interleukin
IR	:	insulin resistance
KD	:	Kawasaki disease
LDL	:	low density lipoprotein
MCP	:	monocyte chemotactic protein
MMP	:	matrix metalloprotease
mRNA	:	messenger ribonucleic acid
OA	:	osteoarthritis
PAN	:	polyarteritis nodosa
PBMC	:	peripheral blood mononuclear cell
PGE2	:	prostaglandin E2
RA	:	rheumatoid arthritis
Ra	:	receptor antagonist
SF	:	synovial fluid
SLE	:	systemic lupus erythematosus
TGF-β	:	transforming growth factor beta
Th	:	T helper cell
TNF-α	:	tumour necrosis factor alpha
VCAM	:	vascular adhesion molecule
WG	:	Wegener's granulomatosis

References

Aguilar-Chavez, E.A. and J.I. Gamez-Nava, M.A. Lopez-Olivo, S. Galvan-Melendres, E.G. Corona-Sanchez, C.A. Loaiza-Cardenas, A. Celis, E.G. Cardona-Munoz, and L. Gonzalez-Lopez. 2009. Circulating leptin and bone mineral density in rheumatoid arthritis. J. Rheumatol. 36: 512-516.

Al, M. and L. Ng, P. Tyrrell, J. Bargman, T. Bradley, and E. Silverman. 2009. Adipokines as novel biomarkers in paediatric systemic lupus erythematosus. Rheumatology (Oxford) 48: 497-501.

Anders, H.J. and M. Rihl, A. Heufelder, O. Loch, and M. Schattenkirchner. 1999. Leptin serum levels are not correlated with disease activity in patients with rheumatoid arthritis. Metabolism 48: 745-748.

Bokarewa, M. and D. Bokarew, O. Hultgren, and A. Tarkowski. 2003. Leptin consumption in the inflamed joints of patients with rheumatoid arthritis. Ann. Rheum. Dis. 62: 952-956.

Bokarewa, M. and I. Nagaev, L. Dahlberg, U. Smith, and A. Tarkowski. 2005. Resistin, an adipokine with potent proinflammatory properties. J. Immunol. 174: 5789-5795.

Brentano, F. and O. Schorr, C. Ospelt, J. Stanczyk, R.E. Gay, S. Gay, and D. Kyburz. 2007. Pre-B cell colony-enhancing factor/visfatin, a new marker of inflammation in rheumatoid arthritis with proinflammatory and matrix-degrading activities. Arthritis Rheum. 56: 2829-2839.

Chen, T.H. and L. Chen, M.S. Hsieh, C.P. Chang, D.T. Chou, and S.H. Tsai. 2006. Evidence for a protective role for adiponectin in osteoarthritis. Biochim. Biophys. Acta. 1762: 711-718.

Chung, C.P. and A.G. Long, J.F. Solus, Y.H. Rho, A. Oeser, P. Raggi, and C.M. Stein. 2009. Adipocytokines in systemic lupus erythematosus: relationship to inflammation, insulin resistance and coronary atherosclerosis. Lupus 18: 799-806.

De Sanctis, J.B. and M. Zabaleta, N.E. Bianco, J.V. Garmendia, and L. Rivas. 2009. Serum adipokine levels in patients with systemic lupus erythematosus. Autoimmunity 42: 272-274.

Dumond, H. and N. Presle, B. Terlain, D. Mainard, D. Loeuille, P. Netter, and P. Pottie. 2003. Evidence for a key role of leptin in osteoarthritis. Arthritis Rheum. 48: 3118-3129.

Ebina, K. and A. Fukuhara, W. Ando, M. Hirao, T. Koga, K. Oshima, M. Matsuda, K. Maeda, T. Nakamura, T. Ochi, I. Shimomura, H. Yoshikawa, and J. Hashimoto. 2009. Serum adiponectin concentrations correlate with severity of rheumatoid arthritis evaluated by extent of joint destruction. Clin. Rheumatol. 28: 445-451.

Ehling, A. and A. Schaffler, H. Herfarth, I.H. Tarner, S. Anders, O. Distler, G. Paul, J. Distler, S. Gay, J. Scholmerich, E. Neumann, and U. Muller-Ladner. 2006. The potential of adiponectin in driving arthritis. J. Immunol. 176: 4468-4478.

Faggioni, R. and K.R. Feingold, and C. Grunfeld. 2001. Leptin regulation of the immune response and the immunodeficiency of malnutrition. FASEB J. 15: 2565-2571.

Figenschau, Y. and G. Knutsen, S. Shahazeydi, O. Johansen, and B. Sveinbjornsson. 2001. Human articular chondrocytes express functional leptin receptors. Biochem. Biophys. Res. Commun. 287: 190-197.

Filkova, M. and M. Liskova, H. Hulejova, M. Haluzik, J. Gatterova, A. Pavelkova, K. Pavelka, S. Gay, U. Muller-Ladner, and L. Senolt. 2009. Increased serum adiponectin levels in female patients with erosive compared with non-erosive osteoarthritis. Ann. Rheum. Dis. 68: 295-296.

Forsblad d'Elia, H. and R. Pullerits, H. Carlsten, and M. Bokarewa. 2008. Resistin in serum is associated with higher levels of IL-1Ra in post-menopausal women with rheumatoid arthritis. Rheumatology (Oxford) 47: 1082-1087.

Garcia-Gonzalez, A. and L. Gonzalez-Lopez, I.C. Valera-Gonzalez, E.G. Cardona-Munoz, M. Salazar-Paramo, M. Gonzalez-Ortiz, E. Martinez-Abundis, and J.I. Gamez-Nava. 2002. Serum leptin levels in women with systemic lupus erythematosus. Rheumatol. Int. 22: 138-141.

Gegout, P.P. and P.J. Francin, D. Mainard, and N. Presle. 2008. Adipokines in osteoarthritis: friends or foes of cartilage homeostasis? Joint Bone Spine 75: 669-671.

Giles, J.T. and M. Allison, C.O. Bingham 3rd, W.M. Scott Jr., and J.M. Bathon. 2009. Adiponectin is a mediator of the inverse association of adiposity with radiographic damage in rheumatoid arthritis. Arthritis Rheum. 61: 1248-1256.

Gonzalez-Gay, M.A. and M.T. Garcia-Unzueta, C. Gonzalez-Juanatey, J.A. Miranda-Filloy, T.R. Vazquez-Rodriguez, J.M. De Matias, J. Martin, P.H. Dessein, and J. Llorca. 2008. Anti-TNF-alpha therapy modulates resistin in patients with rheumatoid arthritis. Clin. Exp. Rheumatol. 26: 311-316.

Gonzalez-Gay, M.A. and J. Llorca, M.T. Garcia-Unzueta, C. Gonzalez-Juanatey, J.M. De Matias, J. Martin, M. Redelinghuys, A.J. Woodiwiss, G.R. Norton, and P.H. Dessein. 2008. High-grade inflammation, circulating adiponectin concentrations and cardiovascular risk factors in severe rheumatoid arthritis. Clin. Exp. Rheumatol. 26: 596-603.

Gonzalez-Gay, M.A. and T.R. Vazquez-Rodriguez, M.T. Garcia-Unzueta, A. Berja, J.A. Miranda-Filloy, J.M. de Matias, C. Gonzalez-Juanatey, and J. Llorca. 2010. Visfatin is not associated with inflammation or metabolic syndrome in patients with severe rheumatoid arthritis undergoing anti-TNF-α therapy. Clin. Exp. Rheumatol. 28: 56-62.

Gosset, M. and F. Berenbaum, C. Salvat, A. Sautet, A. Pigenet, K. Tahiri, and C. Jacques. 2008. Crucial role of visfatin/pre-B cell colony-enhancing factor in matrix degradation and prostaglandin E2 synthesis in chondrocytes: possible influence on osteoarthritis. Arthritis Rheum. 58: 1399-1409.

Gualillo, O. 2007. Further evidence for leptin involvement in cartilage homeostases. Osteoarthritis Cartilage 15: 857-860.

Gunaydin, R. and T. Kaya, A. Atay, N. Olmez, A. Hur, and M. Koseoglu. 2006. Serum leptin levels in rheumatoid arthritis and relationship with disease activity. South. Med. J. 99: 1078-1083.

Hizmetli, S. and M. Kisa, N. Gokalp, and M.Z. Bakici. 2007. Are plasma and synovial fluid leptin levels correlated with disease activity in rheumatoid arthritis ? Rheumatol. Int. 27: 335-338.

Iliopoulos, D. and K.N. Malizos, and A. Tsezou. 2007. Epigenetic regulation of leptin affects MMP-13 expression in osteoarthritic chondrocytes: possible molecular target for osteoarthritis therapeutic intervention. Ann. Rheum. Dis. 66: 1616-1621.

Laurberg, T.B. and J. Frystyk, T. Ellingsen, I.T. Hansen, A. Jorgensen, U. Tarp, M.L. Hetland, K. Horslev-Petersen, N. Hornung, J.H. Poulsen, A. Flyvbjerg, and K. Stengaard-Pedersen.

2009. Plasma adiponectin in patients with active, early, and chronic rheumatoid arthritis who are steroid- and disease-modifying antirheumatic drug-naive compared with patients with osteoarthritis and controls. J. Rheumatol. 36: 1885-1891.

Matsui, H. and A. Tsutsumi, M. Sugihara, T. Suzuki, K. Iwanami, M. Kohno, D. Goto, I. Matsumoto, S. Ito, and T. Sumida. 2008. Visfatin (pre-B cell colony-enhancing factor) gene expression in patients with rheumatoid arthritis. Ann. Rheum. Dis. 67: 571-572.

Migita, K. and Y. Maeda, T. Miyashita, H. Kimura, M. Nakamura, H. Ishibashi, and K. Eguchi. 2006. The serum levels of resistin in rheumatoid arthritis patients. Clin. Exp. Rheumatol. 24: 698-701.

Otero, M. and R. Lago, R. Gomez, F. Lago, C. Dieguez, J.J. Gomez-Reino, and O. Gualillo. 2006. Changes in plasma levels of fat-derived hormones adiponectin, leptin, resistin and visfatin in patients with rheumatoid arthritis. Ann. Rheum. Dis. 65: 1198-1201.

Popa, C. and M.G. Netea, T.R. Radstake, P.L. van Riel, P. Barrera, and J.W. van der Meer. 2005. Markers of inflammation are negatively correlated with serum leptin in rheumatoid arthritis. Ann. Rheum. Dis. 64: 1195-1198.

Presle, N. and P. Pottie, H. Dumond, C. Guillaume, F. Lapicque, S. Pallu, D. Mainard, P. Netter, and B. Terlain. 2006. Differential distribution of adipokines between serum and synovial fluid in patients with osteoarthritis. Contribution of joint tissues to their articular production. Osteoarthritis Cartilage 14: 690-695.

Rho, Y.H. and C.P. Chung, J.F. Solus, P. Raggi, A. Oeser, T. Gebretsadik, A. Shintani, and C.M. Stein. 2010. Adipocytokines, insulin resistance and coronary atherosclerosis in rheumatoid arthritis. Arthritis Rheum.

Rho, Y.H. and J. Solus, T. Sokka, A. Oeser, C.P. Chung, T. Gebretsadik, A. Shintani, T. Pincus, and C.M. Stein. 2009. Adipocytokines are associated with radiographic joint damage in rheumatoid arthritis. Arthritis Rheum. 60: 1906-1914.

Rovin, B.H. and H. Song, L.A. Hebert, T. Nadasdy, G. Nadasdy, D.J. Birmingham, C. Yung Yu, and H.N. Nagaraja. 2005. Plasma, urine, and renal expression of adiponectin in human systemic lupus erythematosus. Kidney Int. 68: 1825-1833.

Sada, K.E. and Y. Yamasaki, M. Maruyama, H. Sugiyama, M. Yamamura, Y. Maeshima, and H. Makino. 2006. Altered levels of adipocytokines in association with insulin resistance in patients with systemic lupus erythematosus. J. Rheumatol. 33: 1545-1552.

Salazar-Paramo, M. and M. Gonzalez-Ortiz, L. Gonzalez-Lopez, A. Sanchez-Ortiz, I.C. Valera-Gonzalez, E. Martinez-Abundis, B.R. Balcazar-Munoz, A. Garcia-Gonzalez, and J.I. Gamez-Nava. 2001. Serum leptin levels in patients with rheumatoid arthritis. J. Clin. Rheumatol. 7: 57-59.

Schaffler, A. and A. Ehling, E. Neumann, H. Herfarth, I. Tarner, J. Scholmerich, U. Muller-Ladner, and S. Gay. 2003. Adipocytokines in synovial fluid. JAMA 290: 1709-1710.

Senolt, L. and D. Housa, Z. Vernerova, T. Jirasek, R. Svobodova, D. Veigl, K. Anderlova, U. Muller-Ladner, K. Pavelka, and M. Haluzik. 2007. Resistin in rheumatoid arthritis synovial tissue, synovial fluid and serum. Ann Rheum Dis 66: 458-463.

Senolt, L. and K. Pavelka, D. Housa, and M. Haluzik. 2006. Increased adiponectin is negatively linked to the local inflammatory process in patients with rheumatoid arthritis. Cytokine 35: 247-252.

Seven, A. and S. Guzel, M. Aslan, and V. Hamuryudan. 2009. Serum and synovial fluid leptin levels and markers of inflammation in rheumatoid arthritis. Rheumatol. Int. 29: 743-747.

Simopoulou, T. and K.N. Malizos, D. Iliopoulos, N. Stefanou, L. Papatheodorou, M. Ioannou, and A. Tsezou. 2007. Differential expression of leptin and leptin's receptor isoform (Ob-Rb) mRNA between advanced and minimally affected osteoarthritic cartilage; effect on cartilage metabolism. Osteoarthritis Cartilage 15: 872-883.

Tan, W. and F. Wang, M. Zhang, D. Guo, Q. Zhang, and S. He. 2009. High adiponectin and adiponectin receptor 1 expression in synovial fluids and synovial tissues of patients with rheumatoid arthritis. Sem. Arthritis Rheum. 38: 420-427.

Targonska-Stepniak, B. and M. Dryglewska, and M. Majdan. 2010. Adiponectin and leptin serum concentrations in patients with rheumatoid arthritis. Rheumatol. Int. 30: 731-737.

Targonska-Stepniak, B. and M. Majdan, and M. Dryglewska. 2008. Leptin serum levels in rheumatoid arthritis patients: relation to disease duration and activity. Rheumatol. Int. 28: 585-591.

Tilg, H. and A.R. Moschen. 2006. Adipocytokines: mediators linking adipose tissue, inflammation and immunity. Nat. Rev. Immunol. 6: 772-783.

Tong, K.M. and D.C. Shieh, C.P. Chen, C.Y. Tzeng, S.P. Wang, K.C. Huang, Y.C. Chiu, Y.C. Fong, and C.H. Tang. 2008. Leptin induces IL-8 expression via leptin receptor, IRS-1, PI3K, Akt cascade and promotion of NF-kappaB/p300 binding in human synovial fibroblasts. Cell Signal. 20: 1478-1488.

Vadacca, M. and D. Margiotta, A. Rigon, F. Cacciapaglia, G. Coppolino, A. Amoroso, and A. Afeltra. 2009. Adipokines and systemic lupus erythematosus: relationship with metabolic syndrome and cardiovascular disease risk factors. J. Rheumatol. 36: 295-297.

Wislowska, M. and M. Rok, B. Jaszczyk, K. Stepien, and M. Cicha. 2007. Serum leptin in rheumatoid arthritis. Rheumatol. Int. 27: 947-954.

CHAPTER 27

Adipokines and Allergy

Giorgio Ciprandi,[1,*] Davide Caimmi,[2] Silvia Caimmi,[2] Alessia
Marseglia,[2] Lucia Bianchi,[2] Anna Maria Castellazzi[2] and
Gian Luigi Marseglia[2]

[1]Department of Internal Medicine, Azienda Ospedaliera Universitaria San
Martino University of Genoa, Genoa, Italy
[2]Department of Pediatric Science, Pediatric Clinic, University of Pavia,
Foundation IRCCS, San Matteo, Pavia, Italy

ABSTRACT

The prevalence of both obesity and allergic sensitization continually increases in parallel in industrialized countries. An association between obesity and allergy-related diseases has been reported in many studies, although considerable debate still exists about the existence and meaning of the association. Nevertheless, in general, obesity and weight gain have been associated with an increased risk of respiratory allergy.

Since both asthma and obesity are characterized by inflammation, a common inflammatory pathway has been proposed as a plausible explanation for the frequent association of the two diseases. Allergic rhinitis is characterized by an inflammatory response consequent to allergen exposure. A Th2 polarization orchestrates a cascade of events that induces the occurrence of symptoms. Actually, an increase in white adipose tissue is characterized by a chronic inflammation that pushes the immune system toward a Th2 polarization as well. A series of studies were performed to investigate the possible role of adipokines in patients with allergic rhinitis. Serum levels of adiponectin, leptin, adipsin, and resistin were assessed in patients with pollen allergy (evaluated both during and outside the pollen season) and in patients treated with sublingual immunotherapy.

*Corresponding author

Findings demonstrated that these adipokines are modulated by allergen exposure and immunotherapy is able to affect them.

INTRODUCTION

The prevalence of both obesity and allergic sensitization continually increases in parallel in industrialized countries. Even though several environmental factors have been hypothesized to be involved in the development of allergic diseases, none may fully explain this rapid increase in prevalence. However, some lifestyle factors, including dietary factors, alcohol consumption, physical inactivity, and obesity, have recently received close attention. Indeed, an increase in affluence, typical of western society, may have contributed to an increased availability of food and decreased physical activity, both of which may promote the prevalence of obesity. As for asthma, it has to be noted that, as a group, subjects with asthma are heavier than subjects without asthma and overweight is often related to worse asthma control. Moreover, greater initial weight and greater weight change among adults increase the risk for developing asthma in later life.

In general, obesity and weight gain have been associated with an increased risk of respiratory allergy.

Many studies have reported that obesity is a risk factor for the development of asthma and that obesity precedes the development of asthma. It has also been reported that patients with allergic rhinitis tend to have higher body mass index (BMI) than normal subjects and that BMI is significantly related to bronchial hyperreactivity. Also, some studies have hypothesized a possible relationship between an increased BMI and respiratory allergic diseases such as asthma and rhinitis, with a stronger association in women, which could be due to the fact that, for any given body weight, women have a higher percentage of body fat or that BMI is a less accurate measure for men.

Moreover, since both asthma and obesity are characterized by inflammation, a common inflammatory pathway has been proposed as a plausible explanation for the frequent association of the two diseases. In fact, inflammatory markers, such as C-reactive protein and interleukin (IL) 6, are increased in obese subjects. The presence of systemic inflammation has been linked to the increased risk of development of cardiovascular disease and type 2 diabetes in obese individuals.

White adipose tissue (WAT) is composed of many cell types, including adipocytes and macrophages. Adipocytes secrete a variety of protein signals, namely adipokines, including leptin and adiponectin. Therefore, adipose tissue is currently viewed as an active secretory organ, sending out and responding to signals that modulate appetite, energy expenditure, insulin sensitivity, endocrine and reproductive systems, bone metabolism, inflammation and immunity.

It has also been reported that serum adiponectin and leptin may be impaired in patients with allergic rhinitis and it has been shown that immunotherapy may affect these adipokines in patients with seasonal allergic rhinitis.

LEPTIN AND ADIPONECTIN: TWO ADIPOKINES INVOLVED IN ALLERGIES

Leptin is an adipokine released by adipocytes. Laboratory studies have indicated that leptin is associated with a T helper cell (Th) 1 response but can also augment ovalbumin-induced IgE production in sensitized mice, indicating its pro-inflammatory potential. In contrast, the expression of adiponectin, another important adipokine, is diminished in obese subjects and adiponectin can attenuate allergen responses in mice. As part of its anti-inflammatory action, adiponectin is capable of reducing tumor necrosis factor alpha (TNF-α)–induced NF-jB activation and lipopolysaccharide-induced TNF-α production. Such immunomodulatory factors are also thought to be involved in the protective effects of farm exposure. These observations raise the question whether the associations between body weight, early childhood exposure and allergy are related to blood levels of leptin and adiponectin, as two partly counteracting hormonal immunomodulatory compounds.

Leptin has an influence on the hypothalamus by inducing satiety and by increasing energy metabolism. The serum concentration of leptin is increased in obese subjects, suggesting a leptin resistance mechanism in obesity. Moreover, leptin protects T lymphocytes from apoptosis, regulates T cell proliferation and activation, recruits and activates monocytes and macrophages, and promotes angiogenesis.

Recently, it was reported that leptin exerts a key role in the control of the proliferation of Tregs (regulatory T lymphocytes). Thus, it has been hypothesized that an increased serum concentration of leptin would decrease the proliferation of Tregs.

In addition, exogenous leptin has been shown to modulate allergic airway responses in mice independently of obesity. Leptin infusion caused an enhancement of airway hyperresponsiveness and an increase in serum IgE after inhaled ovalbumin challenge. Moreover, it has been demonstrated that human eosinophils express leptin surface receptors under *in vitro* and *in vivo* conditions, and leptin delays the apoptosis of mature eosinophils *in vitro*. The antiapoptotic effects of leptin are concentration dependent and blocked by an antileptin receptor monoclonal antibody. The efficacy of leptin in blocking eosinophil apoptosis is similar to that of GM-CSF (granulocyte monocyte colony stimulating factor). Leptin delays the cleavage of Bax, as well as the mitochondrial release of cytochrome c, a second mitochondria-derived activator of caspase, suggesting that it blocks proapoptotic pathways proximal to mitochondria in eosinophils. Leptin is therefore a survival cytokine for human eosinophils; this finding suggests potential pathological relevance in allergic and parasitic diseases.

Leptin could also stimulate the chemokinesis of eosinophils and induce the release of inflammatory cytokines IL-1 and IL-6 and chemokines IL-8, growth-related oncogene and monocyte chemotactic protein 1. In view of these findings and the elevated production of leptin in patients with allergic diseases such as atopic asthma and atopic dermatitis, leptin could play crucial immunopathophysiological roles in allergic inflammation by activation of eosinophils via differential intracellular signaling cascades.

Leptin is therefore a survival cytokine for human eosinophils, and this finding might suggest potential pathological relevance in allergic and parasitic diseases. In fact, serum leptin levels were found to be significantly higher in patients with allergic rhinitis in symptomatic periods. Therefore, leptin might have a role in the inflammatory process of the allergic rhinitis. In this regard, it has very recently been shown that there are significantly higher leptin serum levels in female patients with pollen-induced allergic rhinitis outside the pollen season.

On the other hand, adiponectin is the most prevalent adipokine secreted by adipose tissue and it exerts numerous actions on the immune response as well. Initially, adiponectin was considered an anti-inflammatory adipokine, as it reduces the production and activity of TNF and IL-6, and the expression of some adhesion molecules, such as intracellular adhesion molecule 1 and vascular cellular adhesion molecules, and it also promotes the production of IL-10 and IL-1 receptor antagonist. However, more recent studies have shown that adiponectin may also exert pro-inflammatory effects. Indeed, adiponectin serum levels are elevated in patients with classic chronic inflammatory/autoimmune diseases that are unrelated to increased adipose tissue, such as rheumatoid arthritis, systemic lupus erythematosus, inflammatory bowel disease, type 1 diabetes, and cystic fibrosis. In these patients, adiponectin levels positively correlate with inflammatory markers. Thus, adiponectin may be regulated in opposite directions and may exert different functions in classic versus obesity-associated inflammatory disorders. Recently it was reported that adiponectin serum levels were significantly higher in male patients with pollen-induced allergic rhinitis than in normal male subjects when evaluated outside the pollen season.

THE ROLE OF CYTOKINES IN ALLERGIC DISORDERS

Very few studies have investigated the effects of obesity on allergic rhinitis. A Swedish study, conducted on a large cohort of 1,247,038 military conscripts, reported that obesity was not associated with allergic rhinitis in patients with nasal symptoms only. Another study evidenced that the risk of allergic rhinitis increased with increasing BMI among women, but not among men.

On the other hand, the association between obesity and respiratory allergy may, at least partly, depend on decreased immunological tolerance to the allergen as a consequence of immunological changes induced by adipokines, such as adiponectin and leptin, and some pro-inflammatory cytokines secreted by WAT. Therefore, an increase in WAT is characterized by a chronic inflammation that pushes the immune system toward a Th2 polarization. Allergic rhinitis is characterized by Th2-polarized inflammation, as well.

Th2-derived cytokines, such as IL-4 and IL-13, are the primary pathogenic factors that induce, maintain, and amplify allergic inflammation. IL-4 and IL-13 orchestrate this inflammation by promoting IgE synthesis, up-regulating adhesion molecules selective for eosinophil recruitment, and causing increased mucus production and airway hyperreactivity. Moreover, allergic inflammation is closely dependent on allergen exposure, as it persists until the patient inhales the causal allergen.

A few studies have been conducted, investigating the possible role played by leptin and adiponectin in patients affected by respiratory allergic diseases.

Leptin and Allergic Rhinitis

In a study conducted outside the pollen season, no significant difference was found in serum leptin between allergic males and normal males, while a significant difference was observed between allergic females and normal females, the former showing higher leptin values. Also, BMI and leptin levels were strongly correlated both in males and in females, confirming previous results, suggesting that a resistance to leptin may exist and that it depends upon an increase in adipose tissue.

A trend of increasing leptin levels with age has also been detected in both allergic and normal females only, which may be related to the increased adipose tissue that occurs more manifestly in females as they age.

Moreover, a significant relationship was found between serum leptin levels and some clinical and immunological parameters, including symptom severity, peripheral eosinophils and degree of allergy (assessed by the allergen threshold dose for nasal challenge). These relationships outline the close correlation between leptin levels and the severity of atopy.

Adiponectin and Allergic Rhinitis

As for adiponectin, in another study conducted outside the pollen season, cytokine levels seem not to significantly differ between allergic and normal subjects if considered without gender distinction, even though there seems to be a trend showing higher levels in allergic patients. Nevertheless, there was a significant difference in adiponectin values between healthy males and females: females had higher levels. In the analysis by gender, we observed a significant difference between allergic males and normal males, the former showing higher adiponectin values. On the contrary, there was no significant difference in females, even though there was an inverse trend: allergic females have lower levels than normal females. In addition, BMI and adiponectin levels were not correlated.

Table I   Key features of the relationships between allergic disorders and adipokines

- Subjects with asthma are heavier than subjects without asthma and overweight is often related to worse asthma control.
- Asthma and obesity are characterized by inflammation, and a common inflammatory pathway could explain the frequent association of the two diseases.
- There is a significant relationship between serum leptin levels and some clinical and immunological parameters.
- Adiponectin may exert both anti-inflammatory and pro-inflammatory effects, exerting multifunctional actions depending on the immunological milieu.
- SLIT induces a slightly increasing trend in leptin and adioponectin serum levels.

Since these studies were conducted outside the pollen season, leptin and adiponectin levels were not influenced by the presence of allergic inflammation, but may reflect the atopic status per se. In this regard, a very recent review underlines the concept that adiponectin may also exert pro-inflammatory effects. Indeed, adiponectin levels are elevated in classic chronic inflammatory/autoimmune diseases that are unrelated to increased adipose tissue, such as rheumatoid arthritis, systemic lupus erythematosus, inflammatory bowel disease, type 1 diabetes, and cystic fibrosis. In these patients, adiponectin levels positively correlate with inflammatory markers. Thus, adiponectin is regulated in the opposite direction and may exert different functions in classic versus obesity-associated inflammatory disorders, and allergic diseases are chronic inflammatory disorders. Therefore, and this is true at least for adiponectin, such cytokines may exert multifunctional actions depending on the immunological milieu.

Adiponectin Levels during Pollen Season

In a study designed to investigate the possible role of adiponectin in patients with seasonal allergic rhinitis, during pollen season, a deep difference in adiponectin levels was detected between allergic patients during pollen season and normal controls. Moreover, a significant difference was evident comparing in-season with out-season male patients. In addition, there was a significant difference between in-season allergic females and both out-season and normal females.

These findings underline the relevant effect played by pollen exposure on the inflammatory response, which might implicate the adipokine network. Therefore, adiponectin levels seem to be sincerely influenced by the presence of allergic inflammation.

However, since adiponectin was typically considered as an anti-inflammatory cytokine, this study confirms that this adipokine may exert pro-inflammatory effects as well.

Resistin in Allergic Rhinitis

Resistin is a new adipokine recently included in the "found in inflammatory zone" family. Serum resistin concentration was shown to be higher both in children and in adult patients with persistent allergic rhinitis. Moreover, the levels of serum resistin seem to depend on symptom severity.

Resistin not only may exert pro-inflammatory effects in such patients, contributing to the puzzle of nasal allergic inflammation, it may also be a marker of allergy severity.

SLIT and Cytokines

When our group evaluated the possible connection between SLIT and adipokines, we found that a single pre-seasonal SLIT course tended to increase the serum levels of both leptin and adiponectin and this trend appeared to be more pronounced in women.

SLIT induces a slightly increasing trend in leptin and adioponectin serum levels. This phenomenon could appear to be in conflict with the well-known anti-allergic effects exerted by immunotherapy. However, allergic inflammation represents a very complex cascade of several events involving numerous mediators. For example, TGF-beta released by Tregs exerts many anti-inflammatory actions, but it also induces fibrosis and tissue remodeling, three antithetic effects that may depend, though, on several other factors.

Adipsin, SLIT and Allergic Rhinitis

We also evaluated two groups of patients to investigate whether adipsin might have a role in seasonal allergic rhinitis: patients were treated or not treated with immunotherapy. Since immunotherapy is the only treatment capable of inducing a tolerance toward the causal allergen we tried to detect a possible reduction in adipsin values in patients treated with SLIT.

SLIT-treated patients had less intense inflammation and less severe symptoms and needed a lower dose of drug than untreated patients, being therefore deeply different from untreated patients from both a clinical and an immunological point of view.

Serum adipsin was increased in allergic patients in comparison with normal subjects, but this increase was statistically significant only in SLIT group. As for

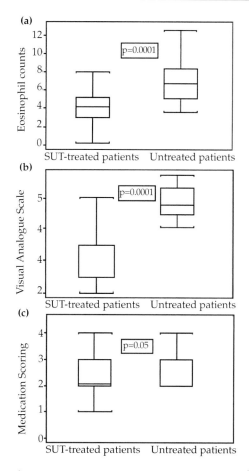

Fig. 1 Comparison between patient treated and untreated with SLIT. (a) Eosinophil counts distribution. (b) Visual analogue scale distribution. (c) Medication scoring distribution. Values are represented with median (white line), quartiles (25th and 75th percentiles, black box), whisker lines, which extend from the highest to the lowest values, and p-values between the groups.

gender distinction, adipsin levels were found to be increased in allergic males only and increased after SLIT therapy, while no difference was found in females. As adiponectin may exert pro-inflammatory action, the hypothesis whether there was a possible relationship between adiponectin and adipsin has been explored and a positive association was observed between the two adipokines, even though only in untreated male patients.

Adipsin, like other adipokines, may exert multiple actions, even antagonistic ones, depending on the existing milieu. In other words, adipsin might be considered both a pro-inflammatory and anti-inflammatory cytokine. In this regard, the level

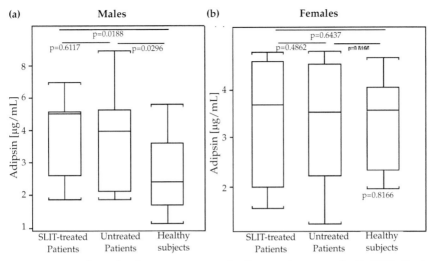

Fig. 2 Serum adipsin distribution in, respectively, allergic patients treated with SLIT, allergic patients not treated with SLIT, and non-allergic patients, divided into (a) males and (b) females. Values are represented with median (white line), quartiles (25th and 75th percentiles, black box), whisker lines, which extend from the highest to the lowest values, and p values between the groups.

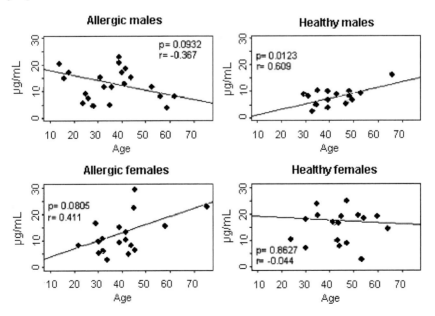

Fig. 3 Serum adiponectin and age in allergic and healthy subjects. Correlation, expressed with Pearson's correlation coefficient (r) and p value, between serum adiponectin and age in allergic and healthy subjects, evaluated separately for males (top) and females (bottom).

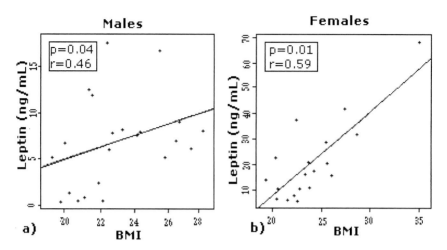

Fig. 4 Serum leptin levels and BMI in allergic subjects treated with SLIT. Correlation, expressed with Pearson's correlation coefficient (r) and p value, between serum leptin and body mass index (BMI) in allergic subjects, after 3 mo of SLIT.

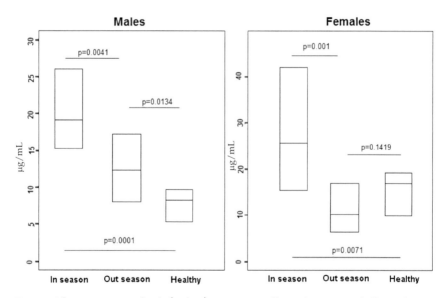

Fig. 5 Adiponectin serum levels (μg/mL) in patients suffering from seasonal allergic rhinitis. Allergic males and females evaluated in and out of pollen season, and in healthy males and females. Values are represented as medians, quartiles, and p values between the groups.

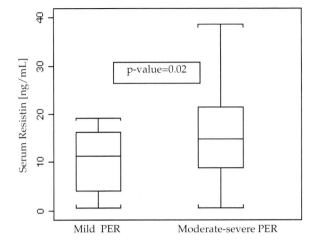

Fig. 6 Serum resistin distribution (ng/mL) in allergic patients (mild PER and moderate-severe PER). Values are represented with median (white line), quartiles (25th and 75th percentiles, black box), whisker lines, which extend from the highest to the lowest values, and p values between the groups.

of allergen exposure might be a factor capable of increasing adipsin values: in fact, immunotherapy is characterized by the assumption of extremely high allergen quantity, such as higher than natural exposure. However, this hypothesis still needs to be confirmed.

CONCLUSIONS

An association between obesity and allergy-related diseases has been reported in many studies, although considerable debate still exists about the existence and meaning of the association. Despite an abundant literature, the ultimate cause of the relationship between high BMI and atopic disorders has not been identified.

Leptin has been shown to be significantly related to BMI in both men and women. Moreover, leptin is significantly related with some clinical and immunological parameters, with gender differences: specifically, it is more evident in women. Leptin serum levels appear to be associated with clinical severity, as suggested by the relationship with symptom severity, medication use, and allergy grade, as well as with allergic inflammation, as shown by a correlation with eosinophil recruitment. On the other hand, adiponectin seems to be less related to clinical events but is associated with eosinophil recruitment.

These findings underline a direct involvement of adipokines in allergic reactions and in their clinical manifestations.

Summary Points

- The prevalence of both obesity and allergic sensitization keeps increasing in parallel in industrialized countries.
- Many studies have reported that obesity is a risk factor for the development of asthma and that obesity precedes the development of asthma.
- Moreover, since both asthma and obesity are characterized by inflammation, a common inflammatory pathway has been proposed as a plausible explanation for the frequent association of the two diseases.
- Serum adiponectin and leptin may be impaired in patients with allergic rhinitis and immunotherapy may affect these adipokines in patients with seasonal allergic rhinitis.
- Exogenous leptin may modulate allergic airway responses independently of obesity and its infusion increases airway hyperresponsiveness IgE serum level.
- Leptin may have a role in the inflammatory process of allergic rhinitis.
- Adiponectin exerts not only anti-inflammatory but even pro-inflammatory effects, its level being elevated in patients with classic chronic inflammatory/autoimmune diseases that are unrelated to increased adipose tissue.
- SLIT induces a slightly increasing trend in leptin and adioponectin serum levels.
- Adpokine serum level is impaired in allergic patients.

Abbreviations

BMI	:	body mass index
SLIT	:	sub-lingual immuno-therapy
WAT	:	white adipose tissue

Definition of Terms

Allergic rhinitis: An allergic disorder triggered by exposure to pollens of specific seasonal plants. It can be divided into seasonal and perennial and its severity into mild, moderate and severe.

Allergy: A disorder of the immune system in which abnormal reactions occour when normally harmless substances get in contact with the body, causing inflammatory responses.

Pollen season: Period of the year during which patients are exposed to the enviromental pollen they are allergic to. It is very important to understand that certain tests run to verify allergy may be affected by the presence of pollen in the air. The pollen season may, therefore, be a bias when evaluating serum levels of cytokines in allergic patients.

SLIT: Sublingual mmunotherapy treatment, a method used to treat allergy and, mostly, allergic rhinitis, by administering to patients a solution given under the tongue. Such treatment aims to reduce sensitivity to allergens, and it has a very good safety profile in both adults and children.

White adipose tissue: One of the two kinds of adipose tissue found in mammals. (The other is brown adipose tissue.) White adipose tissue is as much as 20% of body weight in men and 25% of body weight in women. It is considered a main store of energy. Nevertheless, it has an important endocrinological role as well.

References

Ciprandi, G. and G. Murdaca, G.L. Marseglia, B.M. Colombo, S. Quaglini, and M. De Amici. 2008. Serum adiponectin levels in patients with pollen-induced allergic rhinitis. Int. Immunopharmacol. 8: 945-949.

Ciprandi, G. and M. De Amici, and G.L. Marseglia. 2009. Serum adipsin levels in patients with seasonal allergic rhinitis: preliminary data. Int. Immunopharmacol. 9: 1460-1463.

Ciprandi, G. and M. De Amici, G. Murdaca, G. Filaci, D. Fenoglio, and G.L. Marseglia. 2009. Adipokines and sublingual immunotherapy: preliminary report. Hu. Immunol. 70: 73-78.

Ciprandi, G. and M. De Amici, M.A. Tosca, and G.L. Marseglia. 2009. Serum leptin levels depend on allergen exposure in patients with seasonal allergic rhinitis. Immunol. Invest. 38: 681-689.

Ciprandi, G. and M. De Amici, M.A. Tosca, S. Negrini, G. Murdaca, and G.L. Marseglia. 2009. Two year sublingual immunotherapy affects serum leptin. Int. Immunopharmacol. 9: 1244-1246.

Ciprandi, G. and G. Filaci, S. Negrini, M. De Amici, and M. Fenoglio. 2009. Serum leptin levels in patients with pollen-induced allergic rhinitis. Int. Arch. Allerg. Immunol. 148: 211-218.

Ciprandi, G. and M. De Amici, M.A. Tosca, G.L. Marseglia. In press. Serum resistin in persistent allergic rhinitis: preliminary data in adults. Int. Immunopharmacol.

Fantuzzi, G. 2005. Adipose tissue, adipokines, and inflammation. J. Allerg. Clin. Immunol. 115: 911-919.

Litonjua, A.A. and D.R. Gold. 2008. Asthma and obesity: common early-life influences in the inception of disease. J. Allerg. Clin. Immunol. 121: 1075-1084.

McLachlan, C.R. and R. Poulton, G. Car, J. Cowan, S. Filsell, J.M. Greene, D.R. Taylor, D. Welch, A. Williamson, M.R. Sears, and R.J. Hancox. 2007. Adiposity, asthma, and airway inflammation. J. Allerg. Clin. Immunol. 119: 634-639.

Radon, K. and A. Schulze, R. Schierl, G. Dietrich-Gümperlein, D. Nowak, and R.A. Jörres. 2008. Serum leptin and adiponectin levels and their association with allergic sensitization. Allergy 63: 1448-1454.

CHAPTER 28

Adipokines and Sleep Disorders

Kapil Dhawan,[1] Aneesa M. Das,[1] Shariq I. Sherwani,[1] Sainath R. Kotha,[1] Narasimham L. Parinandi[1,*] and Ulysses J. Magalang[1]

[1]Division of Pulmonary, Allergy, Critical Care, and Sleep Medicine
[1]Department of Internal Medicine, The Ohio State University College of Medicine, Dorothy M. Davis Heart and Lungs Research Institute, 473 W. 12th Avenue Columbus OH 43210, USA

ABSTRACT

Sleep-disordered breathing, a widespread condition, leads to obstructive sleep apnea syndrome, which is associated with severe health consequences including metabolic disorders, cardiovascular diseases, and diabetes. OSA has been recognized more frequently in obese individuals as the intermittent airway obstruction during sleep with frequent arousals, thus establishing links between obesity, sleep-induced airway obstruction, sleep fragmentation, and daytime sleepiness. Adipose tissue secretes pro-inflammatory cytokines and hormone-like molecules called adipokines such as the interleukins (IL), tumor necrosis factor α (TNF-α), leptin, and adiponectin. These adipose-derived cytokines and adipokines exert a profound impact on the neural pathways of respiratory control. Leptin has been shown to act as a respiratory stimulant and is associated with obesity hypoventilation syndrome. Obesity and OSA have been shown to be associated with elevated circulating levels of leptin and impaired leptin signaling appears to contribute to respiratory depression. Leptin is also implicated in nocturnal hypoventilation during sleep. Pro-inflammatory cytokines such as TNF-α, IL-1β, and IL-6 secreted by visceral adipose tissue seem to be involved in physiological sleep regulation. Adiponectin, another important anti-atherogenic adipokine, is secreted exclusively and abundantly by adipose tissue. Although significant reduction in circulating adiponectin levels in obese individuals has

*Corresponding author

been documented, the effect of sleep apnea on circulating adiponectin levels in OSA patients, particularly on the biologically active high molecular weight form, is unclear. The association between adipose-derived adipokines and sleep disorders such as OSA is emerging and more studies in this area are needed.

INTRODUCTION

Sleep-disordered breathing is a highly prevalent condition that leads to obstructive sleep apnea syndrome (OSAS), which affects approximately 5% of the western population and is associated with serious adverse health consequences (Fig. 1) (Young *et al.* 2002). There is growing body of evidence that OSAS is related to increased prevalence of cardiovascular disease (Koskenvuo *et al.* 1987, Hung *et al.* 1990, Shahar *et al.* 2001) and metabolic disorders such as hypertension, dyslipidemia, and diabetes mellitus (Levinson *et al.* 1993, Grunstein *et al.* 1993, Wilcox *et al.* 1998, Lattimore *et al.* 2003, Coughlin *et al.* 2004). Sleep-disordered

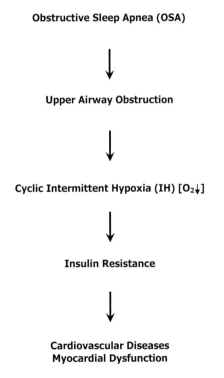

Fig. 1 Obstructive sleep apnea (OSA) results in cyclic intermittent hypoxia (IH) through the upper airway obstruction in humans. OSA is also known to be associated with cardiovascular diseases. IH associated with OSA is also known to cause insulin resistance. Insulin resistance is established to cause cardiovascular and myocardial pathologies.

breathing is secondary to often cyclical cessations in breathing rhythm (apneas) or momentary or sustained reductions in the breath amplitude (hypopneas), which can lead to arousals, arterial hypoxemia and hypercapnia. These apneas and hypopneas are specific to the sleeping state and are accompanied by a compromised, often even completely closed, extrathoracic upper airway (obstructive event) to a marked reduction or cessation of brain stem respiratory motor output (central event) and a combination of obstructive and central events (Table 1).

Table I   Evaluation factors and symptoms of obstructive sleep apnea

Evaluation factors	Symptoms
• Obesity • Male • Craniofacial abnormalities, including large uvula, enlarged tonsils, long soft palate, macroglossia • Aggravated by alcoholism and the use of sedatives • Occurrence of five or more scoreable respiratory events per hour on the polysomnograph. • The respiratory events may include any combination of obstructive apneas, hypopneas, or respiratory-associated arousals	• Snoring • Sleepiness • Sleeplessness • Fatigue • Dry mouth • Sore throat • Headaches • Confusion • Insomnia • Impaired concentration • Impotence

Anatomical predisposition for airway closure is an essential component for the development of OSA. The upper airway is a complex structure required to perform deglutition, vocalization, and respiration. Upper airway obstruction in sleep is most prevalent in humans partly because the hyoid bone, a key anchoring site for pharyngeal dilator muscles, is not rigidly attached to skeletal structures. The oropharynx region is usually where closure occurs in most subjects with OSA, and this region is also smaller in OSA patients versus controls even during wakefulness. Craniofacial and upper airway soft tissue structures also play a significant role in OSA patients. Recent advances in imaging techniques using volumetric time overlapped magnetic resonance imaging or computer tomography (CT) show that the thickness of the lateral pharyngeal walls is a major site of airway compromise. Several studies have shown that most patients with OSA have narrowed airway primarily in the lateral dimension (Pevernagie *et al.* 1995).

OBESITY AND OBSTRUCTIVE SLEEP APNEA

There are many potential causes of lateral wall thickening of the upper airway in OSA patients. As shown in both humans and rodent models, obesity is a major contributor to airway collapse through increased area and volume of pharyngeal fat deposits. Obesity also gives rise to excess fat-free muscular tissue, thereby increasing the size of many upper airway structures (Bradley *et al.* 1986, Schwartz *et al.* 1991) and compressing the lateral airway walls. Obesity also contributes indirectly to upper airway narrowing, especially during sleep, because lung volumes are markedly reduced by a combination of increased abdominal fat mass and the recumbent position. The decreased lung volume (end-expiratory lung volume) in obese subjects, especially in the recumbent position, together with increased tissue oxygen consumption rates, means that lung oxygen stores are more quickly depleted during an apnea, resulting in more severe arterial oxygen desaturation for any given apneic length (Findley *et al.* 1983).

Most patients with OSA do not exhibit daytime hypoventilation. Those obese patients with daytime hypoventilation, not secondary to lung disease, are said to have obesity hypoventilation syndrome. The patients were previously referred to as "Pickwickian" (Bickelmann *et al.* 1956). Most patients with obesity hypoventilation syndrome will have a high apnea-hypopnea index, number of apneas and hypopneas per hour of sleep, and severe arterial oxygen desaturation (Berger *et al.* 2001).

OBESITY AND RESPIRATORY PHYSIOLOGY

Obese patients require a greater effort of breathing in order to move their excess weight (Dempsey *et al.* 1966). The increased metabolic demand is manifested in an increase in resting oxygen consumption (VO_2) and carbon dioxide production (VCO_2) (Zavala and Printen 1984). While obese patients tend to have an increased ventilatory eucapnic state, they have been reported to have an overall increase in minute ventilation (Burki and Baker 1984). Obesity affects both the static and dynamic respiratory parameters. The most prominent changes seen in static lung function with increasing body mass index (BMI) are a reduction in expiratory reserve volume and functional residual capacity (Jones and Nzekwu 2006). The total lung capacity, maximum voluntary ventilation and vital capacity can be reduced with morbid obesity. Residual volume is typically unchanged in obesity, leading to an increased ratio of residual volume to total lung capacity.

Dynamic lung flow is decreased with increasing weight. Forced expiratory volume in one second (FEV1) is inversely correlated with an increasing BMI. In a longitudinal study the weight gain is shown to be associated with a decrease in FEV1 and weight loss is associated with an improvement in FEV1 (Bottai *et al.* 2002). As the BMI increases the decrease in FEV1 is typically in proportion to the change in forced vital capacity. Body fat distribution has an independent effect on ventilatory function (Lazarus *et al.* 1998). When adipose tissue is predominantly accumulated in the subcutaneous and visceral abdominal area, the distribution is referred to as central obesity. Waist circumference is better correlated with metabolic risk than BMI due to its reflection of metabolically active visceral fat. The effect of weight gain on lung function appears to be more significant in men than women. Mild hypoxemia can be seen in obese patients and has been associated with central obesity in the morbidly obese (Zavorsky *et al.* 2007). The diffusing capacity of the lung for carbon monoxide is a parameter to determine the extent to which oxygen passes from the alveoli to the blood.

ADIPOKINES, ADIPOSE TISSUE, OBESITY AND OBSTRUCTIVE SLEEP APNEA

Adipose tissue is now viewed as a crucial center of major endocrine function secreting adipokines with hormonal properties such as leptin, adiponectin and a diverse range of proteins collectively called adipokines (Fig. 2). These humoral mediators and inflammatory cytokines can have an impact on neural pathways associated with respiratory control (Schwartz *et al.* 2008). Perhaps the most intensively studied adipocyte-derived cytokine affecting respiratory control is leptin, which was initially determined to regulate satiety and food intake by binding to receptors in the hypothalamus (Friedman *et al.* 1998). Leptin can also act as a

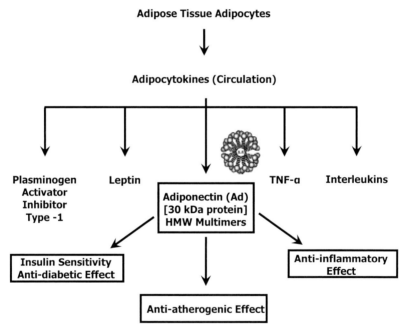

Fig. 2   Adipose tissue synthesizes and secretes a wide variety of bioactive peptides and proteins that are collectively called "adipocytokines". These secretory products include leptin, tumor necrosis factor α (TNF-α), plasminogen activator inhibitor type 1, interleukins, resistin, and adiponectin. Adiponectin has received considerable attention in recent years because of its potent physiological effects and pleiotropic actions, which include its antidiabetic, anti-atherogenic, and anti-inflammatory effects.

respiratory stimulant, and impairment of the leptin signaling pathway, as occurs in leptin-resistant or leptin-deficient states of obesity, causes respiratory depression in mice (O'Donnell *et al.* 1999) and is associated with obesity hypoventilation syndrome in humans (Phipps *et al.* 2002). In addition to its relationship with adiposity, serum leptin levels seem to be also modulated by hypoxia, as shown by increased levels of plasma leptin in mice exposed to intermittent hypoxia (Polotsky *et al.* 2003). In OSA subjects, leptin has been found elevated and, in some cases, leptin levels seem to be more related to obesity than to OSA (Barcelo *et al.* 2005).

Adiponectin is a cytokine-like hormone secreted exclusively and most abundantly by adipose tissue (Yamamoto *et al.* 2005). Adipocytes secrete both the low-molecular-weight (LMW) and high-molecular-weight (HMW) forms of adiponectin along with smaller trimer that are virtually undetectable in serum. Adiponectin concentrations in the serum are affected by changes in adipose tissue mass and, unlike other adipose-derived hormones, its serum levels decrease in obesity. Circulating adiponectin levels and percentage HMW species are higher in females (who have higher levels of body fat for a given BMI) than in males. Plasma levels of adiponectin exhibit strong negative correlations with direct (CT scan) and indirect (waist-to-hip ratio) measures of intra-abdominal fat mass (Cnop *et al.* 2003, Gavrila *et al.* 2003). It is not only the amount of fat but also the distribution of adipose mass that determines adiponectin levels.

Visceral adipose tissue releases many other humoral factors including pro-inflammatory cytokines such as tumor necrosis factor α (TNF-α), interleukin 1β (IL-1β) and interleukin (IL-6) that appear to be involved in physiological sleep regulation (Opp 2005, Kapsimalis *et al.* 2005) and their increased secretion or exogenous administration to humans is associated with sleepiness and fatigue (Table 2) (Mastorakos *et al.* 1993). There are several studies on cytokines and disorders of excessive daytime sleepiness that have shown that both TNF-α and IL-6 plasma concentrations positively correlated with excessive daytime sleepiness (Patel *et al.* 2003, Vgontzas *et al.* 1997, 2004, Ryan *et al.* 2005). Furthermore, in interventional studies, use of continuous positive airway pressure therapy, which acts as a pneumatic airway splint, shows reduction in these inflammatory markers (Minoguchi *et al.* 2004, Yokoe *et al.* 2003).

Table 2 Key features of obstructive sleep apnea

- OSA, which occurs in 90% of sleep apnea patients, affects 5% of the western population.
- OSA is associated with severe health outcomes, including metabolic disorders, cardiovascular diseases and diabetes.
- Obesity, particularly visceral fat obesity, is a potent risk factor for sleep apnea.
- OSA and diabetes are co-existing disorders that have shared risk factors which may be additive or even synergistic in a common environment.
- Obesity and OSA are associated with elevated circulating levels of leptin.
- Adipokines, such as leptin and adiponectin are centered in the adipose tissue.

In OSA patients, in spite of a few conflicting studies reporting unchanged (Makino *et al.* 2006, Tokuda *et al.* 2008) or even augmented adiponectin levels (Wolk *et al.* 2005), most published data support a significant reduction in adiponectin levels that seems independent of the obesity status (Lam *et al.* 2008, Masserini *et al.* 2006, Zhang *et al.* 2004, 2006, 2007). Along with its anti-inflammatory and anti-atherosclerotic effects, adiponectin plays a protective role against insulin resistance and the metabolic syndrome (Invitti *et al.* 2006), and low levels of adiponectin have also been associated with type 2 diabetes in both children and adults (Figs. 1 and 2) (Cruz *et al.* 2004, Spranger *et al.* 2003).

Summary Points

- Sleep-disordered breathing is a prevalent condition that leads to obstructive sleep apnea syndrome and is associated with metabolic disorders, cardiovascular diseases, and diabetes.
- OSA is encountered in obese individuals and manifests as intermittent airway obstruction with frequent arousals during sleep.
- Obesity has been identified to affect the respiratory physiology.
- Adipose tissue, in addition to undertaking fat storage, also functions as an endocrine organ and secretes hormone-like metabolic mediators called "adipokines".
- These adipose-generated adipokines including pro-inflammatory cytokines such as leptin, adiponectin, TNF-α, IL-1β, and IL-6 may influence respiratory control and respiratory physiology.
- Obesity and OSA have been shown to be associated with the altered circulating levels of leptin and adiponectin.
- Adipokines are emerging as important players in OSA and other metabolic disorders.

Abbreviations

BMI	:	body mass index
CT	:	computer tomography
FEV1	:	forced expiratory volume in one second
HMW	:	high molecular weight
IL-1β	:	interleukin 1β
IL-6	:	interleukin 6
LMW	:	low molecular weight
OSA	:	obstructive sleep apnea
OSAS	:	obstructive sleep apnea syndrome
TNF-α	:	tumor necrosis factor α

Definition of Terms

Adipokines: Adipose tissue–secreted hormone-like mediators.
Inflammatory cytokines: Mediators secreted by cells that cause inflammation.
Obesity: Excess fat deposition and overweight.
Obstructive sleep apnea: Intermittent airway obstruction with frequent arousals.
Sleep-disordered breathing: Breathing condition associated with sleep disorder.

References

Berger, K.I. and I. Ayappa, B. Chatr-Amontri, A. Marfatia, I.B. Sorkin, D.M. Rapoport, and R.M. Goldring. 2001. Obesity hypoventilation syndrome as a spectrum of respiratory disturbances during sleep. Chest 120: 1231-1238.

Bottai, M. and F. Pistelli, F. Di Pede, L. Carrozzi, S. Baldacci, G. Matteelli, A. Scognamiglio, and G. Viegi. 2002. Eur. Respir. J. 20: 665-673.

Bradley, T.D and I.G. Brown, R.F. Grossman, N. Zamel, D.Martinez, E.A.Phillipson, and V. Hoffstein. 1986. Pharyngeal size in snorers, nonsnorers, and patients with obstructive sleep apnea. N. Engl. J. Med. 315: 1327-1331.

Burki, N.K. and R.W. Baker. 1984. Ventilatory regulation in eucapnic morbid obesity. Am. Rev. Respir. Dis. 129: 538-543.

Cnop, M. and P.J. Havel, K.M. Utzschneider, D.B. Carr, M.K. Sinha, E.J. Boyko, B.M. Retzlaff, R.H. Knopp, J.D. Brunzell, and S.E. Kahn. 2003. Relationship of adiponectin to body fat distribution, insulin sensitivity and plasma lipoproteins: evidence for independent roles of age and sex. Diabetologia 46: 459-469.

Coughlin, S.R. and L. Mawdsley, J.A. Mugarza, P.M. Calverley, and J.P. Wilding. 2004. Obstructive sleep apnoea is independently associated with an increased prevalence of metabolic syndrome. Eur. Heart J. 25: 735-741.

Cruz, M. and R. García-Macedo, Y. García-Valerio, M. Gutiérrez, R. Medina-Navarro, G. Duran, N. Wacher, and J. Kumate. 2004. Low adiponectin levels predict type 2 diabetes in Mexican children. Diabetes Care 27: 1451-1453.

Dempsey, J.A. and W. Reddan, B. Balke, and J. Rankin. 1966. Work capacity determinants and physiologic cost of weight-supported work in obesity. J. Appl. Physiol. 21: 1815-1820.

Findley, L.J. and A.L. Ries, G.M. Tisi, and P.D. Wagner. 1983. Hypoxemia during apnea in normal subjects: mechanisms and impact of lung volume. J. Appl. Physiol. 55: 1777-1783.

Friedman, J.M. and J.L. Halaas. 1998. Leptin and the regulation of body weight in mammals. Nature 395: 763-770.

Gavrila, A. and J.L. Chan, N. Yiannakouris, M. Kontogianni, L.C. Miller, C. Orlova, and C.S. Mantzoros. 2003. Serum adiponectin levels are inversely associated with overall and central fat distribution but are not directly regulated by acute fasting or leptin administration in humans: cross-sectional and interventional studies. J. Clin. Endocrinol. Metab. 88: 4823-4831.

Grunstein, R. and I. Wilcox, T.S. Yang, Y. Gould, and J. Hedner. 1993. Snoring and sleep apnoea in men: association with central obesity and hypertension. Int. J. Obes. Relat. Metab. Disord. 17: 533-540.

Hung, J. and E.G. Whitford, R.W. Parsons, and D.R. Hillman. 1990. Association of sleep apnoea with myocardial infarction in men. Lancet 336: 261-264.

Invitti, C. and C. Maffeis, L. Gilardini, B. Pontiggia, G. Mazzilli, A. Girola, A. Sartorio, F. Morabito, and G.C. Viberti. 2006. Metabolic syndrome in obese Caucasian children: prevalence using WHO-derived criteria and association with nontraditional cardiovascular risk factors. Int. J. Obes. 30: 627-633.

Jones, R.L. and M.M. Nzekwu. 2006. The effects of body mass index on lung volumes. Chest 130: 827-833.

Kapsimalis, F. and G. Richardson, M.R. Opp, and M. Kryger. 2005. Cytokines and normal sleep. Curr. Opin. Pulm. Med. 11: 481-484.

Koskenvuo, M. and J. Kaprio, K. Heikkilä, S. Sarna, T. Telakivi, and M. Partinen. 1987. Snoring as a risk factor for ischaemic heart disease and stroke in men. Br. Med. J. 294: 643.

Lam, J.C. and A. Xu, S. Tam, P.I. Khong, T.J. Yao, D.C. Lam, A.Y. Lai, B. Lam, K.S. Lam, and S.M. Mary. 2008. Hypoadiponectinemia is related to sympathetic activation and severity of obstructive sleep apnea. Sleep 31: 1721-1727.

Lattimore, J.D. and D.S. Celermajer, and I. Wilcox. 2003. Obstructive sleep apnea and cardiovascular disease. J. Am. Coll. Cardiol. 41: 1429-1437.

Lazarus, R. and C.J. Gore, M. Booth, and N. Owen. 1998. Effects of body composition and fat distribution on ventilator function in adults. Am. J. Clin. Nutr. 68: 35-41.

Levinson, P.D. and S.T. McGarvey, C.C. Carlisle, S.E. Eveloff, P.N. Herbert, and R.P. Millman. 1993. Adiposity and cardiovascular risk factors in men with obstructive sleep apnea. Chest 103: 1336-1342.

Makino, S. and H. Handa, K. Suzukawa, M. Fujiwara, M. Nakamura, S. Muraoka, I. Takasago, Y. Tanaka, K. Hashimoto, and T. Sugimoto. 2006. Obstructive sleep apnoea syndrome, plasma adiponectin levels, and insulin resistance. Clin. Endocrinol. 64: 12-19.

Masserini, B. and P.S. Morpurgo, F. Donadio, C. Baldessari, R. Bossi, P. Beck-Peccoz, and E. Orsi. 2006. Reduced levels of adiponectin in sleep apnea syndrome. J. Endocrinol. Invest. 29: 700-705.

Mastorakos, G. and G.P. Chrousos, and J.S. Weber. 1993. Recombinant interleukin-6 activates the hypothalamic-pituitary-adrenal axis in humans. J. Clin. Endocrinol. Metab. 77: 1690-1694.

Makino, S. and H. Handa, K. Suzukawa, M. Fujiwara, M. Nakamura, S. Muraoka, I. Takasago, Y. Tanaka, K. Hashimoto, and T. Sugimoto. 2006. Obstructive sleep apnoea syndrome, plasma adiponectin levels, and insulin resistance. Clin. Endocrinol. 64: 12-19.

Minoguchi, K. and T. Tazaki, T. Yokoe, H. Minoguchi, Y. Watanabe, M. Yamamoto, and M. Adachi. 2004. Elevated production of tumor necrosis factor-alpha by monocytes in patients with obstructive sleep apnea syndrome. Chest 126: 1473-1479.

O'Donnell C.P. and C.D. Schaub, A.S. Haines, D.E. Berkowitz, C.G. Tankersley, A.R. Schwartz, and P.L. Smith. 1999. Leptin prevents respiratory depression in obesity. Am. J. Respir. Crit. Care Med. 159: 1477-1484.

Opp, M.R. 2005. Cytokines and sleep. Sleep Med. Rev. 9: 355-364.

Patel, S.R. and D.P. White, A. Malhotra, M.L. Stanchina, and N.T. Ayas. 2003. Continuous positive airway pressure therapy for treating sleepiness in a diverse population with obstructive sleep apnea: results of a meta-analysis. Arch. Intern. Med. 163: 565-571.

Pevernagie, D.A. and A.W. Stanson, P.F. Sheedy 2nd, B.K. Daniels, and J.W. Shepard Jr. 1995. Effects of body position on the upper airway of patients with obstructive sleep apnea. Am. J. Respir. Crit. Care Med. 152: 179-185.

Phipps, P.R. and E. Starritt, I. Caterson, and R.R. Grunstein. 2002. Association of serum leptin with hypoventilation in human obesity. Thorax 57: 75-76.

Polotsky, V.Y. and J. Li, N.M. Punjabi, A.E. Rubin, P.L. Smith, A.R. Schwartz, and C.P. O'Donnell. 2003. Intermittent hypoxia increases insulin resistance in genetically obese mice. J. Physiol. 552: 253-264.

Ryan, S. and C.T. Taylor and W.T. McNicholas. 2005. Selective activation of inflammatory pathways by intermittent hypoxia in obstructive sleep apnea syndrome. Circulation 112: 2660-2667.

Schwartz, A.R. and A.R. Gold, N. Schubert, A. Stryzak, R.A. Wise, S. Permutt, and P.L. Smith. 1991. Effect of weight loss on upper airway collapsibility in obstructive sleep apnea. Am. Rev. Respir. Dis. 144: 494-498.

Schwartz, A.R. and S.P. Patil, A.M. Laffan, V. Polotsky, H. Schneider, and P.L. Smith. 2008. Obesity and obstructive sleep apnea: pathogenic mechanisms and therapeutic approaches. Proc. Am. Thorac. Soc. 5: 185-192.

Shahar, E. and C.W. Whitney, S. Redline, E.T. Lee, A.B. Newman, F. Javier Nieto, G.T. O'Connor, L.L. Boland, J.E. Schwartz, and J.M. Samet. 2001. Sleep-disordered breathing and cardiovascular disease: cross-sectional results of the Sleep Heart Health Study. Am. J. Respir. Crit. Care Med. 163: 19-25.

Spranger, J. and A. Kroke, M. Möhlig, M.M. Bergmann, M. Ristow, H. Boeing, and A.F. Pfeiffer 2003. Adiponectin and protection against type 2 diabetes mellitus. Lancet 361: 226-228.

Tokuda, F. and Y. Sando, H. Matsui, H. Koike, and T. Yokoyama. 2008. Serum levels of adipocytokines, adiponectin and leptin, in patients with obstructive sleep apnea syndrome. Intern. Med. 47: 1843-1849.

Vgontzas, A.N. and D.A. Papanicolaou, E.O. Bixler, A. Kales, K. Tyson, and G.P. Chrousos. 1997. Elevation of plasma cytokines in disorders of excessive daytime sleepiness: role of sleep disturbance and obesity. J. Clin. Endocrinol. Metab. 82: 1313-1316.

Vgontzas, A.N. and E. Zoumakis, H.M. Lin, E.O. Bixler, G. Trakada, and G.P. Chrousos. 2004. Marked decrease in sleepiness in patients with sleep apnea by etanercept, a tumor necrosis factor-alpha antagonist. J. Clin. Endocrinol. Metab. 89: 4409-4413.

Wilcox, I. and S.G. McNamara, F.L. Collins, R.R. Grunstein, and C.E. Sullivan. 1998. "Syndrome Z": the interaction of sleep apnoea, vascular risk factors and heart disease. Thorax 53: S25-S28.

Wolk, R. and A. Svatikova, C.A. Nelson, A.S. Gami, K. Govender, M. Winnicki, and V.K. Somers. 2005. Plasma levels of adiponectin, a novel adipocyte-derived hormone, in sleep apnea. Obes. Res. 13: 186-190.

Yamamoto, K. and T. Kiyohara, Y. Murayama, S. Kihara, Y. Okamoto, T. Funahashi, T. Ito, R. Nezu, S. Tsutsui, J.I. Miyagawa, S. Tamura, Y. Matsuzawa, I. Shimomura, and

Y. Shinomura. 2005. Production of adiponectin, an anti-inflammatory protein, in mesenteric adipose tissue in Crohn's disease. Gut 54: 789-796.

Yokoe, T. and K. Minoguchi, H. Matsuo, N. Oda, H. Minoguchi, G. Yoshino, T. Hirano, and M. Adachi. 2003. Elevated levels of C-reactive protein and interleukin-6 in patients with obstructive sleep apnea syndrome are decreased by nasal continuous positive airway pressure. Circulation 107: 1129-1134.

Young, T. and P.E. Peppard, and D.J. Gottlieb. 2002. Epidemiology of obstructive sleep apnea: a population health perspective. Am. J. Respir. Crit. Care Med. 165: 1217-1239.

Zavala, D.C. and K.J. Printen. 1984. Basal and exercise tests on morbidly obese patients and after gastric bypass. Surgery 95: 221-229.

Zavorsky, G.S. and J.M. Murias, J. Kim Do, J. Gow, J.L. Sylvestre, and N.V. Christou. 2007. Waist-to-hip ratio is associated with pulmonary gas exchange in the morbidly obese. Chest 13: 362-367.

Zhang, X.L. and K.S. Yin, C. Li, E.Z. Jia, Y.Q. Li, and Z.F. Gao. 2007. Effect of continuous positive airway pressure treatment on serum adiponectin level and mean arterial pressure in male patients with obstructive sleep apnea syndrome. Chin. Med. J. 120: 1477-1481.

Zhang, X.L. and K.S. Yin, H. Mao, H. Wang, and Y. Yang. 2004. Serum adiponectin level in patients with obstructive sleep apnea hypopnea syndrome. Chin. Med. J. 117: 1603-1606.

Zhang, X.L. and K.S. Yin, H. Wang, and S. Su. 2006. Serum adiponectin levels in adult male patients with obstructive sleep apnea hypopnea syndrome. Respiration 73: 73-77.

CHAPTER 29

Adipokines in Children and Adolescents and Implications for Disease

HuiLing Lu[1] and Katherine Cianflone[2],*

[1]Department of Pediatrics, Tongji Hospital, Tongji Medical College, Huazhong University of Science and Technology, Wuhan 430030, China
[2]Centre de Recherche Institut Universitaire de Cardiologie et Pneumologie de Quebec, Université Laval Y4323, 2725 Chemin Ste-Foy, Québec, QC, Canada, G1V 4G5

ABSTRACT

Adipokines, secreted by adipose tissue, are involved in satiety, energy expenditure and storage, as well as immune functions including cytokine production and macrophage activation. Adipokines are implicated in obesity and related diseases (metabolic syndrome (MetS) and atherosclerosis) in adults and children, as well as inflammatory diseases (rheumatoid arthritis, osteoarthritis, sepsis). Although some adipokines are produced exclusively in adipose tissue, many can be secreted elsewhere, such as myeloid cells. Metabolic and immune crossroads are highlighted by the recent perception that obesity and related diseases are chronic inflammatory conditions. Many adipokines have been examined and implicated in obesity/immune-related diseases in adults. In children there is far less data. Based on literature review, the following adipokines were identified in children/adolescents: leptin, adiponectin, acylation-stimulating protein (ASP), resistin, tumor necrosis factor (TNFα), visfatin and retinol-binding protein 4 (RBP4). The present review outlines data on adipokines in children and their association with (1) obesity, insulin resistance (InsRes), MetS and type 2 diabetes (T2D), (2) obesity-associated liver and vascular disorders, and (3) immune-related disorders.

*Corresponding author

Lastly, the environmental influences that contribute to critical development periods (intra-uterine milieu, post-natal growth, early childhood and puberty-adolescence) are addressed.

INTRODUCTION

Research on adipokines in children has focused on obesity, InsRes and MetS. The NHANES survey (2003-2004) indicated 17% overweight in adolescents, with similar results for children aged 6-11 yr and 2-5 yr. The worldwide trend is similar, with ethnicity, gender, puberty and age influencing it. Often obese children and adolescents become obese adults, but the level of fatness at which morbidity increases is an ongoing question. Nonetheless, 30% of children have MetS, regardless of the MetS definition used (Eyzaguirre and Mericq 2009, Korner *et al.* 2007).

Can adipokines be used as biomarkers for obesity development or, once a subject is obese, to identify risk for InsRes, MetS or T2D (Eyzaguirre and Mericq 2009, Korner *et al.* 2007)? Developing evidence indicates many adipokine alterations in immune-related diseases in children, and adipokine dysregulation as a whole, are likely involved. Associations of adipokines in three areas will be presented: (1) obesity, InsRes, MetS and T2D, (2) obesity-associated liver and vascular disorders, and (3) immune-related disorders. Lastly, environmental influences contributing to critical development periods are addressed.

ADIPOKINES IN OBESITY, InsRes AND MetS

Based on literature review, the following adipokines were identified in children/adolescents: leptin, adiponectin, ASP, resistin, TNFα, visfatin and RBP4 (Table 1). Data are lacking on other novel adipokines, such as apelin, chemerin and vaspin.

Leptin in Children/Adolescents

Leptin is best known for its role in satiety and absence results in severe obesity (Korner *et al.* 2007). Leptin affects reproduction, vascular function and bone mass and targets kidney, bowel, pancreas and muscle. In immune function, leptin stimulates secretion of cytokines and chemotactic factors and macrophage activation. Disease associations in adults are wide-ranging, from obesity (and related disorders) to immune dysfunction (hepatitis, osteoarthritis, sepsis, rheumatoid arthritis). Leptin correlates closely with BMI and fat mass in children (as with adults) and correlates negatively to fat-free mass. The increased leptin in women becomes apparent with puberty and is likely a combination of differences in fat mass and fat-free mass between girls and boys. Gram per gram, subcutaneous fat produces more leptin. Increased leptin in obesity and correlations with metabolic

Table 1   Major metabolic and immune functions of adipokines relevant to studies in children and adolescents with obesity-induced changes

ADIPOKINE	METABOLIC	IMMUNE	CHILDREN	ADULTS
Leptin	Satiety, hormone regulation, energy expenditure	Cytokine induction, chemotaxis, macrophage activation	↑	↑
Adiponectin	Fatty acid oxidation, gene regulation, insulin secretion, hepatic gluconeogenesis	Production of cytokines and adhesion molecules, inflammation	↓	↓
ASP	Triglyceride storage, glucose transport, inhibits hormone-sensitive lipase	Stimulates cytokine secretion	↑	↑
Resistin	Hepatic and/or muscle InsRes and glucose metabolism	Stimulates expression of TNFα and IL-6, associated with inflammatory markers	↑ or →	↑
TNFα	Interferes with insulin signaling	Induces adhesion molecule expression, endothelial cell apoptosis	↑ or ↓ or →	↑
Visfatin	Increases muscle and adipose glucose uptake, decreases hepatic glucose release	Stimulation of B cell growth, delays neutrophil apoptosis, increases TNFα and IL-6 production	↑	↑ or →
RBP4	Increases hepatic gene expression, impairs muscle insulin signaling	Acute phase reactant	↑	↑

ASP, acylation-stimulating protein; TNFα, tumor necrosis factor α; RBP4, retinol-binding protein 4; ↑ increase; ↓ decrease; → no change.

parameters (InsRes and lipids) have been confirmed in many studies and are consistent with the hypothesis of a leptin-resistant state.

Weight loss results in leptin decreases. A comprehensive summary of exercise interventions in obese children/adolescents indicated leptin change was best determined by initial leptin, and obese children with appropriate (neither high nor low) leptin were more environmentally sensitive and more likely to respond to weight-loss intervention.

It has been suggested that decreased soluble leptin receptor (sOB-R), a major leptin-binding protein, may contribute to leptin resistance. sOB-R decreases with puberty as well as in obesity. With significant weight reduction, sOB-R increases and remains constant. Instances of higher sOB-R/leptin ratio in neonates, postnatal weight loss, T1D children, and malnourished children may impact on leptin action.

Adiponectin in Children/Adolescents

Adiponectin influences insulin secretion, food intake and energy expenditure via fatty acid oxidation in liver and skeletal muscle. In adults, adiponectin is decreased in obesity, atherosclerosis, T2D, cardiovascular disease, and rheumatoid arthritis (Eyzaguirre and Mericq 2009, Korner *et al.* 2007). Adiponectin, lower in boys than girls, changes progressively with puberty and is inversely related to androgens; thus, pubertal development and age need to be corrected for in disease assessment. Adiponectin decreases with increasing obesity, especially central adiposity, in children/adolescents. High-molecular-weight adiponectin correlates positively with HDL, but inversely with insulin and triglycerides. The association with MetS parameters (InsRes, hyperinsulinemia, blood pressure), more than BMI and pubertal development, suggests epidemiological and pathophysiological relevance in childhood.

Low adiponectin predicts presence or occurrence of MetS. In very young obese children (2-6 yr), adiponectin decreased, even without lipid changes (Cianflone *et al.* 2005). In pre-pubertal children, fasting and postprandial adiponectin alterations anticipated lipid and TNFα changes. Obese children aged 12-13 yr had a worse metabolic profile than those aged 9-10 yr. Lifestyle interventions in overweight/obese children increased adiponectin regardless of weight changes, with no consistent leptin changes. Adiponectin may be an early biomarker for treatment efficacy and to identify obese children/adolescents at risk of future diabetes and atherosclerosis.

Resistin in Children/Adolescents

Resistin may be increased in obesity, InsRes and T2D (depending on the study). A role in the immune system is apparent in inflammatory bowel disease, sepsis, and coronary heart disease. Resistin, reportedly higher in females (children and adults), with negative age correlation, suggests developmental links (Li *et al.* 2009, Martos-Moreno *et al.* 2006, Stringer *et al.* 2009, Kelly *et al.* 2007). However, in a Spanish pediatric population, there was no gender or pubertal difference. In contrast to adults, obese and non-obese children have reportedly similar resistin and in First Nations youth, obese, non-obese and T2D had equivalent resistin. These findings involved small groups (< 350), limiting potential subgroup analysis. In analysis of a large cohort of children/adolescents (~3500, aged 6-18 yr), resistin was higher in girls vs. boys, with a decreased trend with age. Increased resistin associated with central obesity (boys and girls), adiposity (only girls), and percent fat mass, BMI and lipid parameters, but only weakly with InsRes and MetS (Li *et al.* 2009), suggesting that resistin is not strongly implicated in these pathologies.

Resistin decreased with long-term exercise training in overweight adolescents in concert with decreased body fat and triglycerides. However, without weight loss, exercise training did not improve adipokines (resistin, adiponectin, leptin and TNFα).

ASP in Children/Adolescents

ASP produced via C3-B-adipsin interaction increases triglyceride storage and adipocyte differentiation. In adults, ASP increases with obesity, T2D, cardiovascular disease and nephrotic syndrome, but there is little data in children (Wamba *et al.* 2008, Cianflone *et al.* 2005). A large cohort analysis (~1600, aged 6-18 yr) demonstrated increased ASP in overweight/obese vs. normal weight, while C3 showed little variation. This effect remained throughout early pubertal stages in boys and girls, although age and puberty had no effect. Increased ASP correlated with cholesterol, triglyceride, glucose, insulin and HOMA.

MetS was strongly associated with ASP; a single factor (hypertension, central obesity or hyperglycemia) correlated with increased ASP. Similarly, ASP-dyslipidemia correlations were identified in pediatric proteinuric renal disease, without obesity. While C3 was unchanged or decreased, ASP increased and correlated with triglycerides and apolipoproteinA1. In very young obese children (aged 2-6 yr), increased ASP without dyslipidemia suggests a predisposition to enhanced fat storage further driving the obesity profile.

TNFα in Children/Adolescents

Links between InsRes and inflammation were first proposed based on TNFα interference of insulin signaling. In adults, adipose TNFα expression correlates positively with BMI, body fat and hyperinsulinemia (Eyzaguirre and Mericq 2009, Korner *et al.* 2007, Maffeis *et al.* 2007). In children, some have found increased TNFα in obese children, others have not. By contrast, higher TNFα levels were present among non-obese vs. obese pre-pubertal girls, with no difference in boys. In obese children, correlation between TNFα and subcutaneous adipocyte diameter was stronger than with BMI or fat mass.

Paradoxically, peak oxygen consumption (fitness indicator) was also positively correlated with TNFα and negatively with adiponectin. Pro-inflammatory cytokines, including TNFα, increased with acute exercise in adolescent boys. In more obese children, fat mass may attenuate potentially positive effects of fitness on circulating levels of TNFα.

Visfatin in Children/Adolescents

Visfatin (aka PBEF) is increased in obesity, T2D, MetS, and beta-cell deterioration, although that is not confirmed in all studies. Immune interactions abound with associations in sepsis patients, lung injury, and unstable coronary heart disease. In children, visfatin consistently correlates with age, BMI, weight, waist circumference, insulin and homeostatic model assessment of insulin resistance (HOMA) but not lipid profiles (Araki *et al.* 2008, Dedoussis *et al.* 2009). In surgical samples (aged 7-14 yr), adipose and immune cell visfatin expression correlated with BMI and TNFα (especially in immune cells); plasma visfatin correlated with visceral and subcutaneous adipose mass and plasma leptin.

RBP4 in Children/Adolescents

In adults, RBP4 is increased in obesity, InsRes, MetS and T2D, although not confirmed in other studies. Recent interest in RBP4 as a potential InsRes promoter has resulted in studies in children/adolescents (Santoro *et al.* 2009, Kanaka-Gantenbein *et al.* 2008). Both pre- and post-puberty RBP4 increases in obesity. RBP4 correlates with BMI, HOMA and, in some cases, percent fat mass and triglycerides. There are studies supporting and contradicting increased RBP4 with puberty. By contrast, a decrease of RBP4 in obesity with inverse correlation to BMI and no correlation to HOMA was demonstrated in pre- and post-puberty children, although the expected increase in leptin and decrease in adiponectin were seen.

Over time, 3 yr RBP4 monitoring in post-puberty overweight black adolescents correlated with InsRes, body size and triglycerides. In a one-year weight-loss study (median age 11 yr), decreased weight was associated with decreased RBP4, blood pressure, triglycerides and HOMA; the change in RBP4 correlated with both HOMA and weight status.

Finally, it has been suggested that the RBP4/serum retinol (SR) ratio may be a better biomarker than RBP alone. RBP4/SR is increased in obesity, and decreased with weight loss. Further, the RBP4/SR ratio correlated with BMI and percent body fat. Independent of adiposity, it correlated with triglycerides, insulin and more strongly with MetS indices than RBP4 alone.

OBESITY-ASSOCIATED DISEASES

Liver Disease in Children/Adolescents

Nonalcoholic fatty liver disease (NAFLD) is characterized by visceral obesity, InsRes and MetS. Increasing evidence implicates adipokines in pathogenesis/progression of NAFLD/fibrosis with increased resistin and TNFα, decreased adiponectin, but unchanged leptin in adults (Table 2). In obese children, neither leptin nor resistin was associated with NAFLD. By contrast, adiponectin was decreased, correlating inversely with hepatic fat, BMI and NAFLD presence/severity. Surprisingly, RBP4 correlated inversely with liver damage, necroinflammatory activity, NAFLD score, fibrosis and disease severity in NAFLD children. In hepatitis-C-infected children, RBP4 inversely correlated with liver damage and HOMA, discrepant with RBP4-InsRes association in obesity (Tsochatzis *et al.* 2009, Nobili *et al.* 2009, Zou *et al.* 2005, Pacifico *et al.* 2008).

Table 2   Presented adipokines in obesity-associated disorders in children

FAT-ASSOCIATED DISEASES	ADIPOKINES
InsRes	leptin, adiponectin, resistin, ASP, visfatin, RBP4
MetS	leptin, adiponectin, resistin, ASP, visfatin, RBP4
T2D	leptin, adiponectin
Liver steatosis/ NAFLD	leptin, adiponectin, resistin, RBP4
Hyperlipidemia	leptin, adiponectin, resistin, ASP, visfatin, RBP4
Atherosclerosis	leptin, adiponectin, resistin

ASP, acylation-stimulating protein; MetS, metabolic syndrome; NAFLD, non-alcoholic fatty liver disease; RBP4, retinol-binding protein 4; T2D, type 2 diabetes.

Vascular Disease/Inflammation in Children/Adolescents

Tarquini comprehensively summarized anti-inflammatory/anti-atherosclerotic roles of adiponectin in atherosclerosis, hypertension, and related heart diseases in adults. In healthy non-obese pubertal adolescents, decreased adiponectin was associated with InsRes, with no changes in endothelial-dependent vasodilation or arterial distensibility. However, in obese children, carotid intima-media thickness (IMT), an early marker of atherosclerosis, correlated inversely with adiponectin and positively with resistin, HOMA, BMI and hypertension. Corrected for gender, Tanner stage and relative BMI, only adiponectin retained significant IMT correlation. In obese children with/without NAFLD, IMT also inversely correlated with adiponectin and positively correlated with leptin, BMI, and HOMA (Tarquini *et al.* 2007, Pacifico *et al.* 2008, Beauloye *et al.* 2007).

IMMUNE-RELATED DISORDERS IN CHILDREN/ADOLESCENTS

Blood Diseases

Cure rate and life expectancy of children with cancer has increased significantly; acute lymphoblastic leukemia (ALL), the most common form, has the highest survival rate. Prevention and handling of treatment effects have gained considerable importance. Endocrine and metabolic complications, including osteoporosis and overweight/obesity, are reported in childhood cancer survivors. ALL patients re-examined 20 yr post-treatment indicated no obesity, but one-third were overweight. Differences, primarily in males, were increased percent fat mass, triglycerides and LDL cholesterol, but no change in leptin (normalized to kg fat). In Hodgkin's disease and ALL children, 7 yr post-treatment, no influence of treatment on leptin or BMI was detected, although a higher fat mass trend was observed (Table 3). At time of diagnosis in non-Hodgkin's lymphoma patients, there was no difference in leptin levels. Immediately post-treatment (leukemia and lymphoma patients), decreased leptin correlated with bone mineral density changes, although in multivariate analysis only age contributed. Overall, these studies do not suggest a role for leptin in post-treatment consequences (Petridou *et al.* 2009, Jarfelt *et al.* 2005).

Table 3 Major adipokines associated with changes in immune/inflammation-related disorders

IMMUNE-RELATED DISORDERS	ADIPOKINES
Blood disorders	leptin, adiponectin
Respiratory disorders	leptin, adiponectin, resistin
Type 1 diabetes	leptin, adiponectin
Celiac disease	leptin

Elevated adiponectin levels are independently associated with childhood non-Hodgkin's lymphoma and poor prognosis at time of diagnosis, suggesting that adiponectin may represent a diagnostic biomarker and may be implicated in its pathogenesis.

Respiratory Diseases

Adipokines may contribute to the obesity-asthma link, as evidenced by prospective studies in children in developed countries. In atopic-asthmatic, non-atopic-asthmatic and control children, resistin was decreased in atopic-asthmatics vs. other groups, significantly predicting asthma. While adiponectin and leptin did not differ, both correlated with pulmonary function, suggesting disease-modifying effects. Adiponectin, but not leptin, determined risk of wheezing in early childhood, when measured in cord blood of mothers with atopy history. In pediatric allergic rhinitis, increased resistin correlated with disease severity, while increased leptin and decreased adiponectin were significant predictive factors, correlating with nasal symptom scores (Kim *et al.* 2008, Matricardi *et al.* 2007).

Pancreas-Intestinal Diseases in Children/Adolescents

In children newly diagnosed with celiac disease, with intestinal/immune dysfunction, leptin is decreased. Following a one-year gluten-free diet, leptin increased three-fold, significantly higher vs. pre-treatment and controls, suggesting beneficial effects of leptin on immune function (Ertekin *et al.* 2006).

The incidence of T1D and T2D is increasing worldwide, particularly in young children. Physicians are faced with differentiating between the two at initial diagnosis when clinical/laboratory parameters may overlap (Huerta *et al.* 2006). Although T2D children are usually overweight, obesity can also be present in T1D. Additional markers, such as C-peptide, do not necessarily differentiate early on. Disparate changes in adiponectin and leptin may help reflect the differential pathophysiology, providing diagnostic tools. As reviewed by Huerta, in adults, adiponectin was higher in T1D vs. control and T2D. Limited data exist in children: adiponectin was higher in pediatric T1D vs. control and T2D, although gender differences were not always accounted for. Data regarding leptin are contradictory: most report decreased leptin at T1D diagnosis vs. T2D and controls. It has been hypothesized that leptin, which promotes Th1-type immune response, may play a role in T1D. However, leptin was not higher in auto-antibody-positive relatives vs. gender and BMI matched auto-antibody-negative controls.

EARLY ENVIRONMENTAL INFLUENCES

Prenatal/early-life origins of obesity are supported by epidemiological human and animal studies worldwide (Jasik and Lustig 2008, Saleh *et al.* 2008, Smith *et al.* 2009, Savino and Liguori 2008, Muhlhausler and Smith 2009). Inappropriately high/low nutrition during crucial development associates with obesity/disease risk. Limited adult adipogenesis and slow adipocyte turnover support fetal/early post-natal periods as key windows in adipose development. Both low birth weight (small-for-gestational-age, SGA) and high birth weight relate to future fatness, cardiovascular disease, InsRes and MetS (Table 4, Fig. 1).

Table 4 Critical development stages in which environment, obesity and adipokines may influence development

DEVELOPMENT STAGES	CONSIDERATIONS
Intrauterine environment	Low birth weight; high birth weight; maternal influence; cord blood levels as biomarkers
Post-natal growth	Catch-up effect in children small for gestational age; effect of breast-feeding
Infants and young children	Adipokine changes in obesity prior to lipid and insulin/glucose changes; adipokines as biomarkers for future risk
Early adolescence	Obesity and early puberty; PCOS; breast cancer risk

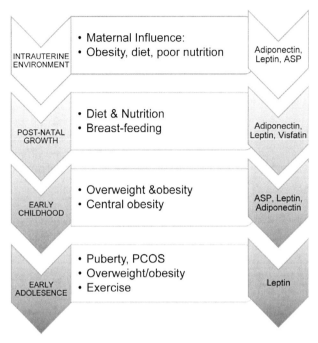

Fig. 1 Adipokine involvement in metabolic and related diseases and immune diseases in children and adolescents. Adipokines shown to be altered in the specific diseases are indicated. These changes may have potential use as biomarkers or may be implicated in pathological disturbances. INSR, insulin resistance; METS, metabolic syndrome; T2D, type 2 diabetes.

Adiponectin and leptin are produced by placental/fetal tissue; blood levels decrease over early months/years. Cord blood leptin and sOB-R correlated with birth weight and higher sOB-R may protect the fetus/infant from energy waste. In contrast to adults, cord blood adiponectin correlated positively with birth weight, reflecting developing adipose mass. Cord blood ASP was lower than maternal serum and correlated with fetal birth weight, maternal triglycerides and fatty acids. By contrast, cord blood RBP4 did not correlate with birth weight.

In SGA newborns, adiponectin was decreased vs. appropriate-for-gestational-age (AGA) newborns. In AGA, fat and lean mass were gender specific, although adiponectin and visfatin were not. In SGA, visfatin and adiponectin were gender specific (higher in girls), whereas body adiposity was not. With matching of SGA and AGA groups, leptin and visfatin were comparable, but adiponectin was still lower. By contrast, high birth weight and early weight gain, specifically associated with higher adiponectin, suggested protection of obese children and adolescents from central obesity.

Recently, a focus on breast feeding and childhood obesity indicates that adiponectin and leptin are in breast milk. Adiponectin declines throughout lactation due to prolactin inhibition. Leptin transfers from maternal circulation to breast milk to neonatal blood, potentially exerting biological effects on the infant. Milk leptin levels correlated with maternal leptin and with infant leptin, and was higher in exclusively breast-fed infants vs. formula-fed.

Longer-term effects of birth parameters are an important area of research. Lower cord blood leptin was associated with smaller birth weight but higher BMI at age 3. Cord blood adiponectin was predictive of central adiposity at age 3. Children/ adolescents from post-bariatric-surgery mothers had lower birth weight, three-fold lower prevalence of severe obesity, greater insulin sensitivity, improved lipid profile and lower leptin vs. children from pre-bariatric-surgery mothers. A positive correlation of ASP in children (6-18 yr) with mothers' BMI as well as birth weight was noted.

Early adolescence is another particularly high-risk time with synergy of metabolic and behavioral changes. A problem with early weight gain is precocious puberty, with potential influence on future polycystic ovary syndrome or breast cancer. Analysis of NHLBI data in girls aged 9-10 yr indicates the greatest BMI increases between thelarche and menarche, pointing to puberty as a high-risk time. Leptin, which impacts puberty timing and is permissive for normal reproductive function, may be of particular importance.

CONCLUSION

The perception of obesity and related diseases as chronic inflammatory conditions underscores metabolic and immune interactions. The cross-boundary functions of adipokines exemplify this and demonstrate implications in children for diseases, including obesity, obesity-related diseases, inflammation/immune-related disorders as well as life-long environmental influences (Fig. 2).

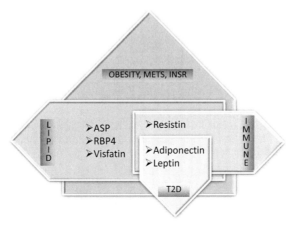

Fig. 2 Potential impact of adipokines on critical development periods from intrauterine growth to adolescence for adiposity and related environmental influences.

Summary Points

- Adipokines play multiple roles in metabolism and immune function.
- In adults, adipokine changes are associated with obesity, obesity-related diseases and inflammation/immune disorders.
- In children, alterations in leptin, adiponectin, ASP, resistin, TNFα, visfatin and RBP4 have been identified in relation to obesity, insulin resistance, metabolic syndrome and type 2 diabetes as well as liver and vascular disease.
- In children, changes in adipokines are associated with inflammation/immune disorders in blood, respiratory, pancreatic and intestinal disorders.
- Early environmental influences such as intrauterine, post-natal, early childhood and puberty impact future obesity and metabolic risk.

Abbreviations

AGA	:	appropriate-for-gestational age
ALL	:	acute lymphoblastic leukemia
ASP	:	acylation stimulating protein
BMI	:	body mass index
CRP	:	C-reactive protein
HOMA	:	homeostatic model assessment of insulin resistance
IL	:	interleukin
IMT	:	intima-media thickness
InsRes	:	insulin resistance
IRS	:	insulin receptor substrate
MetS	:	metabolic syndrome
NAFLD	:	non-alcoholic fatty liver disease
PBEF	:	pre-B-cell colony-enhancing factor
PMN	:	peripheral mononuclear
RBP4	:	retinol binding protein 4
SGA	:	small for gestational age
sOB-R	:	soluble leptin receptor
SR	:	serum retinol
T1D	:	type 1 diabetes
T2D	:	type 2 diabetes
TNFα	:	tumor necrosis factor α

Definition of Terms

Adipokine: Hormone secreted partly or exclusively by adipose tissue.

Atherosclerosis: Vascular dysfunction associated with endothelial damage, smooth muscle proliferation, macrophage infiltration, lipid accumulation, blood clot and calcification.

Hyperlipidemia: Increased circulating plasma lipids, measured as increased triglycerides, cholesterol or apolipoprotein B.

Inflammation: Condition associated with increased circulating levels of inflammatory factors such as cytokines and C-reactive protein.

InsRes: Metabolic state characterized by diminished insulin response, identified as increased fasting insulin with/without increased glucose or following an oral glucose tolerance test.

Intrauterine: Fetal environment affected by placental and maternal conditions.

Lipodystrophy: Disturbance in fat deposition associated with redistribution, partial or total loss of specific fat depots.

MetS: Syndrome in adults and children characterized by a combination of central obesity, increased triglycerides, decreased HDL or increased blood pressure.

Steatosis: Increased accumulation of fat (triglycerides) in the liver.

Acknowledgements

This study was supported by grants from CIHR/NSERC (to KC) and National Natural Science Foundation of China (#30600213 to HLL). K. Cianflone holds a Canada Research Chair in Adipose Tissue. We appreciate the work of Mélanie Cianflone in manuscript preparation, and critical reading by Jessica Smith.

References

Araki, S. and K. Dobashi, K. Kubo, R. Kawagoe, Y. Yamamoto, Y. Kawada, K. Asayama, and A. Shirahata. 2008. Plasma visfatin concentration as a surrogate marker for visceral fat accumulation in obese children. Obesity (Silver Spring) 16: 384-388.

Beauloye, V. and F. Zech, H.T. Tran, P. Clapuyt, M. Maes, and S.M. Brichard. 2007. Determinants of early atherosclerosis in obese children and adolescents. J. Clin. Endocrinol. Metab. 92: 3025-3032.

Cianflone, K. and H. Lu, J. Smith, W. Yu, and H. Wang. 2005. Adiponectin, acylation stimulating protein and complement C3 are altered in obesity in very young children. Clin. Endocrinol. (Oxf.) 62: 567-572.

Dedoussis, G.V. and A. Kapiri, A. Samara, D. Dimitriadis, D. Lambert, M. Pfister, G. Siest, and S. Visvikis-Siest. 2009. Visfatin: the link between inflammation and childhood obesity. Diabetes Care 32: e71.

Ertekin, V. and Z. Orbak, M.A. Selimoglu, and L. Yildiz. 2006. Serum leptin levels in childhood celiac disease. J. Clin. Gastroenterol. 40: 906-909.

Eyzaguirre, F. and V. Mericq V. 2009. Insulin resistance markers in children. Horm. Res. 71: 65-74.

Huerta, M.G. 2006. Adiponectin and leptin: potential tools in the differential diagnosis of pediatric diabetes? Rev. Endocr. Metab. Disord. 7: 187-196.

Jarfelt, M. and B. Lannering, I. Bosaeus, G. Johannsson, and R. Bjarnason. 2005. Body composition in young adult survivors of childhood acute lymphoblastic leukaemia. Eur. J. Endocrinol. 153: 81-89.

Jasik, C.B. and R.H. Lustig. 2008. Adolescent obesity and puberty: the "perfect storm". Ann. NY Acad. Sci. 1135: 265-279.

Kanaka-Gantenbein, C. and A. Margeli, P. Pervanidou, S. Sakka, G. Mastorakos, G.P. Chrousos, and I. Papassotiriou. 2008. Retinol-binding protein 4 and lipocalin-2 in childhood and adolescent obesity: when children are not just "small adults". Clin Chem. 54: 1176-1182.

Kelly, A.S. and J. Steinberger, T.P. Olson, and D.R. Dengel. 2007. In the absence of weight loss, exercise training does not improve adipokines or oxidative stress in overweight children. Metabolism 56: 1005-1009.

Kim, K.W. and Y.H. Shin, K.E. Lee, E.S. Kim, M.H. Sohn, and K.E. Kim. 2008. Relationship between adipokines and manifestations of childhood asthma. Pediatr. Allergy Immunol. 19: 535-540.

Körner, A. and J. Kratzsch, R. Gausche, M. Schaab, S. Erbs, and W. Kiess. 2007. New predictors of the metabolic syndrome in children—role of adipocytokines. Pediatr. Res. 61: 640-645.

Li, M. and A. Fisette, X.Y. Zhao, J.Y. Deng, J. Mi, and K. Cianflone. 2009. Serum resistin correlates with central obesity but weakly with insulin resistance in Chinese children and adolescents. Int. J. Obes. (Lond). 33: 424-439.

Maffeis, C. and D. Silvagni, R. Bonadonna, A. Grezzani, C. Banzato, and L. Tatò. 2007. Fat cell size, insulin sensitivity, and inflammation in obese children. J. Pediatr. 151: 647-652.

Martos-Moreno, G.A. and V. Barrios, and J. Argente. 2006. Normative data for adiponectin, resistin, interleukin 6, and leptin/receptor ratio in a healthy Spanish pediatric population: relationship with sex steroids. Eur. J. Endocrinol. 155: 429-434.

Matricardi, P.M. and C. Grüber, U. Wahn, and S. Lau. 2007. The asthma-obesity link in childhood: open questions, complex evidence, a few answers only. Clin. Exp. Allergy 37: 476-484.

Muhlhausler, B. and S.R. Smith. 2009. Early-life origins of metabolic dysfunction: role of the adipocyte. Trends Endocrinol. Metab. 20: 51-57.

Nobili, V. and N. Alkhouri, A. Alisi, S. Ottino, R. Lopez, M. Manco, and A.E. Feldstein. 2009. Retinol-binding protein 4: a promising circulating marker of liver damage in pediatric nonalcoholic fatty liver disease. Clin. Gastroenterol. Hepatol. 7: 575-579.

Pacifico, L. and V. Cantisani, P. Ricci, J.F. Osborn, E. Schiavo, C. Anania, E. Ferrara, G. Dvisic, and C. Chiesa. 2008. Nonalcoholic fatty liver disease and carotid atherosclerosis in children. Pediatr. Res. 63: 423-427.

Petridou, E.T. and T.N. Sergentanis, N. Dessypris, I.T. Vlachantoni, S. Tseleni-Balafouta, A. Pourtsidis, M. Moschovi, S. Polychronopoulou, F. Athanasiadou-Piperopoulou, M.

Kalmanti, and C.S. Mantzoros. 2009. Serum adiponectin as a predictor of childhood non-Hodgkin's lymphoma: a nationwide case-control study. J. Clin. Oncol. 27: 5049-5055.

Saleh, J. and H.D. Al-Riyami, T.A. Chaudhary, and K. Cianflone. 2008. Cord blood ASP is predicted by maternal lipids and correlates with fetal birth weight. Obesity (Silver Spring) 16: 1193-1198.

Santoro, N. and L. Perrone, G. Cirillo, C. Brienza, A. Grandone, N. Cresta, and E. Miraglia del Giudice. 2009. Variations of retinol binding protein 4 levels are not associated with changes in insulin resistance during puberty. J. Endocrinol. Invest. 32: 411-414.

Savino, F. and S.A. Liguori. 2008. Update on breast milk hormones: leptin, ghrelin and adiponectin. Clin. Nutr. 27: 42-47.

Smith, J. and K. Cianflone, S. Biron, F.S. Hould, S. Lebel, S. Marceau, O. Lescelleur, L. Biertho, S. Simard, J.G. Kral, and P. Marceau. 2009. Effects of maternal surgical weight loss in mothers on intergenerational transmission of obesity. J. Clin. Endocrinol. Metab. 94: 4275-4283.

Stringer, D.M. and E.A. Sellers, L.L. Burr, and C.G. Taylor. 2009. Altered plasma adipokines and markers of oxidative stress suggest increased risk of cardiovascular disease in First Nation youth with obesity or type 2 diabetes mellitus. Pediatr. Diabetes 10: 269-277.

Tarquini, R. and C. Lazzeri, G. Laffi, and G.F. Gensini. 2007. Adiponectin and the cardiovascular system: from risk to disease. Intern. Emerg. Med. 2: 165-176.

Tsochatzis, E.A. and G.V. Papatheodoridis, and A.J. Archimandritis. 2009. Adipokines in nonalcoholic steatohepatitis: from pathogenesis to implications in diagnosis and therapy. Mediators Inflamm. Rev.

Wamba, P.C. and J. Mi, X.Y. Zhao, M.X. Zhang, Y. Wen, H. Cheng, D.Q. Hou, and K. Cianflone. 2008. Acylation stimulating protein but not complement C3 associates with metabolic syndrome components in Chinese children and adolescents. Eur. J. Endocrinol. 159: 781-790.

Zou, C.C. and L. Liang, F. Hong, J.F. Fu, and Z.Y. Zhao. 2005. Serum adiponectin, resistin levels and non-alcoholic fatty liver disease in obese children. Endocr. J. 52: 519-524.

Section IV
Specific Adipokines in Disease

CHAPTER 30

Role of Leptin in Immune Cell Development and Innate Immunity

Pamela J. Fraker[1,*] and Afia Naaz[1]

[1]Department of Biochemistry and Molecular Biology, 419 Biochemistry Building, Michigan State University, East Lansing, MI 48824-1319, USA

ABSTRACT

Leptin is an adipokine with structural and signaling similarity to the helical cytokines and therefore plays an essential role in the modulation of hematopoiesis and innate immune cells. Obese individuals or mice are characterized by leptin resistance, on the contrary; conditions such as starvation and malnutrition are characterized by decreased leptin. Both of these conditions create a defective immune system and increased susceptibility to infection. The leptin-deficient ob/ob mouse and the mouse lacking functional leptin receptor, the db/db mouse, are used as effective models to study these effects of leptin on the immune system. There is thymic atrophy and a decrease in the bone marrow cell population in the ob/ob mouse. All circulating and tissue-resident innate immune cells such as the monocytes, macrophages, neutrophils, dendritic cells and natural killer cells express the leptin receptor and are able to respond to leptin and transduce leptin signal. Indeed, obesity or lack of leptin disrupts signaling though the leptin receptor and its downstream JAK/STAT pathway. Leptin is either essential or enhances the functional capability of innate immune cells. It induces innate immune cells to produce pro-inflammatory cytokines and promotes phagocytosis, chemotaxis, cytotoxicity and effective antigen presentation, in both *in vivo* and *in vitro* conditions. The decreased functional capacity of the innate immune cells in the absence of leptin could account for the increased susceptibility to infection in the obese and malnourished. This review discusses the various roles of leptin in

---

*Corresponding author

hematopoiesis and its ability to modify the functional capacity of innate immune cells and their development in the marrow.

INTRODUCTION

Leptin is a peptide hormone, produced predominantly by white adipose tissue and to a lesser extent by the placenta, fetal tissues, brown adipose tissue, and intestines. It is designated as an adipokine, based on its structural similarity to helical cytokines such as interleukin 6 (IL-6) and granulocyte-macrophage colony-stimulating factor. Various recent studies have shown that, in addition to being an important player in energy metabolism and appetite suppression, leptin is also a major player in the modulation of immune function (La Cava and Matarese 2004). Here we review the role of leptin in the development of immune cells and the function of the innate immune system (Fig. 1 summarizes the sites of leptin production and immune activity).

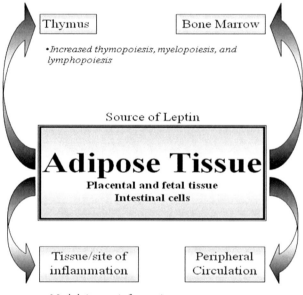

Fig. 1 Sites of leptin production and immune action. Leptin is produced primarily from white adipose tissue and to a lesser extent by brown fat, placental and fetal tissue, and intestinal cells. Leptin's sites of immune activity are (a) the thymus and bone marrow, where it promotes hematopoiesis, and (b) immune cells in the peripheral circulation, and sites of injury and infection, where it modulates the production of pro-inflammatory cytokines, chemotaxis, phagocytosis and antigen presentation.

Human conditions of energy imbalance are also associated with changes in leptin levels. Obesity is associated with high circulating leptin and peripheral leptin resistance; conversely, malnourished or starved individuals have low circulating leptin, a decrease in immune response and increased susceptibility to infections (Fig. 2). The potential therapeutic use of leptin depends on our better understanding of the contribution of leptin to the immune aspect of these conditions.

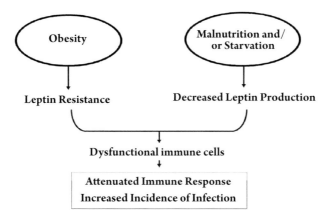

Fig. 2 Immune effects of leptin dysfunction in human conditions. Common human conditions of leptin insufficiency are malnutrition or starvation and obesity. In malnutrition and starvation there is decreased circulating leptin and in obesity there is increase in circulating leptin, but leptin resistance at the cellular level. In both conditions individuals are prone to a decreased immune function and susceptibility to infection.

Our existing knowledge of leptin's activity can be attributed to the availability of rodent models lacking leptin function, i.e., the ob/ob mouse lacking functional leptin, the db/db mouse lacking a functional leptin receptor and the fa/fa rat deficient in cellular leptin receptors. In addition to transgenic models, conditions such as malnutrition, starvation and diet-induced obesity in mice also provide good models of human relevance to ascertain the impact of leptin activities on the immune system.

LEPTIN STRUCTURE AND SIGNALING PATHWAY

Leptin is structurally similar to the long chain helical cytokines such as IL6 and the leptin receptor belongs to the class I cytokine receptor family. There are six splice variants of the leptin receptor, designated as OBR-a, -b, -c, -d, -e, and -f. The long form of the receptor (OBR-b) is the known functional form and has a transmembrane region. The other splice variants lack the transmembrane region.

The most common short form of the receptor is OBR-a. The main function of the short forms of the receptor could be the transport and degradation of leptin, rather than transducing signal (La Cava and Matarese 2004). Both OBR-b and OBR-a are present in most immune and non-immune cells (Table 1) and are up-regulated in response to leptin. Leptin acts by binding to its functional receptor. The leptin/receptor complex recruits JAK2 and activates the JAK/STAT pathway. Downstream of JAK2 recruitment, the AKT/PI3K and ERK1/2 pathways may also be activated; these pathways are involved in various cellular functions such as cell growth, differentiation and migration that are an important part of development and functional role of immune cells (Shuai and Liu 2003). Absence of leptin or leptin receptor in the ob/ob and db/db mouse or the presence of leptin resistance as in obesity is associated with the disruption of JAK/STAT and AKT/PI3K pathway. This could explain why these conditions of leptin insufficiency are also associated with a deficiency of immune function (La Cava and Matarese 2004).

Table I Presence of leptin receptor in innate immune cells and impact of leptin deficiency on their function

Cells of innate immunity	Presence of leptin receptor		Changes in innate immune function in ob/ob and/ or db/db mice compared to WT
	OBR-b	OBR-a	
Neutrophils	No	Yes	↓ in phagocytosis, chemotaxis, superoxide production
Monocytes	Yes	Yes	↓ in activation marker, chemotaxis
Macrophages	Yes	Yes	↓ in phagocytosis, activation markers, chemotaxis, oxidative burst
Dendritic cells	Yes	Yes	↓ in inflammatory cytokine and ↑ in anti-inflammatory cytokine production, ↓ in maturation, ↓ proliferation of allogenic CD4 + cells
NK cells	Yes	Yes	↑ in cell number but decrease in function

DEPENDENCY OF LYMPHOPOIESIS, THYMOPOIESIS, AND MYELOPOIESIS ON LEPTIN

The bone marrow is one of the largest tissues of the body. The marrow of the legs, arms, fingers, ribs, skull, etc. produce billions of new cells each day. In spite of its size and high activity rate, the role of the marrow in metabolism and disease is often overlooked. How the marrow responds to neuroendocrine changes has also been neglected. Thus the impact of leptin on hematopoiesis and other marrow functions remains to be better defined. What is known to date regarding the impact of leptin on lymphopoiesis (the production of lymphocytes), myelopoiesis (the production of monocytes/macrophages and granulocytes, especially neutrophils), and thymopoiesis is fascinating and will be the focus of this section.

The marrow is endlessly complex. The hematopoietic stem cells present in the marrow are a part of a percentage of the total population giving rise to all the precursors of B cells, very early T cell lineages, myeloids, erythrocytes, etc. needed to replenish the blood. Each early progenitor requires a plethora of growth factors in order to differentiate into mature lineages. Some of these factors are provided by so-called stromal cells. Likewise, osteoblasts needed for development of the bone are present in the marrow along with osteoclasts that promote bone resorption. Perhaps more amazing is that adipocyte precursors also develop in the marrow from so-called mesenchymal stem cells. In general, the marrow contains substantial numbers of adipocytes that increase substantially in the obese. Indeed, one question that remains to be answered is, how much leptin is being produced within the marrow by adipocytes and what impact does it have on lymphopoiesis and myelopoiesis, especially in the obese?

Leptin clearly enhances lymphopoiesis and myelopoiesis. Perhaps the best evidence that leptin supports lymphopoiesis and may indeed be an essential factor in the development of lymphocytes in the marrow and thymus came from studies of the ob/ob mouse that can not produce functional leptin. When WT controls were compared to the ob/ob mutant, it was found that the marrow of the leptin-deficient mice contained 40% fewer cells in their femurs and tibias. Given the size of the marrow, this represents a huge loss in cellularity that suggests a dependency on leptin. Flow cytometric analysis of the major hematopoietic lineages of the marrow noted a remarkable loss of 70% of the developing B cells with only 20% of the normal number of pre-B cells remaining in the ob/ob mouse. Obviously, lymphopoiesis in the marrow had been altered. Myelopoiesis was also impaired with a 25% decline in granulocyte lineages and a 40% decline in monocytic cells. Both WT and ob/ob mice were injected with recombinant leptin for 7 to 12 d. By day 12 myeloid lineages were nearly normal both in proportion and absolute numbers in the ob/ob mice. The B cell compartment doubled in size by day 7, being 75% of normal by day 12 (Claycombe *et al.* 2008).

Analogies to the marrow were noted by several labs examining changes in the thymus in the ob/ob mouse (Hick *et al.* 2006). The thymus is a primary immune tissue where T cells mature. The CD4+CD8+ immature thymocytes changed from 76% of the cells of the thymus in control mice to 49% in the ob/ob mouse. This loss in pre-T cells is analogous to the loss of pre-B cells observed in the marrow (Claycombe *et al.* 2008). Leptin supplementation also substantially increased thymopoiesis and the proportion of pre-T cells (Hick *et al.* 2006). As was the case for the marrow, provision of additional leptin did not increase thymopoiesis in normal strains of mice (Hick *et al.* 2006). Collectively, these experiments strongly support a role for leptin in lymphopoiesis and myelopoiesis.

LEPTIN SIGNALING IN THE CELLS OF INNATE IMMUNITY

The initial immune response to infection, trauma or wound healing involves the cells of innate immunity that produce cytokines (monocytes, macrophages, dendritic cells), migrate to site of infection (neutophils, monocytes), engulf and kill microbes (neutrophils, macrophages, dendritic cells) and present antigens to cells of adaptive immune system, to convert them from naïve cells to responsive cells. Monocytes, macrophages, mature and immature dendritic cells and natural killer cells express both OBR-a and OBR-b form of the receptor (Table1). The level of the non-signal tranducing OBR-a receptor expression is higher than OBR-b, i.e., the functional receptor in circulating monocytes and dendritic cells (Tsiotra *et al.* 2000). In response to leptin these cells activate the JAK/STAT pathway; loss of leptin signaling causes a decrease in STAT phosphorylation and interruption in the signaling cascade. Polymorphonuclear cells such as neutrophils and eosinophils have been shown to express only OBR-a, which is up-regulated in response to leptin (Zarkesh-Esfahani *et al.* 2001). Zarkesh-Esfahani *et al.* showed that the leptin activated neutrophils in whole blood but had no effect on purified neutrophils; they also showed that the TNF-alpha produced by the leptin activated monocytes is responsible for indirectly activating neutrophils (Zarkesh-Esfahani *et al.* 2004).

The presence of the leptin receptor and the ability of the innate immune cells to respond to leptin indicates that leptin might have an impact on the functional capabilities of these cells. The following sections describe the role of leptin on the functions of innate immunity (Fig. 3 summarizes the effect of leptin on the functions of innate immune cells).

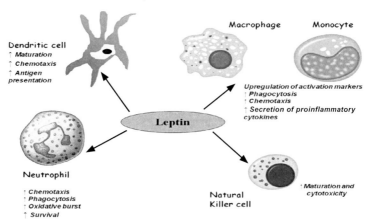

Fig. 3 Leptin modulates the functions of innate immune cells. Leptin acts as a pro-inflammatory cytokine on innate immune cells such as neutrophils, monocytes, macrophages, dendritic cells and natural killer cells. Leptin increases chemotaxis, phagocytosis in neutrophils; increases maturation, chemotaxis, cytokine production and phagocytosis in monocytes and macrophages; increases maturation and antigen presentation in dendritic cells; and increases the maturation and cytotoxicity of natural killer cells.

FUNCTIONAL ASPECT OF INNATE IMMUNITY

Cytokine Production

Cytokines are produced by immune cells in response to pathogens or pathogen-associated molecules such as LPS and act on other immune cells to produce an immune response. Pretreatment of leptin or leptin stimulation increased the production of cytokines in response to bacterial toxins such as LPS. Leptin alone did not increase cytokine production in peritoneal macrophages, neutrophils or dendritic cells, but leptin stimulation or pretreatment of LPS-treated cells increased the production of cytokines such as IL6, TNF-alpha, IL1beta, IL10 and IL12 (Loffreda *et al.* 1998, Mattioli *et al.* 2005). Cytokine production in response to LPS from peritoneal macrophages and dendritic cells was lower in ob/ob mice and fa/fa rats than in WT mice. db/db mice produced low levels of circulating pro-inflammatory cytokines such as IL-12 and TNF-alpha while producing more of the anti-inflammatory cytokine such as IL-10 (Lam *et al.* 2007). These studies show that leptin might have a priming or additive effect on cytokine production in response to pathogenic molecules and elicit a pro-inflammatory function from immune cells.

Table 2 Key features of immune cell function

- Innate immune cells are involved in the initial response to infection or trauma.
- They are involved in cytokine and chemokine production in response to infections.
- They migrate to the site of infection or inflammation through a process called chemotaxis.
- They engulf and kill pathogens by a process called phagocytosis.
- They process antigens and present them to adaptive immune cells such as T cells, in a process known as antigen presentation, which leads to proliferation and priming of the adaptive immune cells.
- Innate immune cells such as NK cells are also involved in cytotoxicity, a process by which they secrete proteins such as perforin and granzyme that can kill virus-infected cells and tumors.

Chemotaxis

Chemotaxis is the process by which cells of innate immunity are attracted and migrate to the site of infection and inflammation; this involves the secretion of chemokines/chemoattractants such as monocyte chemotactic protein 1 (MCP-1), RANTES and monocyte inflammatory protein 1 alpha that induce morphological changes and mobilize immune cells such as monocytes and neutrophils to these sites.

Leptin induces chemotactic migration in monocytic cell lines and mouse peritoneal macrophages; this process is blocked by inhibitors of leptin signaling. Leptin's chemotactic properties are weaker and not additive for MCP-1, which is a major monocyte chemoattractant. Leptin also causes an intracellular Ca++ flux, which stimulates motility towards chemoattractants in both monocytes (Gruen *et al.* 2007) and neutrophils (Montecucco *et al.* 2006).

Ob/ob mice and db/db mice treated with LPS showed a decreased recruitment of neutrophils to the brain when compared to WT; this recruitment became more robust after treatment of leptin in ob/ob mice. The decreased neutrophil recruitment is attributed to decreased expression of intercellular adhesion molecule 1 (ICAM-1) in response to LPS treatment in the brain of mice lacking leptin signaling; the expression of ICAM-1 was comparable to that of WT in ob/ob mice treated with leptin (Rummel *et al.* 2009). Conversely, another study using WT and db/db mice and a *Staphylococcus aureus* hind paw infection as a model showed a more robust neutrophil infiltration to the site of infection in db/db mice than in WT mice, but while the WT infection cleared in 10 d, the bacterial load was still high in the db/db. The *in vivo* studies are in agreement with the results from the *in vitro* study, indicating that leptin has chemotactic properties, but of lower strength than chemokines such as MCP-1 (Park *et al.* 2009). Leptin might not be essential for adequate chemotaxis in all situations.

Phagocytosis

Phagocytosis is a process by which opsonized bacteria or foreign bodies are engulfed and destroyed by phagocytes such as neutrophils and macrophages. A group from the University of Michigan conducted some valuable studies in ob/ob mice. They showed that phagocytosis and removal of *Klebsiella pneumoniae* by circulating neutrophils was significantly decreased in the absence of leptin in ob/ob mice compared to WT mice, which could be overcome with leptin replacement. The deficiency in phagocytic function in the ob/ob was due to the decreased expression of CD11b, a part of complement receptor 3 essential for the recognition of opsonized bacteria (Moore *et al.* 2003). Another study by Loffreda *et al.* showed that fewer peritoneal macrophages from ob/ob and db/db showed phagocytic activity in the presence of *Candida parasilopsis* compared to WT (Loffreda *et al.* 1998). Phagocytosis increased in both ob/ob and WT, but not in the macrophages of db/db mice when treated with leptin, suggesting that presence of leptin signaling as well as higher concentration of leptin could promote and increase phagocytosis.

Neutrophils also show a decreased respiratory burst in response to *Staphylococcus aureus* and *Staphylococcus pneumoniae* challenge in ob/ob and db/db compared to WT mice (Hsu *et al.* 2007).

Collectively these studies indicate that leptin facilitates phagocytosis and clearance of pathogens.

Antigen Presentation

Antigen-presenting cells such as dendritic cells engulf, process and present antigen attached to MHC class II receptor and co-stimulatory proteins. These stimulatory and co-stimulatory signals communicate with the T cell receptor of the T-helper cells and facilitate their proliferation and priming, which in turn can activate or regulate adaptive immune cells.

Lam *et al.* showed that leptin facilitates maturation of bone marrow–derived dendritic cells, which in turn makes them more capable of presenting antigens and increasing T cell proliferation in co-culture (Lam *et al.* 2007). Mattioli *et al.* conducted another dendritic cell study using human naïve T cells and heterologous dendritic cells pretreated with leptin, LPS or LPS + leptin. There was increased proliferation in the T cells in contact with dendritic cells treated with leptin over control and leptin + LPS over LPS (Mattioli *et al.* 2005). Dendritic cells from ob/ob mice are less potent in stimulating T cells *in vitro* compared to WT (Macia *et al.* 2006), indicating leptin is an essential component in the maturation and antigen-presentation capacity of antigen-presenting cells.

Cytotoxic Activity

Cytotoxicity is an innate immune function by which cells, i.e., natural killer cells, secrete proteins such as perforin and granzyme that can kill virus-infected cells and tumors. Obese individuals have decreased circulating NK cells and a decreased ability to destroy tumor cells *in vitro* compared to lean subjects. In db/db mice, the NK portion of mononuclear cells was significantly decreased in the circulation, spleen, lung, and liver compared to WT mice. NK cell activation and cytotoxity were also retarded in db/db mice. Leptin treatment dose-dependently increases the ability of mouse NK cells to kill tumor cells *in vitro* (Elinav *et al.* 2006). In another study, decreased functional capacity of NK cells in obese rats is reversed by adoptive transfer into lean rats, with an improvement in downstream leptin signaling (Lautenbach *et al.* 2009). These studies indicate that leptin may have a functionally significant role in NK cell development and cytotoxicity. Leptin resistance in the NK cells (Nave *et al.* 2008) might explain the proclivity of the obese to develop tumors (Calle *et al.* 2003).

Apoptosis and Cell Survival

Leptin is an anti-apoptotic factor and promotes cell survival in immune cells. The pathways involved in leptin's anti-apoptotic activity have been studied in peripheral immune cells such as monocytes (Najib and Sanchez-Margalet 2002), dendritic cells (Mattioli *et al.* 2005), neutrophils (Bruno *et al.* 2005) and eosinophils (Conus *et al.* 2005). A study by Najib *et al.* showed that leptin decreases apoptosis in peripheral monocytes through MAPKK (Najib and Sanchez-Margalet 2002). Another study with circulating neutrophils showed that leptin decreased apoptosis by inhibition of caspase-3 through the activation of MAPK as well as PI3K pathways of leptin signaling. Leptin also increased dendritic cell survival by activating the

NF-κb pathway after spontaneous and UV-induced apoptosis (Mattioli *et al.* 2009). Overall leptin has an important role in promoting innate immune cell survival, by its ability to activate multiple anti-apoptotic pathways.

CONDITIONS RELATED TO LEPTIN INSUFFICIENCY

Obesity

Some studies have shown a positive correlation with obesity and decreased immune function. Smith *et al.* showed that mice with diet-induced obesity have decreased immune response to influenza virus and increased mortality compared to control mice (Smith *et al.* 2007). Amar *et al.* 2007 showed a decreased immune response to Porphyromonas gingivalis infection associated with hypo-responsive macrophages in mice with diet-induced obesity (Amar *et al.* 2007). Although, very few human studies provide conclusive evidence on decreased immunity with obesity, some studies have shown that obese individuals have increased post-operative infections, increased infection and delayed recovery after burn injury, and increased susceptibility to some bacterial and viral agents (reviewed in Falagas and Kompoti 2006, Marti *et al.* 2001).

Obesity is characterized by increased leptin levels and leptin resistance. The central leptin resistance is also followed by leptin resistance at the cellular level of innate immune cells (Nave *et al.* 2008). This leptin resistance at the cellular level could decrease the functions of the innate immune cells explaining the increased susceptibility to infection in both human and mouse models of obesity.

Malnutrition

Malnutrition is notorious for causing thymic atrophy, losses of lymphocytes, and impaired cell and antibody-mediated responses (Fraker and King 2004). Acute forms of nutritional deficiencies initiate stress in the mice that induces increased levels of glucocorticoids, but also reduces serum leptin. In zinc-deficient adult mice, leptin declines 30% to 50% after a 30 d period of deficiency. The thymus is atrophied and lymphopoiesis declines in these mice (Fraker and King 2004). In another stress model, mice were starved for 2 d. Thymocytes were reduced by almost 90% in these mice (Howard *et al.* 1999). Injection of leptin during the starvation period provided substantial protection for thymopoiesis (Howard *et al.* 1999). Of course, malnutrition causes many metabolic and endocrine changes, so that discerning a specific role for leptin in the changes in myelopoiesis and lymphopoiesis would be difficult. Nevertheless, reduced levels of leptin undoubtedly contribute to the plethora of changes in immune function caused by suboptimal nutrition (Hick *et al.* 2006).

CONCLUSION

To sum up, leptin plays a significant role in both hematopoiesis and the innate immune system. Leptin could be absolutely essential or act as an enhancer. This opens doors for the use of leptin as an immunotherapy agent.

Summary Points

- Leptin is an adipokine with structural similarity to helical cytokines. Its receptor and signaling pathways are similar to type I cytokines.
- Leptin is an essential factor in the development of immune cells. The ob/ob mouse has thymic atrophy as well as a significant decrease in bone marrow immune cells and progenitors. Leptin treatment reverses this deficit.
- Cells of innate immunity such as neutrophils, monocytes, macrophages, dendritic cells, and natural killer cells express leptin receptor.
- Leptin has a stimulatory effect on innate immune functions such as chemotaxis, phagocytosis cytotoxicity and antigen presentation.
- Presence of leptin promotes the survival of immune cells.
- Leptin produces a pro-inflammatory effect inducing secretion of pro-inflammatory cytokines from innate immune cells. It also is capable of priming T helper cells to the Th1 phenotype.
- Conditions of leptin insufficiency are characterized by a deficient innate immune function and increased susceptibility to infection.

Abbreviations

CD	:	cluster of differentiation
IL	:	interleukin
JAK/STAT	:	Janus kinases/signal transducers and activators of transcription
LPS	:	lipopolysaccharide
MAPK	:	mitogen-activated protein kinase
NF-κB	:	nuclear factor-kappa B
NK cell	:	natural killer cell
OBR-a and OBR-b	:	leptin receptor a and leptin receptor b isoforms
PI3K	:	phosphoinositide 3-kinase
RANTES	:	Regulated upon activation, normal T cell expressed and secreted
TNF-alpha	:	tumor necrosis factor alpha
WT	:	wild type

References

Amar, S. and Q. Zhou, Y. Shaik-Dasthagirisaheb, and S. Leeman. 2007. Diet-induced obesity in mice causes changes in immune responses and bone loss manifested by bacterial challenge. Proc. Natl. Acad. Sci. USA 104: 20466-20471.

Bruno, A. and S. Conus, I. Schmid, and H.U. Simon. 2005. Apoptotic pathways are inhibited by leptin receptor activation in neutrophils. J. Immunol. 174: 8090-8096.

Calle, E.E. and C. Rodriguez, K. Walker-Thurmond, and M.J. Thun. 2003. Overweight, obesity, and mortality from cancer in a prospectively studied cohort of U.S. adults. N. Engl. J. Med. 348: 1625-1638.

Claycombe, K. and L.E. King, and P.J. Fraker. 2008. A role for leptin in sustaining lymphopoiesis and myelopoiesis. Proc. Natl. Acad. Sci. USA 105: 2017-2021.

Conus, S. and A. Bruno, and H.U. Simon. 2005. Leptin is an eosinophil survival factor. J. Allergy Clin. Immunol. 116: 1228-1234.

Elinav, E. and A. Abd-Elnabi, O. Pappo, I. Bernstein, A. Klein, D. Engelhardt, E. Rabbani, and Y. Ilan. 2006. Suppression of hepatocellular carcinoma growth in mice via leptin, is associated with inhibition of tumor cell growth and natural killer cell activation. J. Hepatol. 44: 529-536.

Falagas, M.E. and M. Kompoti. 2006. Obesity and infection. Lancet Infect. Dis. 6: 438-446.

Fraker, P.J. and L.E. King. 2004. Reprogramming of the immune system during zinc deficiency. Annu. Rev. Nutr. 24: 277-298.

Gruen, M.L. and M. Hao, D.W. Piston, and A.H. Hasty. 2007. Leptin requires canonical migratory signaling pathways for induction of monocyte and macrophage chemotaxis. Am. J. Physiol. Cell Physiol. 293: C1481-1488.

Hick, R.W. and A.L. Gruver, M.S. Ventevogel, B.F. Haynes, and G.D. Sempowski. 2006. Leptin selectively augments thymopoiesis in leptin deficiency and lipopolysaccharide-induced thymic atrophy. J. Immunol. 177: 169-176.

Howard, J.K. and G.M. Lord, G. Matarese, S. Vendetti, M.A. Ghatei, M.A. Ritter, R.I. Lechler, and S.R. Bloom. 1999. Leptin protects mice from starvation-induced lymphoid atrophy and increases thymic cellularity in ob/ob mice. J. Clin. Invest. 104: 1051-1059.

Hsu, A. and D.M. Aronoff, J. Phipps, D. Goel, and P. Mancuso. 2007. Leptin improves pulmonary bacterial clearance and survival in ob/ob mice during pneumococcal pneumonia. Clin. Exp. Immunol. 150: 332-339.

La Cava, A. and G. Matarese. 2004. The weight of leptin in immunity. Nat. Rev. Immunol. 4: 371-379.

Lam, Q.L. and B.J. Zheng, D.Y. Jin, X. Cao, and L. Lu. 2007. Leptin induces CD40 expression through the activation of Akt in murine dendritic cells. J. Biol. Chem. 282: 27587-27597.

Lautenbach, A. and C.D. Wrann, R. Jacobs, G. Muller, G. Brabant, and H. Nave. 2009. Altered phenotype of NK cells from obese rats can be normalized by transfer into lean animals. Obesity (Silver Spring) 17: 1848-1855.

Loffreda, S. and S.Q. Yang, H.Z. Lin, C.L. Karp, M.L. Brengman, D.J. Wang, A.S. Klein, G.B. Bulkley, C. Bao, P.W. Noble, M.D. Lane, and A.M. Diehl. 1998. Leptin regulates proinflammatory immune responses. Faseb J. 12: 57-65.

Macia, L. and M. Delacre, G. Abboud, T.S. Ouk, A. Delanoye, C. Verwaerde, P. Saule, and I. Wolowczuk. 2006. Impairment of dendritic cell functionality and steady-state number in obese mice. J. Immunol. 177: 5997-6006.

Marti, A. and A. Marcos, and J.A. Martinez. 2001. Obesity and immune function relationships. Obes. Rev. 2: 131-140.

Mattioli, B. and L. Giordani, M.G. Quaranta, and M. Viora. 2009. Leptin exerts an anti-apoptotic effect on human dendritic cells via the PI3K-Akt signaling pathway. FEBS Lett. 583: 1102-1106.

Mattioli, B. and E. Straface, M.G. Quaranta, L. Giordani, and M. Viora. 2005. Leptin promotes differentiation and survival of human dendritic cells and licenses them for Th1 priming. J. Immunol. 174: 6820-6828.

Montecucco, F. and G. Bianchi, P. Gnerre, M. Bertolotto, F. Dallegri, and L. Ottonello. 2006. Induction of neutrophil chemotaxis by leptin: crucial role for p38 and Src kinases. Ann. NY Acad. Sci. 1069: 463-471.

Moore, S.I. and G.B. Huffnagle, G.H. Chen, E.S. White, and P. Mancuso. 2003. Leptin modulates neutrophil phagocytosis of Klebsiella pneumoniae. Infect. Immun. 71: 4182-4185.

Najib, S. and V. Sanchez-Margalet. 2002. Human leptin promotes survival of human circulating blood monocytes prone to apoptosis by activation of p42/44 MAPK pathway. Cell Immunol. 220: 143-149.

Nave, H. and G. Mueller, B. Siegmund, R. Jacobs, T. Stroh, U. Schueler, M. Hopfe, P. Behrendt, T. Buchenauer, R. Pabst, and G. Brabant. 2008. Resistance of Janus kinase-2 dependent leptin signaling in natural killer (NK) cells: a novel mechanism of NK cell dysfunction in diet-induced obesity. Endocrinology 149: 3370-3378.

Park, S. and J. Rich, F. Hanses, and J.C. Lee. 2009. Defects in innate immunity predispose C57BL/6J-Leprdb/Leprdb mice to infection by Staphylococcus aureus. Infect. Immun. 77: 1008-1014.

Rummel, C. and W. Inoue, S. Poole, and G.N. Luheshi. 2009. Leptin regulates leukocyte recruitment into the brain following systemic LPS-induced inflammation. Mol. Psychiatry.

Shuai, K. and B. Liu. 2003. Regulation of JAK-STAT signalling in the immune system. Nat. Rev. Immunol. 3: 900-911.

Smith, A.G. and P.A. Sheridan, J.B. Harp, and M.A. Beck. 2007. Diet-induced obese mice have increased mortality and altered immune responses when infected with influenza virus. J. Nutr. 137: 1236-1243.

Tsiotra, P.C. and V. Pappa, S.A. Raptis, and C. Tsigos. 2000. Expression of the long and short leptin receptor isoforms in peripheral blood mononuclear cells: implications for leptin's actions. Metabolism 49: 1537-1541.

Zarkesh-Esfahani, H. and A.G. Pockley, Z. Wu, P.G. Hellewell, A.P. Weetman, and R.J. Ross. 2004. Leptin indirectly activates human neutrophils via induction of TNF-alpha. J. Immunol. 172: 1809-1814.

Zarkesh-Esfahani, H. and A.G. Pockley, R.A. Metcalfe, M. Bidlingmaier, Z. Wu, A. Ajami, A.P. Weetman, C.J. Strasburger, and R.J. Ross. 2001. High-dose leptin activates human leukocytes via receptor expression on monocytes. J. Immunol. 167: 4593-4599.

CHAPTER 31

Visfatin and Hypocaloric Diets

D.A. de Luis[1,*] **and R. Aller**[1]

[1]Center of Investigation of Endocrinology and Clinical Nutrition, Medicine School, Valladolid University, C/Los perales 16 (Urb Las Aceñas), Simancas 47130 Valladolid, Spain

ABSTRACT

Visfatin was originally identified as a 52 kD protein that is primarily expressed in bone marrow, muscle and liver. Visfatin has many pleiotropic actions in tissues other than adipose tissue. Several features of visfatin suggest that this molecule might be important for understanding the biological differences between subcutaneous and intra-abdominal adipose tissue, and their relation with metabolic syndrome and other related diseases such as gestational diabetes mellitus, non-alcoholic fatty liver disease and polycystic ovary syndrome. After a diet-induced weight reduction, a marked positive correlation between visfatin mRNA expression in subcutaneous adipose tissue and a negative correlation with body mass index were detected. Diet-induced weight reduction and an increase of visfatin expression in subcutaneous adipose tissue with a decrease in circulating visfatin. An interesting investigation showed that aerobic exercise is associated with this reduction in circulating visfatin in young type 2 diabetes patients, too. In another interventional study, it was demonstrated that weight reduction after a hypocaloric diet is associated with a significant decrease in circulating concentrations of visfatin in mildly obese subjects. These changes in visfatin levels, a hypothetical insulin-mimetic adipokine, could have relevant implications for treatment of obesity, polycystic ovary syndrome, diabetes mellitus and other manifestations of metabolic syndrome.

*Corresponding author

INTRODUCTION

The incidence of obesity and co-morbidities is dramatically increasing worldwide in both children and adults. This obese state is characterized by what has been called low-grade systemic inflammation. In fact, inflammatory markers, such as C-reactive protein and interleukin 6, are increased in obese individuals compared with lean subjects (de Luis *et al.* 2002). This systemic inflammation has been linked to the increased risk of development of cardiovascular disease and diabetes mellitus type 2. This evidence has led, in the last decade, to an increase of research of adipose tissue as an active organ. The current view of adipose tissue is that of an active secretory organ, sending out and responding signals that modulate energy expenditure, appetite, endocrine system, insulin sensitivity, reproductive system, bone metabolism and inflammation.

Adipokines are proteins produced mainly by adipocytes. Adipose tissue secretes a variety of these proteins such as leptin, adiponectin, resistin, adipsin and visfatin (Table 1).

Table I   Adipokines from fat tissue

Leptin
Adiponectin
Resistin
Visfatin
Plasminogen activator inhibitor 1
Interleukin 6
TNF-alpha
Angiotensinogen

VISFATIN AS A PLEIOTROPIC PROTEIN

Ten years after the identification of a novel protein secreted from lymphocytes (Samal *et al.* 1994), it was shown that this protein is also secreted from adipose tissue (Fukuhara *et al.* 2005). In the past years, several groups have studied the functions of pre-B cell colony-enhancing factor (PBEF). These studies reveal that PBEF has some interesting roles in several tissues. The discovery that PBEF, also termed visfatin, is highly expressed in visceral fat and that circulating levels correlate with obesity is of great interest to many researchers.

Visfatin was originally identified as a 52 kD protein that is primarily expressed in bone marrow, muscle and liver. Visfatin has many pleiotropic actions in tissues other than adipose tissue. For example, visfatin is also regulated in neutrophils by interleukin beta and functions as a novel inhibitor of apoptosis in response to inflammatory stimuli (Jia *et al.* 2004). Other studies showed that visfatin is constitutively expressed by fetal membranes during pregnancy and is expressed in the amniotic epithelium, mesenchymal cells and chorionic cytotrophoblast (Ognjanovic *et al.* 2002). Visfatin has also been shown to regulate nicotinamide adenine dinucleotide (NAD+)-dependent protein deacetylase activity and promote vascular smooth muscle cell maturation (Van der Veer *et al.* 2005).

VISFATIN AS AN ADIPOKINE

The original observation that visfatin levels were shown to be increased in visceral adipose tissue of mice and in human female participants compared with its levels in subcutaneous adipose tissue (Chen *et al.* 2005) has been supported by a study that demonstrated that visfatin levels are altered in patients with type 2 diabetes mellitus. A decrease in adiponectin and an increase in visfatin were observed in type 2 diabetic patients and the authors suggested that the increasing visfatin levels were associated with type 2 diabetes mellitus. Another study (Berndt *et al.* 2005) in which visceral fat mass was calculated from computed tomography scans indicated that there was no correlation between visceral fat mass and visfatin concentrations. However, there was a significant correlation between visceral visfatin gene expression and percent body fat, but no significant association between percent body fat or body mass index and subcutaneous visfatin mRNA expression. This study concluded that serum visfatin concentrations or visceral visfatin mRNA expression correlated with measurements of obesity but not with waist to hip ratio of visceral fat mass.

In another study (Tan *et al.* 2006), there was significant up-regulation of visfatin mRNA in both subcutaneous and visceral adipose tissue of women with polycystic ovary syndrome (PCOS), when compared to normal controls. The precise reason for the up-regulation of visfatin seen in PCOS, a pro-inflammatory state, is unknown.

The age of the population could be an important factor in these contradictory results. For example, in a population of obese adolescents, visfatin levels were negatively correlated with age and Tanner stage, and positively correlated with high-density cholesterol in this population (Jin *et al.* 2007).

Gestational diabetes mellitus is other controversial area of visfatin physiology. In one study, a decreased concentration of plasma visfatin was detected in patients with gestational diabetes mellitus (9.2 ng/ml vs. 12.6 ng/ml). This may indicate that the hormone plays a role in the pathogenesis of gestational diabetes mellitus (Chan *et al.* 2006). However, another study evaluated the role of visfatin in women with gestational diabetes mellitus. Visfatin concentrations were investigated longitudinally during pregnancy and after delivery. In the analysis, median visfatin levels were elevated in women with gestational diabetes mellitus (64 ng/ml vs. 46 ng/ml). According to the longitudinal analysis, visfatin increased during pregnancy and rose further after delivery (Krzyzanowska *et al.* 2006), and there is no explanation.

Visfatin is related with another disease called non-alcoholic fatty liver disease (NAFLD: the liver manifestation of metabolic syndrome). Our group demonstrated that probabilities of portal inflammation increased 1.11 (CI95%: 1.03-1.50) with each increment of 1 ng/ml of visfatin concentration. Portal inflammation frequencies were different between groups (low visfatin group 13.07 < ng/ml: 37.5% vs. high visfatin group 13.07 > ng/ml: 62.5%; p < 0.05). Perhaps visfatin plasma concentrations could predict the presence of portal inflammation in NAFLD patients (Aller *et al.* 2009).

Moreover, genetic variation in the visfatin gene may have an effect on visceral and subcutaneous visfatin mRNA expression profiles or serum visfatin concentrations. In a study, seven single nucleotide polymorphisms (SNPs) and one insertion/ deletion were identified. Three SNPS (rs9770242, -948G→T and rs4730153) that were representative of their linkage disequilibrium groups were genotyped in Caucasians from Germany. The ratio of visceral/subcutaneous visfatin mRNA expression was associated with all three polymorphisms (Böttcher *et al.* 2006).

In another study, the 11 exons and a promoter region of visfatin gene were screened for SNPs (Tokunaga *et al.* 2008). The -15355T/T genotype was associated with lower serum triglyceride and higher HDL cholesterol levels in non-diabetic subjects.

As we can see, several features of visfatin suggest that this molecule might be important for understanding the biological differences between subcutaneous and intra-abdominal adipose tissue (Table 2), and their relation with metabolic syndrome and other related diseases such as gestational diabetes mellitus, NAFLD and PCOS.

Table 2 Adipokine differences between subcutaneous and visceral adipose tissue

Visceral adipose tissue	Subcutaneous visceral tissue
Plasminogen activator inhibitor	Leptin
Adiponectin	
IL6	
Visfatin	
TNF-alpha	

First, visfatin has been detected (Tokunaga *et al.* 2008) in the plasma and its concentration correlates with intrabdominal fat mass but not with subcutaneous. Second, visfatin changes in plasma following diet, drug treatment and exercise (see next section). Although an endocrine role for visfatin has been suggested, it cannot be ruled out that visfatin might also have a paracrine effect on the visceral adipose tissue, facilitating the differentiation of the adipose tissue through the activation of glucose transport and lipogenesis.

MODULATION OF VISFATIN EXPRESSION BY DRUGS AND OTHER SITUATIONS

Some studies have investigated the regulation of visfatin by thiazolidinediones and other cytokines. The expression of some adipocytokines in visceral fat tissue of Otsuka Long-Evans Tokushima fatty (OLEFT) rats has been investigated (Choi *et al.* 2005). As we can expect, serum insulin and glucose concentrations decreased in rosiglitazone-treated OLEFT rats. In addition, rosiglitazone increased serum adiponectin concentration in these rats. The expression of both adiponectin and visfatin mRNA in visceral fat deposits was elevated by rosiglitazone. If the hypothesis is that visfatin can act as an insulin-mimetic, then it makes some sense that an insulin-sensitizing drug such as rosiglitazone might increase the expression levels of this hormone. It has also been shown that dexamethasone treatment of 3T3-L1 adipocytes increased visfatin mRNA, whereas TNF-alpha, isoproterenol and growth hormone down-regulated visfatin mRNA in a dose- and time-dependent manner (Kralisch *et al.* 2005a). However, insulin did not influence the synthesis of this hormone in this type of cell.

In culture adipocytes, TNF-alpha and IL6 (Kralisch *et al.* 2005b) resulted in a decrease of visfatin mRNA levels. Treatment with pioglitazone (30-45 mg/d) did not alter gene expression or circulating levels of visfatin in either diabetic or non-diabetic individuals (Hammarstedt *et al.* 2006). These last studies do not enhance our understanding of the modulation of visfatin expression by drugs and other situations.

MODULATION OF VISFATIN BY DIETS

As our understanding of the complex brain–adipose tissue axis expands, it becomes increasingly clear that although obesity can occur as a direct result of genetic mutations, the current obesity epidemic may largely be a consequence of modern lifestyle. From an evolutionary perspective, the adipose hormonal responses to short- and long-term energy deficits that act centrally to induce food intake were advantageous in environments of limited food availability. However, the current environment in industrialized nations generally provides unlimited access to food. Unfortunately, the few human studies that have been conducted to date examining the effects of diets on adipokines have contradictory results, and specific studies on visfatin effects are limited.

After a diet-induced weight reduction (Kovacikova *et al.* 2008), a marked positive correlation with visfatin mRNA expression in subcutaneous adipose tissue and a negative correlation with body mass index were detected. There was a clear positive relationship between the magnitude of weight loss and the increase of visfatin mRNA levels. Surprisingly, the response of visfatin levels to the diet was in the opposite direction than gene expression in subcutaneous adipose tissue. A hypothesis is that increased visfatin expression could be a compensatory response to decreased circulating levels. These results are partly in line with the results published on reduction of visfatin after bariatric surgery (Haider *et al.* 2006) or on increase of visfatin (Manco *et al.* 2007) in morbidly obese women after weight loss induced by gastroplasty.

In other protocol (Sun *et al.* 2007), 61 healthy young men were overfed for 7 d. Visfatin decreased 19% overall, -23% in lean, 9% in overweight, and 18% in obese subjects after this overfeeding protocol. This protocol resulted in significant increases in body weight in all subjects that was determined to be a combination of body fat and lean body mass. However, the reduction of visfatin in this study does not appear to be linked to the increase of body fat. Perhaps many other adipokines not addressed in the study may potentially influence the response of visfatin to a positive energy balance. Another potential confounding factor could be the distribution of macronutrients in the diet. For example, male Wistar rats fed on a standard diet or a high-fat diet were treated by oral administration of EPA. In this protocol (Perez Echarri *et al.* 2009), a significant decrease in white adipose tissue visfatin mRNA levels was found, which was reversed by EPA administration.

Glycemic control could be another confounding factor (Table 3). For example, visfatin concentrations reduced from 25 ng/ml at baseline to 20.3 ng/ml after 3 mo of intensive glycemic control, while HbA1c decreased from 9% to 6.2%. Changes of visfatin concentration may be a compensatory mechanism to ameliorate insulin deficiency, too (Zhu *et al.* 2008).

Table 3 Potential confounding fators in visfatin levels

Weight reduction
Bariatric surgery
Overfeeding
Other adipokines
Nutrients
Exercise
Glycemic control

An interesting investigation (Brema *et al.* 2008) showed that aerobic exercise is associated with this reduction in circulating visfatin in young type 2 diabetes patients. Surprisingly, this visfatin reduction was not accompanied with a reduction of fat mass or insulin resistance. These results should be interpreted with caution; further studies on larger samples and with a long-term exercise period will be needed.

Our group has published an interventional study with a hypocaloric diet (de Luis *et al.* 2008). After hypocaloric diet, body mass index, weight, fat mass, waist circumference, systolic blood pressure, glucose, total cholesterol and LDL cholesterol decreased (Tables 4 and 5). Seric visfatin concentrations decreased in a significant weight (112.14 ± 70.2 ng/ml vs. 99.4 ± 58.1 ng/ml: P < 0.05). Basal levels of visfatin were higher in men than in women and the change after weight

Table 4 Changes in anthropometric variables after a hypoclaoric diet (de Luis *et al.* 2008)

Characteristics	Baseline	3 mo
Body mass index	34.1±4.8	32.9±4.9 *
Weight (kg)	87.8±15.2	84.8±15 *
Fat free mass (kg)	47.7±14.9	47.1±13.3
Fat mass (kg)	36.2±11.9	35±12.6 *
Waist circumference	107.7±14.8	104.2±14.4 *
Waist to hip ratio	0.93±0.1	0.91±0.09
Systolic BP (mmHg)	131.7±14.3	124±12.8 *
Diastolic BP (mmHg)	79.1±12.8	82.6±14.8
RMR (kcal/day)	1693±417	1707±489
VO$_2$C (ml/min)	250±73.3	259±72

R, resting metabolic rate; O$_2$C, oxygen consumption; BP, blood pressure. Student's *t* test and Wilcoxon test were used as statistical methods.
*p<0.05, in each group with basal values.

Table 5 Classical cardiovascular risk factors after a hypocaloric diet (de Luis *et al.* 2008)

Characteristics	Baseline	3 mo
Glucose (mg/dl)	101.8±21	98.8±19 *
Total ch. (mg/dl)	214±45	204.5±37 *
LDL-ch. (mg/dl)	132±49	112±51 *
HDL-ch. (mg/dl)	54.1±13.8	53.4±13.1
TG (mg/dl)	125±55	121±70
Insulin (mUI/L)	13.5±7.8	12.6±9.2
HOMA	2.3±1.5	2.1±1.6
CRP (mg/dl)	5.8±7.1	6±7.2
Visfatin (ng/ml)	112.14±70.2	99.4±58.1 *

LDL-ch, low-density lipoprotein; HDL, high-density lipoprotein; CRP, C reactive protein; Ch., cholesterol; TG, triglycerides. Student's *t* test and Wilcoxon test were used as statistical methods.
*p<0.05, in each group with basal values.

loss was significant in women (105 ± 81 ng/ml vs. 90.9 ± 40.1 ng/ml: P < 0.05) and not significant in men (131.9 ± 91 ng/ml vs. 124.9 ± 89 ng/ml: ns). In the multivariate analysis with age- and sex-adjusted basal visfatin concentrations as a dependent variable, only age remained as an independent predictor in the model, with a decrease of 4.1 ng/ml (CI95%: 1.6-6.4) basal visfatin concentrations with each increase of 1 yr. In the second multivariate analysis with age- and sex-adjusted basal post-treatment visfatin concentrations as a dependent variable, only age remained as an independent predictor in the model, with a decrease of 3.7 ng/ml (CI95%: 1.2-6.1) post-treatment visfatin concentrations with each increase of 1 yr. In the third multivariate analysis with age- and sex-adjusted changes of visfatin as a dependent variable, only age remained as an independent predictor in the model, with a decrease of 3.8 ng/ml (CI95%: 0.3-7.7) visfatin change with each increase of 1 yr. The major finding of our study was that weight reduction after a hypocaloric diet is associated with a significant decrease in circulating concentrations of visfatin in mildly obese subjects.

In another of our studies (de Luis *et al.* 2009), 228 obese patients were divided into three groups by visfatin tertile values: (group I < 16.06 ng/ml), (group II 16.06 ng/ml to 60.55 ng/ml) and (group III > 60.55 ng/ml). Table 6 shows dietary intakes in these three groups. Patients in group III had lower intakes of calories, carbohydrate, total fat, monounsaturated fat, polyunsaturated fat, saturated fat, total cholesterol and proteins than group I. Patients in group II had lower intakes of calories, total fat, monounsaturated fat, saturated fat, total cholesterol and proteins than group I. In the multivariate analysis with age-, sex-, weight, fat mass, insulin and HOMA-adjusted *basal* visfatin concentration as a dependent variable, only monounsaturated fat intake remained as an independent predictor in the model ($F = 1.7$; $p < 0.05$), with an inverse correlation. Visfatin concentration decreased -3.71 ng/ml (CI95%: -0.45 to -6.96) for each gram of monounsaturated fat intakes. The relation of fatty acids with visfatin levels has been described. For example, polyunsaturated fatty acids stimulated AMP-activated protein kinase and increased visfatin secretion in cultured murine adipocytes (Lorente Cebrian *et al.* 2009). In a randomized, double-blind clinical trial, the presence of an infusion of free fatty acids antagonize rosiglitazone-induced visfatin release (Haider *et al.* 2006). Visfatin secretion is apparently influenced by oxidation products of free fatty acids since the presence of beta-hydroxibutirate, which blocks fatty acid oxidation, partly reversed the inhibitory effect of these acids on visfatin release by human adipocytes. This data suggests a potential nutritional sensor role for visfatin.

Table 6 Dietary intake of study population by tertile visfatin value (de Luis *et al.* 2009)

Parameters	Group I n=76	Group II n=76	Group III n=76	p
Energy (kcal/day)	2050.4±691.3	1747.8±810.2[#]	1618.1±443.1[*]	0.001
CH (g/day)	204.1±81.4	179.5±91.5	162.1±63.7[*]	0.011
Fat (g/day)	90.3±41.2	75.2±42.1[#]	69.2±22.3[*]	0.003
Protein (g/day)	101.3±27.1	84.1±28.1[#]	81.8±23.7[*]	0.001
S-fat (g/day)	25.4±16.1	22.2±16.7	18.9±8.2[*,$]	0.032
M-fat (g/day)	39.7±18.7	33.8±17.8[#]	32.9±10.5[*]	0.038
P-fat (g/day)	8.9±4.8	7.3±4.1	6.7±3.5[*]	0.007
Cholesterol (mg/day)	405.3±244.7	407.6±210	362.4±16[*]0	0.055
Fiber (g/day)	15.8±6.2	14.2±7.1	13.4±5.3	0.100
Duration of exercise (h/wk)	3.2±2.1	1.5±3.1[#]	0.7±1.7[*,$]	0.001

Visfatin levels: group I < 16.06 ng/ml, group II 16.06 to 60.55 ng/ml and group III > 60.55 ng/ml. CH, carbohydrate; S-fat, saturated fat; M-fat, monosaturated fat; P-fat, polyunsaturated fat.
p, p value in ANOVA test.
[*]Statistical difference between groups III and I.
[$]Statistical difference between groups III and II.
[#]Statistical difference between groups II and I.

It remains unknown whether this dietary factor influences circulating visfatin levels indirectly or directly. One hypothesis is that dietary factors are associated with a decrease in serum concentrations through decreased visfatin production. Alternatively, they may increase visfatin sensitivity, leading in turn to a subsequent decline in visfatin production through unknown feedback mechanisms. However, our study has several limitations. First, the cross-sectional design does not permit the assessment of causality. Second, we used a dietary questionnaire for dietary data collection. Third, residual confounding factors could remain unexplored in our multivariate analysis. Fourth, single measurement of serum visfatin concentration may represent short-term status only and introduce random errors.

These above-mentioned changes in visfatin levels, a hypothetical insulin-mimetic adipokine, could have implications for treatment of obesity, polycystic ovary syndrome, diabetes mellitus and other manifestations of metabolic syndrome.

APLICATIONS TO OTHER AREAS OF HEALTH AND INTEREST

The study of visfatin has many points of interest for other health specialists. For gynecologists, the role of visfatin could be important in potential treatment of PCOS. For endocrinologists, visfatin physiology could be important to learn about metabolic syndrome. For dietitians, visfatin changes could be important in the design of potential hypocaloric diets for obese patients. For gastroenterologists, visfatin may plan an important role in the classification of NAFLD or in the treatment of some kinds of liver damage.

Summary Points

- Visfatin is an adipokine. However, visfatin has many pleiotropic actions in tissues other than adipose tissue. Visfatin is constitutively expressed by fetal membranes during pregnancy and is expressed in the amniotic epithelium, mesenchymal cells, and chorionic cytotrophoblast.
- Several features of visfatin suggest that this molecule might be important for understanding the biological differences between subcutaneous and intra-abdominal adipose tissue, and their relation with metabolic syndrome and other related diseases such as gestational diabetes mellitus, non-alcoholic fatty liver disease and polycystic ovary syndrome.
- After a diet-induced weight reduction, there is a marked decrease in circulating concentrations of visfatin in obese subjects. Gastroplasty, glycemic control and thiazolidinediones have shown a significant influence on visfatin levels.

- Distribution of macronutrients in hypocaloric diets could be an explanation for response of visfatin levels after weight loss.
- These changes in visfatin levels, a hypothetical insulinmimetic adipokine, could have relevant implications in treatment of obesity, polycystic ovary syndrome, diabetes mellitus and other manifestations of metabolic syndrome.

Abbreviations

kD	:	kilodalton
NAFLD	:	non-alcoholic fatty liver disease
OLEFT	:	Otsuka Long-Evans Tokushima fatty
PBEF	:	pre-B cell colony-enhancing factor
PCOS	:	polycystic ovary syndrome
SNP	:	single nucleotide polymorphism

Definition of Terms

Adipokine: Polypeptides produced by the adipocytes. They include leptin, adiponectin, resistin and many cytokines of the immune system, such as tumor necrosis factor alpha, interleukin 6 and complement factor D (also known as adipsin). They have potent autocrine, paracrine and endocrine functions.

Diabetes mellitus type 2: A subclass of diabetes mellitus that is not insulin-responsive or -dependent. It is characterized initially by insulin resistance and hyperinsulinemia and eventually by glucose intolerance, hyperglycemia and overt diabetes. Type 2 diabetes mellitus is no longer considered a disease exclusively found in adults. Patients seldom develop ketosis but often exhibit obesity.

Gestational diabetes mellitus: Diabetes mellitus induced by pregnancy but resolved at the end of pregnancy. It does not include previously diagnosed diabetics who become pregnant (pregnancy in diabetics). Gestational diabetes usually develops in late pregnancy when insulin-antagonistic hormones peak, leading to insulin resistance, glucose intolerance and hyperglycemia.

Hypocaloric diet: Regular course of eating and drinking adopted by a person or animal. This does not include diet therapy, a specific diet prescribed in the treatment of a disease.

Insulin: A 51 amino acid pancreatic hormone that plays a major role in the regulation of glucose metabolism, directly by suppressing endogenous glucose production (glycogenolysis, gluconeogenesis) and indirectly by suppressing glucagon secretion and lipolysis. Native insulin is a globular protein comprising a zinc-coordinated hexamer. Each insulin monomer contains two chains, *a* (21 residues) and *b* (30 residues), linked by two disulfide bonds. Insulin is used as a drug to control insulin-dependent diabetes mellitus.

Obesity: A status with body weight that is grossly above the acceptable or desirable weight, usually due to accumulation of excess fats in the body. The standards may vary with age, sex, genetic or cultural background. A body mass index greater than 30.0 kg/m^2 is considered obese, and a body mass index greater than 40.0 kg/m^2 is considered morbidly obese.

Polycystic ovary syndrome: A complex disorder characterized by infertility, hirsutism, obesity, and various menstrual disturbances such as oligomenorrhea, amenorrhea and anovulation. Polycystic ovary syndrome is usually associated with bilateral enlarged ovaries studded with atretic follicles, not with cysts.

Single nucleotide polymorphism: A single nucleotide variation in a genetic sequence that occurs at appreciable frequency in the population.

Thiazolidinediones: Thiazoles with two keto oxygens. Members are insulin-sensitizing agents that overcome insulin resistance by activation of the peroxisome proliferator activated receptor gamma (PPAR-gamma).

Visfatin: Originally identified as a 52 kD protein that is primarily expressed in bone marrow, muscle and liver. Visfatin has many pleiotropic actions in tissues other than adipose tissue.

References

Aller, R. and D. de Luis, O. Izaola, R. Conde, M. Gonzalez Sagrado, M. Velasco, T. Alvarez, D. Pacheco, and M. González. 2009. Influence of visfatin on histopathological changes on non-alcoholic fatty liver disease. Dig. Dis. Sci. 54: 1772-1777.

B Zhu, J. and M. Schott, R. Liu, C. Liu, B. Shen, Q. Wang, X. Mao, K. Xu, X. Wu, S. Schinner, C. Papewalis, W.A. Scherbaum, and C. Liu. 2008. Intensive glycemic control lowers plasma visfatin levels in patients with type 2 diabetes. Horm. Metab. Res. 11: 801-805.

Berndt, J. and N. Kloting, and S. Kralish. 2005 Plasma visfatin concentrations and fat depot specific mRNA expression in humans. Diabetes 54: 2911-2916.

Böttcher, Y. and D. Teupser, B. Enigk, J. Berndt, N. Kloting, M.R. Schon, and J. Thiery. 2006. Genetic variation in the visfatin gene (PBEF1) and its relation to glucose metabolism and fat-depot-specific messenger ribonucleic acid expression in humans. J. Clin. Endocrinol. Metab. 91: 2721-2731.

Brema, I. and M. Hatunic, F. Finucane, N. Burns, and J. Nolan. 2008. Plasma visfatin is reduced after aerobic exercise in early onset type 2 diabetes mellitus. Diabetes Obes. Metab. 10: 593-602.

Chan, T.F. and Y.L. Chen, H.C. Lee, F.H. Chou, L.C. Wu, S.B. Jong, and E.M. Tsai. 2006. Decreased plasma visfatin concentrations in women with gestational diabetes mellitus. J. Soc. Gynecol. Invest. 13: 364-367.

Chen, M.P. and F.M. Chung, and D.M. Chang. 2005. Elevated plasma level of visfatin/pre-B cell colony-enhancing factor in patients with type 2 diabetes mellitus. J. Clin. Endocrinol. Metab. 60: 876-897.

Choi, K.C. and O.H. Ryu, and K.W. Lee. 2005. Effect of PPAR-alpha and gamma agonist on the expression of visfatin, adiponectin and TNF alpha in visceral fat of OLETF rats. Biochem. Biophys. Res. Commun. 336: 747-753.

de Luis, D.A. and M. Gonzalez Sagrado, R. Conde, R. Aller, and O. Izaola. 2007. Circulating adipocytokines in obese non diabetic patients, relation with cardiovascular risk factors, anthropometry and resting energy expenditure. Ann. Nutr. Metab. 51: 374-378.

de Luis, D.A. and M. Gonzalez Sagrado, R. Conde, R. Aller, O. Izaola, and E. Romero. 2008. Effect of a hypocaloric diet on serum visfatin in obese non-diabetic patients. Nutrition 24: 517-521.

de Luis, D.A. and M. Gonzalez Sagrado, R. Conde, R. Aller, O. Izaola, and E. Romero. 2009. Effects of dietary intakes on visfatin levels in obese patients. Ann. Nutr. Metab. Epub ahead of print.

Fukuhara, A. and M. Matsuda, and N. Nishizawa. 2005. Visfatin: a protein secreted by visceral fat that mimics the effects of insulin. Science 307: 426-430.

Haider, D.G. and F. Mittermayer, G. Schaller, M. Artwohl, S. Baumgartner Parzer, G. Prager, M. Roden, and M. Wolzt. 2006. Free fatty acids normalize a rosiglitazone-infused visfatin release. Am. J. Physiol. Endocrinol. Metab. 291: e885-e890.

Haider, D.G. and K. Schindler, G. Schaller, G. Prager, M. Woltzt, and B. Ludvik. 2006. Increased plasma visfatin concentrations in morbidly obese subjects are reduced after gastric banding. J. Clin. Endocrinol. Metab. 91: 1578-1581.

Hammarstedt, A. and J. Pihlajamaki, V.R. Sopasakis, S. Gogg, P.A. Jansson, M. Laakso, and U. Smith. 2006. Visfatin is an adipokine, but it is not regulated by thiazolidinediones. J. Clin. Endocrinol. Metab. 91: 1181-1184.

Jia, S.H. and Y. Li, and J. Parodo. 2004. Pre-B cell colony enhancing factor inhibits neutrophil apoptosis in experimental inflammation and clinical sepsis. J. Clin. Invest. 113: 1318-1327.

Jinn, H. and B. Jiang, J. Tang, W. Lu, W. Wang, L. Zhou, W. Shang, F. Li, Q. Ma, Y. Yang, and M. Chen. 2008. Serum visfatin conenctrations in obese adolescents and its correlation with age and high density lipoprotein cholesterol. Diab. Res. Clin. Pract. 79: 412-418.

Kovacikova, M. and M. Vitkova, E. Klimcakova, J. Polak, J. Hejnova, M. Bajzova, Z. Kovacova, N. Viguerie, D. Langin, and V. Stich. 2008. Visfatin expression in subcutaneous adipose tissue of pre-menopausal women: relation to hormones and weight reduction. Eur. J. Clin. Invest. 38: 516-522.

Kralisch, S. and J. Klein, and U. Lossner. 2005a. Interleukin-6 is negatively regulator of visfatin gene expression in 3T3L1 adipocytes. Am. J. Physiol. Endocrinol. Metab. 289: E586-E590.

Kralisch, S. and J. Klein, and U. Lossner. 2005b. Isoproterenol, TNF alpha and insulin downregulate adipose triglyceride lipase in 3T3-L1 adipocytes. Mol. Cell Endocrinol. 240: 43-49.

Krzyzanowska, K. and W. Krugluger, F. Mittermayer, R. Rahman, D. Haider, N. Shnawa, and G. Schernthaner. 2006. Increased visfatin concentrations in women with gestational diabetes mellitus. Clin. Sci. 1: 605-609.

Lorente-Cebrian, S. and M. Bustos, A. Marti, J.A. Martinez, and M.J. Moreno-Aliaga. 2009. EPA stimulates AMP activated protein kinase and increases visfatin secretion in cultures murine adipocytes. Clin. Sci. 18 Mar. Epub ahead of print.

Manco, M. and J.M. Fernandez Real, F. Equitani, J. Vendrell, M.E. Varela Mora, and G. Nanni. 2007. Effect of massive weight loss on inflammatory adipocytokines and the innate immune system in morbidly obese women. J. Clin. Endocrinol. Metab. 92: 483-490.

Ognjanovic, S. and G.D. Bryant-Greenwood. 2002. Pre-B cell colony enhancing factor, a novel cytokine of human fetal membranes. Am. J. Obstet. Gynecol. 187: 1051-1058.

Perez Echarri, N. and P. Perez Matute, B. Marcos Gomez, J.A. Martinez, and M.J. Moreno Aliaga 2009. Effects of eicosapentaenoic acid ethyl ester on visfatin and apelin in lean and overweight (cafeteria diet-fed) rats. J. Nutr. 101: 1059-1067.

Samal, B. and Y. Sun, and G. Stearns. 1994. Cloning and characterization of the cDNA encoding a novel human pre-B cell colony-enhancing factor. Mol. Cell Biol. 14: 1431-1437.

Sun, G. and J. Bishop, S. Khalili, S. Vasdev, V. Gill, D. Pace, D. Fitzpatrick, E. Randell, Y.G. Xie, and H. Zhang. 2007. Serum visfatin concentrations are positively correlated with serum triacylglycerols and down-regulated by overfeeding in healthy young men. Am. J. Clin. Nutr. 85: 399-404.

Tan, B.K. and J. Chen, J.E. Digby, S. Keay, R. Kennedy, and H. Randeva. 2006. Increased visfatin mRNA and protein levels in adipose tissue and adipocytes in women with PCOS: parallel increase in plasma visfatin. J. Clin. Endocrinol. Metab. doi 10.1210/jc2006-0936.

Tokunaga, A. and A. Miura, Y. Okauchi, K. Segawa, A. Fukuhara, K. Okita, M. Takahashi, T. Funahashi, J. Miyagawa, I. Shimomura, and K. Yamagata. 2008. The -1535 promoter variant of the visfatin gene is associated with serum triglyceride and HDl cholesterol levels in Japanese subjects. Endocr. J. 55(1): 205-212.

Van der Veer, E. and Z. Nong, and C. ONeil. 2005. Pre B cell colony enhancing factor regulates NAD+dependent protein deacetylase activity and promotes vascular smooth muscle cell maturation. Cir. Res. 1: 25-34.

CHAPTER 32

Apelin in Energy Metabolism

Isabelle Castan-Laurell[1] and Philippe Valet[1,*]
[1]INSERM U858, Université de Toulouse, IFR150
[1]Department of Obesity and Metabolic Disorders, 1 ave J.Pouhlès, BP 84225, 31432 Toulouse cedex 4, France

ABSTRACT

Apelin is a peptide known as the ligand of the G protein-coupled receptor APJ. Several active apelin isoforms exist in humans such as apelin-36, apelin-13 and the pyroglutaminated form of apelin-13 ([Pyr1] apelin-13). Apelin and APJ are expressed in the central nervous system, particularly in the hypothalamus and many peripheral tissues. Apelin has been shown to be involved in the regulation of cardiovascular and fluid homeostasis, food intake, cell proliferation and angiogenesis. Apelinergic neurons have also been identified in different nuclei of the hypothalamus, areas known to be important in the regulation of food intake. However, the different studies performed have yielded conflicting results. In addition to being a ubiquitous peptide, apelin is also produced and secreted by adipocytes. The expression and secretion of apelin by adipocytes is mainly up-regulated by insulin in both humans and rodents. Moreover, in obesity and associated diseases such as type 2 diabetes, plasma apelin levels are increased. The link between apelin as an adipokine and metabolic disorders is an emerging field of investigation. The effects of apelin on energy metabolism are increasingly well described and are the focus of this chapter. Recent data highlighted important roles for apelin not only in glucose but also in lipid metabolism. Apelin appears as a beneficial adipokine with anti-obesity and anti-diabetic properties and thus as a promising therapeutic target in metabolic disorders.

*Corresponding author

INTRODUCTION

Apelin is a peptide, identified as the endogenous ligand of APJ, a G protein-coupled receptor ubiquitously expressed. Different isoforms of apelin exist, resulting from several proteolytic cleavage sites, such as apelin-36, apelin-17, apelin-13 and the pyroglutaminated form of apelin-13 ([Pyr1] apelin-13), which is more stable (preventing enzyme degradation). The shorter isoforms seem to be more efficient and apelin-13 or [Pyr1] apelin-13 is often used to test functional effects. Besides, in humans, [Pyr1] apelin-13 was identified as the predominant isoform in the heart (Maguire *et al.* 2009). Apelin is a circulating peptide (around 150 pg/ml in humans in basal conditions) and its levels are generally considered lower than expected for a hormone (Kleinz and Davenport 2005, Maguire *et al.* 2009).

Before it was found to be an adipocyte-secreted factor (adipokine), apelin was known to exert several central and peripheral effects in different tissues (Kleinz and Davenport 2005). Apelin is involved in the regulation of the cardiovascular, immune and gastrointestinal functions but also in fluid homeostasis, angiogenesis, proliferation of different cell types and embryonal development (Fig. 1). There is also compelling evidence that apelin/APJ system appears as new therapeutic target of interest (Kleinz and Davenport 2005) in view of the prevalence of different pathogenesis related to the cardiovascular system and metabolic disorders. The role of apelin in energy metabolism in relation to obesity and type 2 diabetes is emerging and will be the focus of this chapter.

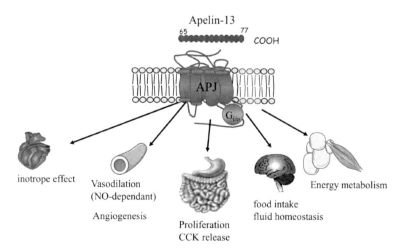

Fig. 1 Principal central and peripheral effects of apelin. Apelin peptides (such as apelin-13) bind to membrane APJ receptor and exert different effects in the cardiovascular and the gastrointestinal systems but also in the central nervous system. Recently, apelin was shown to play a role in energy metabolism.

Color image of this figure appears in the color plate section at the end of the book.

APELIN REGULATION IN OBESITY-ASSOCIATED DISORDERS

The work of Boucher *et al.* 2005 in our team demonstrated a close regulation of the expression and secretion of apelin in adipocytes by insulin both *in vivo* and *in vitro* in mice (Boucher *et al.* 2005). The expression of apelin in adipocytes is increased in various mouse models of obesity associated with hyperinsulinemia. During fasting and after re-feeding in mice, the pattern of apelin expression in adipocytes parallels the plasma levels of insulin. Similarly, the absence of insulin (streptozotocin-treated mice) induces a decrease of apelin mRNAs. In cultured 3T3F442A adipocytes, treatment with insulin also results in increased expression and secretion of apelin. Other factors regulate the expression of apelin in adipocytes either positively, as do TNFα, growth hormone, peroxisome proliferator-activated receptor gamma (PPARγ) coactivator-1 α (PGC1α), or negatively, as do glucocorticoids (Carpéné *et al.* 2007).

In humans, insulin and TNFα also up-regulate the expression of apelin in adipocytes and apelin plasma concentrations are increased in obese and hyperinsulinemic subjects (Boucher *et al.* 2005, Daviaud *et al.* 2006). Different groups found also increased plasma apelin levels in morbidly obese subjects (Heinonen *et al.* 2005), in patients without severe obesity (BMI 23.5 ± 2.4 kg/m^2) but with impaired glucose tolerance or with type 2 diabetes (Li *et al.* 2006). In a recent study in which patients were morbidly obese with or without diabetes, apelin levels were higher only in the morbidly obese diabetic subjects (Soriguer *et al.* 2009). However, reduced plasma apelin levels were described in obese subjects with untreated type 2 diabetes, compared to non-diabetic subjects (BMI 28.9 ± 3.3 kg/m^2) (Erdem *et al.* 2008). In women with gestational diabetes, no significant differences in plasma apelin levels were observed compared to women with normal glucose tolerance (Telejko *et al.* 2009).

Changes in apelin levels after diet-induced weight loss or bariatric surgery in obese individuals were also investigated. Diet-induced weight loss decreases apelin levels in moderately obese women (Castan-Laurell *et al.* 2008) but not significantly in patients with the metabolic syndrome (Heinonen *et al.* 2009). Bariatric surgery resulted in a significant decrease in apelin levels only in morbidly obese patients who had impaired fasting glucose or type 2 diabetes before surgery (Soriguer *et al.* 2009).

Taken together, these studies underline that obesity per se is not the main determinant of increased apelin levels. Circulating apelin is not necessary correlated to BMI in all the studies published. However, especially in obese subjects, adipose tissue might contribute to plasma apelin levels. In addition, apelin levels or changes in apelin levels were found to correlate significantly with serum triglycerides, glucose (Soriguer *et al.* 2009) or HOMA-IR, TNFα (Homeostasis Assessment Model of Insulin Resistance) (Li *et al.* 2006, Castan-Laurell 2008). Thus, a link between apelin and the pathogenesis of insulin resistance or type 2 diabetes might exist and a role for apelin in energy metabolism could be expected.

ROLE OF APELIN IN CARBOHYDRATE METABOLISM

Effect on Insulin Secretion

The first evidence of involvement of apelin in glucose metabolism came from the study of Sorhede Winzell *et al.* showing that apelin inhibits insulin secretion stimulated by glucose *in vivo* in mice and *in vitro* in isolated islets of Langerhans (Sorhede Winzell *et al.* 2005). Apelin-36 was used in this study and had no glucose-lowering effect by itself. More recently, apelin-13 was also shown to inhibit insulin secretion stimulated by high glucose concentrations (10 mM) or potentiated by GLP-1 in INS-1 cells (Guo *et al.* 2009). The mechanism of apelin involves a decrease of cAMP and GLP-1-stimulated cAMP production in the β cells by a PI3-kinase-dependent activation of phosphodiesterase 3B (PDE3B) rather than inhibition of adenylyl cyclase. Interestingly, the dose effect of apelin was biphasic and maximal stimulation was reached at 100 nM. It is tempting to speculate that in hyperinsulinemic obese subjects, the high levels of plasma apelin failed to decrease insulinemia.

Effect on Glucose Uptake in Mice with Standard Body Weight

Apelin administered intravenously at low concentration (200 pmol/kg) decreased blood glucose in mice and improved glucose tolerance (Dray *et al.* 2008). Furthermore, during a hyperinsulinemic-euglycemic clamp, when the hepatic glucose production is totally inhibited, apelin increases glucose utilization throughout the entire organism mainly because of glucose uptake by skeletal muscles and adipose tissues. The liver does not seem to be a direct target of apelin because APJ is not expressed in basal conditions. Apelin also stimulates glucose transport in isolated skeletal muscle (soleus) and its effect is additive to that of insulin (Dray *et al.* 2008).

Apelin Signaling Pathway Involved in Muscle Glucose Transport

It is known that apelin can produce NO by activating endothelial NO synthase (eNOS) in endothelial cells (Tatemoto *et al.* 2001) and that eNOS could be involved in muscle glucose transport (Roy *et al.* 1998, Higaki *et al.* 2001). In the Dray *et al.* 2008 study, apelin was shown to phosphorylate eNOS in the soleus muscle *in vivo* and *in vitro* and this activation was essential in apelin-induced glucose transport (Dray *et al.* 2008). This did not rule out that apelin could also stimulate glucose utilization by altering vascular tone. However, the low dose of apelin used (200 pmol/kg) did not induce changes in mean arterial blood pressure in mice. Furthermore, on isolated muscles of eNOS-deficient mice (eNOS-/-), apelin did not stimulate glucose transport, whereas insulin-induced glucose transport was maintained.

As an upstream target of the eNOS, several studies have described the involvement of AMPK (Barnes and Zierath 2005). AMPK is a key enzyme of energy metabolism activated during ATP depletion in the cell. It is involved in various metabolic processes that stimulate the production of energy such as glucose transport (Barnes and Zierath 2005). We were the first to demonstrate by both *in vivo* and *in vitro* experiments that AMPK was phosphorylated by apelin in soleus muscle and involved in apelin-stimulated glucose transport. More recently, in cultured C2C12 myotubes, apelin-induced glucose uptake was also shown to be dependent on AMPK activation (Yue *et al.* 2009).

In addition, like insulin, apelin phosphorylates Akt and its activation is necessary for glucose transport both *ex vivo* in soleus muscle and in C2C12 myotubes. Moreover, the activation of Akt is AMPK-dependent (Dray *et al.* 2008, Yue *et al.* 2009). The signaling pathway of apelin is thus partly independent of insulin signaling (Fig. 2). This is of interest, particularly in a situation of insulin resistance where insulin receptor is inactivated.

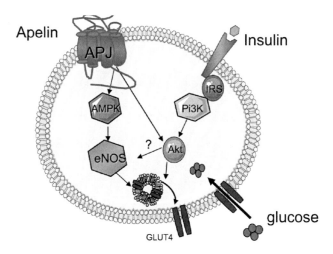

Fig. 2 Apelin signaling pathway involved in muscle glucose transport. Apelin stimulates AMPK, eNOS and Akt activation and then GLUT4 transporters are translocated to the membrane allowing glucose uptake in muscle. It is not known whether apelin can activate Akt independently of AMPK activation. Apelin pathway is partly independent of the first steps of insulin signaling requiring PI3K-associated IRS proteins.

Color image of this figure appears in the color plate section at the end of the book.

Effect of Apelin on Glucose Uptake in Obese and Insulin-Resistant Mice

Mice fed a high-fat diet for several weeks become obese, hyperinsulinemic and insulin-resistant. When receiving an intravenous injection of apelin before an oral glucose tolerance test, their glucose tolerance is significantly improved. During a euglycemic-hyperinsulinemic clamp, the insulin-resistant mice have decreased use of glucose by insulin-sensitive tissues such as muscle and adipose tissue but infusion of apelin during the clamp increases glucose utilization in the same tissues (Dray *et al.* 2008). Thus, apelin in acute treatment is effective in obese insulin-resistant mice and improves the disturbed glucose metabolism. Very recently, chronic apelin treatment (2 wk infusion by osmotic pumps) in insulin-resistant mice was shown to improve insulin sensitivity (Yue *et al.* 2009). The role of apelin in glucose homeostasis was also confirmed by the phenotype of apelin null mice that are hyperinsulinemic and insulin-resistant. The loss of insulin sensitivity in apelin-/- mice was exacerbated by a high fat/high sucrose diet (Yue *et al.* 2009).

Thus, apelin is involved in carbohydrate metabolism and displays beneficial properties such as glucose-lowering effects. Although plasma apelin levels are elevated in obese insulin-resistant mice, exogenous apelin is efficient and thus apelin resistance is unexpected. The increased levels of apelin might constitute a compensatory mechanism to delay the onset of insulin resistance.

EFFECT OF APELIN IN LIPID METABOLISM

Apelin, as an adipocyte-secreted factor and like other adipokines (e.g., leptin, adiponectin), might exert a local effect on adipose tissue in addition to skeletal muscle. The studies that revealed an effect of apelin on lipid metabolism are associated with chronic peripheral administration of apelin (2 wk) in rodents (Higuchi *et al.* 2007, Frier *et al.* 2009). First, treatment of apelin (daily ip injection of apelin) was shown to have an effect on adipose tissue by decreasing the triglycerides content and the weight of different fat depots in chow-fed mice (Higuchi *et al.* 2007). A decreased adiposity was also found in obese mice. Plasma triglycerides were also decreased in both normal and obese mice. Secondly, the treatment did not affect average food intake but did increase rectal temperature and O_2 consumption. An increased expression of mitochondrial uncoupling protein 1 (UCP1) was observed in brown adipose tissue (BAT) (Higuchi *et al.* 2007). All together, the authors suggest that apelin increases energy expenditure through UCP1 activation. In addition to changes in BAT UCP1, apelin treatment also increased UCP3 expression in skeletal muscle but no functional studies were associated (Higuchi *et al.* 2007). Since increase of UCP3 content in muscle could result in an increase of mitochondrial biogenesis, a recent study was conducted in rat in order to know whether apelin treatment would lead to an increase in mitochondrial enzyme activity and protein content in skeletal muscle (Frier *et al.* 2009). Accordingly, enzyme activities of the citrate synthase (involved in the citric acid cycle), the cytochrome C oxidase (involved in the respiratory chain) and β-HAD (involved in the mitochondrial oxidative capacities) were increased in response to apelin treatment in rat triceps. Surprisingly, the increase of mitochondrial markers was independent of increased PGC1α expression, identified as a key player in the regulation of mitochondrial biogenesis. However, PGC1β was up-regulated but its physiological relevance remains to be established. The authors hypothesized that apelin increases skeletal muscle mitochondrial content through a pathway distinct from robust physiological stressors such as exercise (involving PGC1α).

Since chronic apelin treatment decreased adiposity and plasma triglycerides, a question to be addressed is the outcome of these lipids. Modification in the lipid metabolism leading to excess fatty acid accumulation in non-adipose tissues is a hallmark of metabolic diseases. Our group is presently investigating the oxidative capacities of skeletal muscles in response to apelin treatment since apelin activates AMPK in muscle (Dray *et al.* 2008). Interestingly, peripheral apelin treatment increases the β-oxidation in soleus muscle especially in mice fed a high-fat diet (Attané *et al.* 2009). These data are in agreement with the fact that apelin treatment ameliorates insulin sensitivity. Thus, apelin appears as a promising new adipokine involved in whole body energy metabolism (Fig. 3).

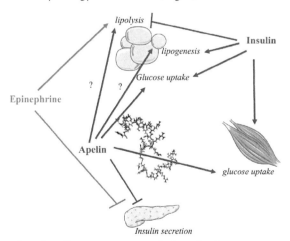

Fig. 3 Apelin and insulin effects on energy metabolism. Insulin increases energy storage by stimulating lipogenesis and inhibiting lipolysis (hydrolysis of triglycerides) in adipose tissue and by stimulating glucose uptake mainly in skeletal muscle. Insulin effects are opposite to those of epinephrine, which known to stimulate lipolysis and to inhibit insulin secretion. Apelin shares effects with both insulin and epinephrine, since apelin stimulates glucose uptake (like insulin) and inhibits insulin secretion (like epinephrine). However, the effects of apelin on lipolysis and lipogenesis are not known but apelin seems rather to stimulate lipid utilization.

Color image of this figure appears in the color plate section at the end of the book.

CENTRAL EFFECT OF APELIN ON ENERGY METABOLISM

Both apelin and its receptor APJ have been detected throughout the central nervous system, particularly in the hypothalamus. Apelin mRNAs are present in different nuclei including the paraventricular, arcuate and supraoptic nuclei that are involved in the control of behavioral, endocrine processes and energy homeostasis (Reaux *et al.* 2001). Apelin-positive nerve fibers in the hypothalamus imply the existence of apelinergic neurons and thus a dual action of apelin as a circulating peptide and a neurotransmitter. So far, it is not known whether peripheral plasma apelin crosses the blood-brain barrier and could modulate apelin levels in the hypothalamus. Higuchi *et al.* mentioned that apelin concentration in the hypothalamus is increased after apelin ip injection (Higuchi *et al.* 2007).

Table I   Key features of energy metabolism

Main processes involved in glucose metabolism:

> *Glucose transport:* Upon insulin action, in skeletal muscle and adipose tissue, the glucose transporter GLUT4 is translocated to plasma membrane allowing glucose entry. Once in the cell, glucose can be directed toward glycogenogenesis or glycolysis.

> *Glycogenogenesis:* It takes places during the post-prandial state when glucose uptake generally exceeds the immediate needs of the resting muscle.

> *Glycolysis:* Cellular degradation of glucose yielding to pyruvate and ATP as an energy source.

Main processes involved in lipid metabolism:

> *Lipolysis:* Breakdown of triglycerides stored in adipocytes into fatty acids (released into the bloodstream) and glycerol.

> *Lipogenesis:* Process by which sugar (glucose) is converted into fatty acids.

> *β-oxidation:* Process by which fatty acids are broken down in the mitochondria to generate Acetyl-CoA. The acetyl-CoA is then ultimately converted into ATP, CO_2, and H_2O using the citric acid cycle and the electron transport chain.

Concerning the effect of acute icv apelin administration on energy expenditure, it was reported that apelin (10 µg) in rat increases core body temperature and locomotor activity (Jaszbereny *et al.* 2004). The same effects as well as food intake were also measured after a chronic 10 d intracerobroventricular (icv) infusion of apelin-13 (1 µg/day) into the third ventricle in mice. Apelin treatment increased food intake (on day 3 to 7), locomotor activity especially during the nocturnal period when feeding occurs, and body temperature only during the period of activity (Valle *et al.* 2008). Several previous studies have determined the effect of acute icv apelin on feeding behavior in rats but the results are contradictory. Apelin icv was shown to decrease food intake in fed and fasting rats (Sunter *et al.* 2003) and during nocturnal administration of apelin, whereas during day-time apelin stimulates feeding (O'Shea *et al.* 2003). No significant effect was reported on the accumulated 24 h food intake in rat (Taheri *et al.* 2002). More recently, Clark *et al.* 2009 showed that apelin icv decreased food and water intake and respiratory exchange ratio in control rats, but had no effect in rats fed a high-fat diet (Clark *et al.* 2009). Moreover, apelin induced a down-regulation of central APJ receptor only in rats fed a high-fat diet, suggesting that a decreased central response to apelin could induce obesity (Clark *et al.* 2009). It is important to better characterize the role of apelin in feeding behavior, since most of the peptides that have been discovered to date exhibit inhibitory effects on food intake, the only exception being ghrelin. To determine the phenotype (orexigen or anorexigen) of the neuron expressing apelin and its receptor APJ (in the arcuate nucleus, for example) and the relation of apelin with other neuropeptides involved in appetite regulation will be very helpful and will allow us to better define the central role of apelin in energy metabolism.

CONCLUSION

The roles of apelin and APJ in energy metabolism should be under continuing expansion in the coming years. Apelin appears as a beneficial adipokine with anti-obesity and anti-diabetic properties and thus as a promising therapeutic target in metabolic disorders. One principal aim will be also to have an integrative view of the apelin/APJ system with regard to other homeostasic functions such as cardiovascular and gastrointestinal functions. Most of these studies have so far been conducted in mice and an approach in humans is essential to confirm the role of apelin in energy metabolism. There is also a need to develop selective APJ ligands in order to facilitate basic research and in view of future clinical investigations.

Summary Points

- Apelin is a peptide known as the ligand of the G protein-coupled receptor APJ that exists in different isoforms. The more active are apelin-36, apelin-17, apelin-13 and the pyroglutaminated form of apelin-13 ([Pyr1] apelin-13).
- Apelin is produced and secreted by adipocytes in humans and rodents. The mRNA expression of apelin in adipocytes is up-regulated by different hormones and factors such as insulin, growth hormone, TNFa and PGC1α and down-regulated by glucocorticoids.
- Apelin is found in the circulation. Plasma apelin levels are increased in obese and in diabetic subjects. However, in newly diagnosed and untreated type 2 diabetic patients, plasma apelin levels are lower.
- Apelin is involved in glucose metabolism. In standard mice, apelin (given by intravenous injection) has a glucose-lowering effect and thus ameliorates glucose tolerance and allows glucose utilization by skeletal muscles and adipose tissue. The effect of apelin is maintained in obese and insulin-resistant mice.

- Apelin is involved in lipid metabolism. Chronic treatment of apelin (14 d) decreases adiposity in standard and obese mice. In addition, apelin treatment increased UCP1 expression in brown adipose tissue, suggesting an increase in energy expenditure, and increased UCP3 in muscle, suggesting an increased in mitochondrial biogenesis.
- Apelin and its receptor APJ have been detected throughout the central nervous system particularly in the hypothalamus. Apelin mRNAs are present in different nuclei including the paraventricular, arcuate and supraoptic nuclei that are involved in the control of behavioral, endocrine processes and energy homeostasis. Apelin is able to regulate food intake and fluid homeostasis.

Abbreviations

AMPK	:	5'AMP activated protein kinase
BAT	:	brown adipose tissue
β-HAD	:	β-hydroxy-acetyl-deshydrogenase
BMI	:	body mass index
eNOS	:	endothelial NO synthase
GLP-1	:	glucagon-like peptide
GLUT4	:	glucose transporter 4
HOMA-IR	:	homeostasis assessment model of insulin resistance
icv	:	intracerebroventricular
PDE3B	:	phosphodiesterase 3B
PGC1α	:	PPARγ coactivator-1 α
PPAR	:	peroxisome proliferator-activated receptor
Pyr 1 apelin-13	:	apelin-13 pyroglutamined
TNFα	:	tumor necrosis factor α
UCP	:	uncoupling protein

Definition of Terms

Hyperinsulinemic euglycemic clamp: It measures the amount of glucose necessary to compensate for increased insulin levels without causing hypoglycemia.

Insulin resistance: A decreased sensitivity to the action of insulin. Conditions in which the increased amount of insulin is inadequate to induce normal insulin responses in insulin-sensitive tissues (liver, skeletal muscles, adipose tissues).

Streptozotocin: A chemical selectively toxic in insulin-producing β cells in mammals and thus used to obtain an animal model of type 1 diabetes.

References

Attané, C. and R. Guzman-Riuz, S. Le Gonidec, V. Bézaire, D. Daviaud, C. Dray, M. Ruiz-Gayo, P. Valet, and I. Castan-Laurell. 2009. Chronic apelin treatment effects on lipid metabolism in wild type and insulin-resistant mice. Diabetologia 52: S 20.

Barnes, B.R. and J.R. Zierath. 2005. Role of AMP-Activated Protein Kinase in the control of glucose homeostasis. Curr. Mol. Med. 5: 341-348.

Boucher, J. and B. Masri, D. Daviaud, S. Gesta, C. Guigné, A. Mazzucotelli, I. Castan-Laurell, I. Tack, B. Knibiehler, C. Carpene, Y. Audigier, J.S. Saulnier-Blache, and P. Valet. 2005. Apelin, a newly identified adipokine up-regulated by insulin and obesity. Endocrinology 146: 1764-1771.

Carpéné, C. and C. Dray, C. Attané, P. Valet, M.P. Portillo, I. Churruca, F.I. Milagro, and I. Castan-Laurell. 2007. Expanding role for the apelin/APJ system in physiopathology. J. Physiol. Biochem. 63: 359-373.

Castan-Laurell, I. and M. Vítkova, D. Daviaud, C. Dray, M. Kováciková, Z. Kovacova, J. Hejnova, V. Stich, and P. Valet. 2008. Effect of hypocaloric diet-induced weight loss in obese women on plasma apelin and adipose tissue expression of apelin and APJ. Eur. J. Endocrinol. 158: 905-910.

Clarke, K.J. and K.W. Whitaker, and T.M. Reyes. 2009. Diminished metabolic responses to centrally-administered apelin-13 in diet-induced obese rats fed a high-fat diet. J. Neuroendocrinol. 21: 83-89.

Daviaud, D. and J. Boucher, S. Gesta, C. Dray, C. Guigne, D. Quilliot, A. Ayav, O. Ziegler, C. Carpene, J.S. Saulnier-Blache, P. Valet, and I. Castan-Laurell. 2006. TNFa up-regulates apelin expression in human and mouse adipose tissue. Faseb J. 20: 1538-1530.

Dray, C. and C. Knauf, D. Daviaud, A. Waget, J. Boucher, M. Buléon, P.D. Cani, C. Attané, C. Guigné, C. Carpéné, R. Burcelin, I. Castan-Laurell, and P. Valet. 2008. Apelin stimulates glucose utilization in normal and obese insulin-resistant mice. Cell Metab. 8: 437-445.

Erdem, G. and T. Dogru, I. Tasci, A. Sonmez, and S. Tapan. 2008. Low plasma apelin levels in newly diagnosed type 2 diabetes mellitus. Exp. Clin. Endocrinol. Diabetes 116: 289-292.

Frier, B.C. and D.B. Williams, and D.C. Wright. 2009. The effect of apelin treatment on skeletal muscle mitochondrial content. Am. J. Physiol. 297: R1761-1768.

Guo, L. and Q. Li, W. Wang, P. Yu, H. Pan, P. Li, Y. Sun, and J. Zhang. 2009. Apelin inhibits insulin secretion in pancreatic b-cells by activation of PI3-Kinase-Phosphodiesterase 3B. Endocrine Res. 34: 142-154.

Heinonen, M.V and A.K. Purhonen, P. Miettinen, M. Paakkonen, E. Pirinen, E. Alhava, K. Akerman, and K.H. Herzig. 2005. Apelin, orexin-A and leptin plasma levels in morbid obesity and effect of gastric banding. Regul. Pept. 130: 7-13.

Heinonen, M.V. and D.E. Laaksonen, T. Karhu, L. Karhunen, T. Laitinen, S. Kainulainen, A. Rissanen, L. Niskanen, and K.H. Herzig. 2009. Effect of diet-induced weight loss on plasma apelin and cytokine levels in individuals with the metabolic syndrome. Nutr. Metab. Cardiovasc. Dis. 19: 626-633.

Higaki, Y. and M.F. Hirshman, N. Fujii, and L.J. Goodyear. 2001. Nitric oxide increases glucose uptake through a mechanism that is distinct from the insulin and contraction pathways in rat skeletal muscle. Diabetes 50: 241-247.

Higuchi, K. and T. Masaki, K. Gotoh, S. Chiba, I. Katsuragi, K. Tanaka, T. Kakuma, and H. Yoshimatsu. 2007. Apelin, an APJ receptor ligand, regulates body adiposity and favors the messenger ribonucleic acid expression of uncoupling proteins in mice. Endocrinology 148: 2690-2697.

Jaszberenyi, M. and E. Bujdoso, and G. Telegdy. 2004. Behavorial, neuroendocrine and thermoregulatory actions of apelin-13. Neuroscience 129: 811-816.

Kleinz, M.J. and A.P. Davenport. 2005. Emerging roles of apelin in biology and medicine. Pharmacol. Therapeutics 107: 198-211.

Li, L. and G. Yang, Q. Li, Y. Tang, M. Yang, H. Yang, and K. Li. 2006. Changes and relations of circulating visfatin, apelin, and resistin levels in normal, impaired glucose tolerance, and type 2 diabetic subjects. Exp. Clin. Endocrinol. Diabetes 114: 544-548.

Maguire, J.J. and M.J. Kleinz, S.L. Pitkin, and A.P. Davenport. 2009. [Pyr1]apelin-13 identified as the predominant apelin isoform in the human heart. Hypertension 54: 598-604.

O'Shea, M. and M.J. Hansen, K. Tatemoto, and M.J. Morris. 2003. Inhibitory effect of apelin-12 on nocturnal food intake in the rat. Nutr. Neurosci. 6: 163-167.

Reaux, A. and K. Gallatz, M. Palkovits, and C. Llorens-Cortes. 2002. Distribution of apelin-synthesizing neurons in the adult rat brain. Neuroscience 113: 653-662.

Roy, D. and M. Perreault, and A. Marette. 1998. Insulin stimulation of glucose uptake in skeletal muscles and adipose tissues *in vivo* is NO dependent. Am. J. Physiol. 274: E692-699.

Soriguer, F. and L. Garrido-Sanchez, S. Garcia-Serrano, J.M. Garcia-Almeida, J. Garcia-Arnes, F.J. Tinahones, and E. Garcia-Fuentes. 2009 Apelin levels are increased in morbidly obese subjects with Type 2 Diabetes Mellitus. Obes. Surg. 2009 (in press).

Sorhede Winzell, M. and C. Magnusson, and B. Ahren. 2005. The apj receptor is expressed in pancreatic islets and its ligand, apelin, inhibits insulin secretion in mice. Regul. Pept. 131: 12-17.

Sunter, D. and A.K. Hewson, and S.L. Dickson. 2003. Intracerebroventricular injection of apelin-13 reduces food intake in the rat. Neurosci. Lett. 353: 1-4.

Taheri, S. and K. Murphy, M. Cohen, E. Sujkovic, A. Kennedy, W. Dhillo, C. Dakin, A. Sajedi, M. Ghatei, and S. Bloom. 2002. The effects of centrally administered apelin-13 on food intake, water intake and pituitary hormone release in rats. Biochem. Biophys. Res. Commun. 291: 1208-1212.

Tatemoto, K. and K. Takayama, M.X. Zou, I. Kumaki, W. Zhang, K. Kumano, and M. Fujimiya. 2001. The novel peptide apelin lowers blood pressure via a nitric oxide-dependent mechanism. Regul. Pept. 99: 87-92.

Telejko, B. and M. Kuzmicki, N. Wawrusiewicz-Kurylonek, J. Szamatowicz, A. Nikolajuk, A. Zonenberg, D. Zwierz-Gugala, W. Jelski, P. Laudanski, A. Kretowski, and M. Gorska. 2009. Plasma apelin levels and apelin/APJ mRNA expression in patients with gestational diabetes mellitus. Diabetes Res. Clin. Pract. (in press).

Valle, A. and N. Hoggard, A.C. Adams, P. Roca, and J.R. Speakman. 2008. Chronic central administration of apelin-13 over 10 days increases food intake, body weight, locomotor activity and body temperature in C57BL/6 mice. J. Neuroendocrinol. 20: 79-84.

Yue, P. and H. Jin, M. Aillaud-Manzanera, A.C. Deng, J. Azuma, T. Asagami, R.K. Kundu, G.M. Reaven, T. Quertermous, and P.S. Tsao. 2009. Apelin is necessary for the maintenance of insulin sensitivity. Am. J. Physiol. Endocrinol. Metab. (in press).

CHAPTER **33**

Adiponectin Deficiency

Wayne Bond Lau,[1] **Ling Tao,**[1] **Yajing Wang**[1] **and Xin L. Ma**[1,*]

[1]Department of Emergency Medicine, 1020 Sansom Street, Thompson Building, Room 239, Philadelphia, PA 19107, USA

ABSTRACT

Adiponectin (Ad) is an abundant protein hormone that regulates numerous metabolic processes. The 30 kDa protein mainly originates from adipose tissue, with both full-length and globular domain forms existing in the bloodstream. A collagenous domain within adiponectin leads to its spontaneous self-assembly into various oligomeric isoforms, including trimers, hexamers, and high molecular weight multimers. Two membrane-spanning receptors for adiponectin have been identified, with differing concentration distribution in various body tissues. The major intracellular pathway activated by adiponectin includes phosphorylation of AMP-activated protein kinase, responsible for many of adiponectin's downstream actions, including increased glucose utilization, fatty acid oxidation, increased muscle glucose uptake, and decreased liver gluconeogenesis.

Since its discovery in 1995, adiponectin has garnered considerable attention for its role in diabetes and cardiovascular pathology. Clinical observations have demonstrated the association of hypoadiponectinemia in patients with obesity, cardiovascular disease, insulin resistance, diabetes, and the metabolic syndrome. In this review, we elaborate currently known information about adiponectin deficiency in relationship to diabetes, cardiovascular disease (including atherosclerosis, endothelial dysfunction, and cardiac injury), and other body system disorders (including the immune, liver, pulmonary, renal, and oncological systems), and finally provide perspective on the future of adiponectin research and the protein's potential therapeutic benefits.

---

*Corresponding author

INTRODUCTION

Adiponectin is a protein hormone responsible for the modulation of numerous metabolic processes. In this review, we elaborate currently known information about adiponectin deficiency in pathological disease processes and provide perspective on the future of research related to adiponectin and its potential therapeutic benefits.

ADIPONECTIN AND ADIPONECTIN RECEPTORS

The cDNA encoding adiponectin was first described by Scherer and colleagues in 1995. Because of its similarity to complement factor C1q, the protein was termed Acrp30, as it was an adipocyte complement-related protein of 30 kDa size. Three other groups isolated both mouse and human forms of adiponectin, and termed the adipocytokine AdipoQ (Hu *et al.* 1996), apM1 (Maeda *et al.* 1996), and GBP28 (Nakano *et al.* 1996) respectively. The human adiponectin gene exists on chromosome 3q26, a region associated with type-2 diabetes mellitus and metabolic syndrome susceptibility.

Adiponectin has a primary sequence 244 amino acids long and contains a signal sequence (which targets the protein for extracellular section and is cleaved in the mature peptide) and a non-conserved N-terminal domain, followed by 22 collagen repeats, and a C-terminal globular domain that has topological similarities to TNFα, despite dissimilar amino-acid sequences (Fig. 1). Full-length adiponectin (fAd) requires post-translational modification for biological activity (e.g., hydroxylation

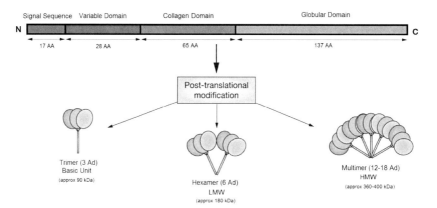

Fig. 1 Structure of adiponectin. Full-length adiponectin (fAd) requires post-translational modifications (e.g., hydroxylation and glycosylation) for activity. Adiponectin molecules (Ad) are secreted from adipocytes as trimers (~90 kDa; the basic unit), LMW hexamers (~180 kDa) and HMW isoforms (12-18-mers; >400 kDa). AA, amino acid (length); N, amino-terminus; C, carboxy-terminus; LMW, low molecular weight; HMW, high molecular weight.

and glycosylation) and is secreted from adipocytes in three major size classes—trimers (~90 kDa; the basic unit), low molecular weight (LMW) hexamers (~180 kDa), and high molecular weight (HMW) isoforms consisting of 12-mers to 18-mers (which can exceed 400 kDa in size) (Fig. 1). Monomeric adiponectin not incorporated into trimers is predicted to be thermodynamically unstable and has not been observed under native conditions, but a proteolytic cleavage product of full-length adiponectin containing purely globular C-terminal domain has been postulated to exist *in vivo*. Recently, studies have shown that leukocyte elastase (secreted by activated monocytes and neutrophils) cleave full-length adiponectin (fAd) and generate globular adiponectin (gAd).

Adiponectin regulates cellular function via binding and activation of its specific receptors, adiponectin receptor 1 (AdipoR1) and adiponectin receptor 2 (AdipoR2), each encoded by their respectively named genes. The adiponectin receptors belong to a new family of membrane receptors structurally predicted to contain seven transmembrane domains, but are topologically distinct from G-coupled receptor proteins. Adiponectin binds to the C-terminal extracellular domain of AdipoR, whereas the N-terminal cytoplasmic domain interacts with an adaptor protein, APPL1 (Fig. 2). AdipoR1 is a receptor with high affinity for globular adiponectin, whereas AdipoR2 is a receptor with intermediate affinity for both full-length and globular adiponectin. AdipoR1 is abundantly expressed in skeletal muscle and endothelial cells, and AdipoR2 is predominantly expressed in the liver. Both AdipoR1 and AdipoR2 are constitutively expressed in adult cardiomyocytes, but studies resolving each receptor's relative contribution to adiponectin cardioprotective signaling remain ongoing. In addition to the two

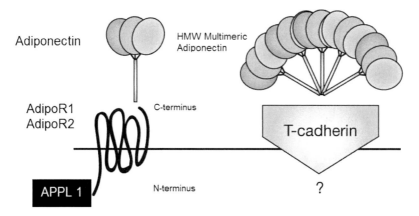

Fig. 2 Adiponectin and the adiponectin receptor. Adiponectin interacts with the extracellular C-terminus region of the adiponectin receptor, which spans the membrane in seven domains, with its N-terminus intracellular, interacting with APPL1. T-cadherin is postulated to be a receptor for multimeric HMW adiponectin isoforms, with yet unknown biological function. AdipoR1/R2, adiponectin receptor 1/2; HMW, high molecular weight; APPL1, adapter protein.

adiponectin receptors, T-cadherin has been proposed to be a receptor for the hexameric and HMW adiponectin forms and has been found to be capable of binding adiponectin in C2C12 myoblasts, but not in hepatocytes (Hug *et al.* 2004). However, as T-cadherin lacks an intracellular domain, the specific biological function of any adiponectin/T-cadherin interaction remains unknown.

ADIPONECTIN DEFICIENCY AND DIABETES

Large epidemiological studies have confirmed that adiponectin levels are greater in lean than in obese populations (Arita *et al.* 1999). Circulating adiponectin is inversely correlated with body fat mass, insulin resistance, and type-2 diabetes mellitus. Adiponectin is unrelated to impaired insulin action in type-1 diabetes. Hypoadiponectinemia has been linked to reduced insulin receptor tyrosine phosphorylation in skeletal muscle, preceding full-body impaired insulin response in humans. Recent meta-analysis of over 14,000 patients revealed an association between greater adiponectin level and decreased risk for type-2 diabetes (Li *et al.* 2009), while numerous ethnic and geographic sources report hypoadiponectinemia predictive of future insulin resistance and diabetes onset. Additionally, adiponectin's specific isoform prevalence and distribution are critically important, as deficient HMW isoform concentration has been linked to poor insulin sensitivity and low basal lipid oxidation, and impaired multimerization of human adiponectin mutants is associated with both hypoadiponectinemia and diabetes.

Many clinical genetic studies have demonstrated the link between hypoadiponectinemia and diabetes. Studies of the human genome suggest the role of adiponectin in insulin resistance and type-2 diabetes susceptibility. Analysis of single nucleotide polymorphisms of the adiponectin gene and its promoter region in various ethnic groups, including the Japanese, German Caucasian, American Caucasian, and French Caucasian, reveal a predisposition for the development of type-2 diabetes, obesity, and insulin resistance. Similarly, polymorphisms of the adiponectin receptor genes AdipoR1 and AdipoR2 have been found to be associated with insulin resistance and type-2 diabetes, but have not been replicated widely across populations. A clinical report reveals healthy Mexican Americans with a family history of diabetes had reduced AdipoR1 and AdipoR2 expression in their muscle tissue compared to individuals with no diabetic history, but the nature of the impairment remains unexplained (Civitarese *et al.* 2004). Multivariate analysis in Caucasians and the Pima Indians (a population with elevated incidence of obesity and type-2 diabetes) reveals that hypoadiponectinemia is more closely related to the degree of insulin resistance and hyperinsulinemia than the magnitude of adiposity and glucose intolerance (Weyer *et al.* 2001). Another case-control cohort study of the Pima Indians revealed that, after controlling for body mass index, higher plasma adiponectin concentration affords a 40% risk reduction in type-2 diabetes development, suggesting the utility of plasma adiponectin level as a future type-2 diabetes risk biomarker (Lindsay *et al.* 2002).

The causative role of adiponectin deficiency in diabetes has been demonstrated in animal models. AdipoR1 and R2 expression is reduced in the skeletal muscle of *ob/ob* mice, correlating with both decreased adiponectin binding to skeletal muscle and consequent signal transduction. Adiponectin administration in the diabetic *db/db* mouse attenuates insulin resistance (Tsuchida *et al.* 2004). Kubota and colleagues demonstrated homozygous adiponectin knockout mice manifest impaired insulin sensitivity (Kubota *et al.* 2002), and Maeda confirmed this finding in knockout animals after 2 wk high-fat diet duress (Maeda *et al.* 2002). Moreover, in a prospective longitudinal study involving rhesus monkeys genetically predisposed to develop insulin resistance, plasma adiponectin levels declined as insulin resistance progressed (Hotta *et al.* 2001). A negative correlation between body weight, body fat, and resting insulin levels existed in association with adiponectin levels. Finally, thiazolidinediones, a group of insulin-sensitizing drugs, have been shown to increase circulating adiponectin levels in insulin-resistant humans and rodents (Pajvani *et al.* 2004). We have recently demonstrated that rosiglitazone's cardioprotective effects are critically dependent on its adiponectin stimulatory action, suggesting that under pathological conditions where adiponectin expression is impaired (such as advanced type-2 diabetes), the harmful cardiovascular effects of rosiglitazone may outweigh its cardioprotective benefits (Tao *et al.* 2010).

In summary, the association between hypoadiponectinemia and insulin resistance and diabetes has been demonstrated abundantly in both basic science and clinical studies. The high correlation shared between adiponectin level and insulin resistance, together with adiponectin's consistency in varying body states (e.g., metabolism, infection) and durability against degradation, has earned the adipocytokine great appeal as a potential diabetic risk biomarker. Adiponectin assays in the face of diabetes mellitus have clinical utility (1) as a predictor of diabetic onset/insulin resistance and (2) as a monitor, via changes in level or isoform prevalence, gauging metabolic progress over time after weight loss or insulin-sensitizing treatment (such as thiazolidinediones) (Kusminski and Scherer 2009). Its promise as a therapeutic agent against insulin resistance progression remains the object of intensive investigation (Table 1).

Table 1   Key features of adiponectin deficiency and diabetes

- Circulating adiponectin is inversely correlated with body fat mass, insulin resistance, type-2 diabetes mellitus, and the metabolic syndrome.
- Clinical aggregate data reveals an association between decreased risk for type-2 diabetes and elevated adiponectin level.
- Multiple basic science studies using transgenic models of adiponectin deficiency and obesity demonstrate the link between altered lipid metabolism and impaired insulin sensitivity in adiponectin-deficient states.
- Thiazolidinediones, pharmaceutical agents that have been shown to increase adiponectin plasma levels, are efficacious insulin-sensitizing agents in both animal and human studies.
- Increasing evidence suggests that oligomeric distribution imbalance and deficiency of high molecular weight isoforms of adiponectin, not only mere broad adiponectin deficiency, are specifically linked to poor insulin sensitivity and diabetes.
- Adiponectin's appeal as a diabetic risk biomarker or gauge of metabolic progress during diabetic treatment is supported by both basic science and clinical studies, but its use as a direct therapeutic agent against the diabetic condition remains the topic of ongoing investigation.

ADIPONECTIN DEFICIENCY AND CARDIOVASCULAR DISEASE

Adiponectin and Atherosclerosis

Basic science studies support adiponectin as an anti-atherogenic molecule relevant for the prevention of atherosclerotic plaques. Epidemiological studies reveal that adiponectin levels are especially low in subjects with coronary artery disease. Carotid intima-media thickness, an early marker for atherosclerosis, is inversely related to adiponectin levels in men. *In vitro* studies have demonstrated that adiponectin is inhibitory upon monocyte/endothelial adhesion, myeloid differentiation, and macrophage cytokine production—processes intimately related with atherosclerotic plaque genesis (Ouchi *et al.* 1999). Adiponectin directly inhibits various adhesion molecules such as intracellular adhesion molecule 1, vascular cellular adhesion molecule 1, and E-selectin; by attenuating expression of scavenger receptor class A-1 macrophages and consequent low density lipoprotein uptake, adiponectin decreases foam cell formation (Ouchi *et al.* 1999). Adiponectin inhibits various growth factors in cultured smooth muscle cells, including platelet-derived growth factor, basic fibroblast growth factor, heparin-binding epidermal growth factor (EGF)-like growth factor, and EGF, as well as the cell proliferation and migration induced by these elements (Arita *et al.* 2002). Additionally, adiponectin exerts anti-inflammatory effects on macrophages. Most importantly, in response to external vascular cuff injury, adiponectin knock-out mice have increased neointimal formation (Kubota *et al.* 2002). Even in mice with pro-atherogenic apoE-/- background, over-expression of globular adiponectin results in decreased atherosclerotic lesions (Arita *et al.* 2002).

Adiponectin and Endothelial Dysfunction

Endothelial dysfunction, the earliest alteration in vascular pathology, plays a critical role in atherosclerosis development. Clinical observations have demonstrated that hypoadiponectinemia is closely related to endothelial dysfunction in peripheral arteries. Our *in vitro* studies demonstrated that aortic vascular rings from adiponectin knockout mice manifest endothelial dysfunction in terms of increased superoxide production, decreased eNOS phosphorylation, and aberrant vasodilatory response to acetylcholine (Cao *et al.* 2009). Exogenous globular adiponectin supplementation to the adiponectin$^{-/-}$ mice *in vivo* attenuated the endothelial dysfunction measured by these parameters. Moreover, we demonstrated that acute adiponectin treatment significantly attenuated hyperlipidemia-induced endothelial dysfunction in rats via reduction of oxidative and nitrative stress, and by differential regulation of eNOS and iNOS activity (Li *et al.* 2007), giving further evidence that reduced adiponectin in metabolic disorders disrupts endothelial function, a pivotal event in macroangiopathy genesis and progression, a significant cause of mortality in patients with metabolic syndrome and type-2 diabetes (Fig. 3).

Adiponectin and Cardiac Injury

Numerous epidemiological studies have shown that reduced adiponectin levels correlate with increased risk of cardiovascular disease, and the relationship persists even after adjustment for diabetes, dyslipidemia, hypertension, smoking, and body mass index (Kumada *et al.* 2003). High plasma adiponectin concentrations are associated with a lower risk of myocardial ischemia (MI) in men (Kumada *et al.* 2003). Ethnic disparities in adiponectin levels might explain predilection in African-American and South Asian populations for increased type-2 diabetes and coronary artery disease (Hulver *et al.* 2004, Retnakaran *et al.* 2004). Persistently low plasma adiponectin concentration after acute myocardial infarction is predictive of future adverse cardiac events, and recent clinical observations have also demonstrated that post-MI plasma adiponectin levels correlated positively with myocardial salvage index and ejection fraction recovery (Shibata *et al.* 2008). As such, reduced adiponectin production has been recognized as a risk factor of cardiovascular disease.

Experimental studies have generated more direct evidence supporting the link between hypoadiponectinemia and cardiomyocyte injury. *In vitro* studies have demonstrated that adiponectin promotes cell survival and inhibits cell death, suggesting that adiponectin may have direct cardioprotective effects atop its vasculo-protective properties, which indirectly shield cardiomyocytes from myocardial ischemia/reperfusion (MI/R) injury. Indeed, we and others have recently demonstrated that, in addition to microvascular defects, adiponectin knockout (AdKO) mice manifest worse MI/R injury than control mice, in

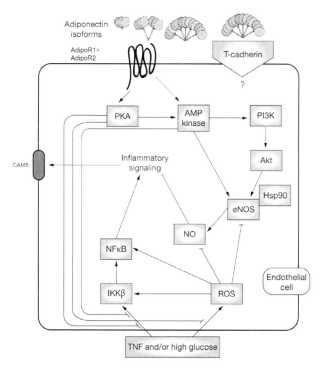

Fig. 3 Adiponectin signaling in endothelial cells. Adiponectin signaling transduction pathways suppress endothelial cell activation elicited by high glucose levels and agonists such as TNF, and suppress inflammatory responses (IKKβ and NFκβ activation). Adiponectin enhances NO generation via AMP kinase cascade activation, and the cAMP-PKA pathway has been shown to mediate adiponectin's anti-oxidative and anti-inflammatory protective effects. AdipoR1 or AdipoR2, adiponectin receptor protein 1 or 2; AMP kinase, AMP-activated protein kinase; cAMP, cyclic AMP; CAMs, cell adhesion molecules; eNOS, endothelial nitric oxide synthase; Hsp90, heat shock protein 90; IKKβ, Iκβ kinase; NFκβ, nuclear factor κβ; NO, nitric oxide; PI3K, phosphatidylinositol 3 kinase; PKA, protein kinase A; ROS, reactive oxygen species; TNF, tumor necrosis factor.

measures such as increased myocardial apoptosis and infarct size, and decreased cardiac function (Shibata *et al.* 2005, Tao *et al.* 2007). Compared to controls, there is increased MI/R injury in heterozygous adiponectin knockout mice (+/-), in which circulating adiponectin levels are reduced by approximately half, but the severity was significantly less than that seen in homozygous knockout animals (Tao *et al.* 2007). This finding suggests that adiponectin levels not only correlate inversely with the risk of development of ischemic heart disease (as indicated by the human epidemiological data), but also are inversely related to the severity of MI/R insult. Exacerbated MI/R injury in AdKO mice was ameliorated by provision of fAd or gAd in a dose-dependent manner (Shibata *et al.* 2005, Tao *et al.* 2007). These exciting results suggest that adiponectin might eventually be used as a novel therapeutic agent in the treatment of ischemic cardiac injury.

As stated earlier, adiponectin is known to stimulate NO production via AMPK-mediated eNOS phosphorylation. Lack of adiponectin, therefore, might conceivably reduce NO production, thus rendering cardiomyocytes more susceptible to MI/R injury. However, this hypothesis is not supported by experimental findings revealing markedly increased, rather than decreased, total NO production in ischemic-reperfused tissue obtained from AdKO mice (Tao *et al.* 2007). Interestingly, inducible nitric oxide synthase (iNOS) expression is increased in AdKO mice compared to controls, and gAd treatment effectively blocks iNOS expression in the AdKO animals (Tao *et al.* 2007). These novel results indicate that adiponectin differentially regulates NO production by both eNOS and iNOS. Under physiological conditions, adiponectin stimulates NO production via eNOS phosphorylation and thus contributes to its vasodilatory, anti-inflammatory, and vascular-protective actions through an AMPK-dependent pathway. In contrast, under pathological conditions when iNOS is induced, adiponectin prevents excess NO generation by inhibiting iNOS expression and this anti-nitrative action of adiponectin is largely AMPK-independent (Wang *et al.* 2009a, b) (Fig. 4).

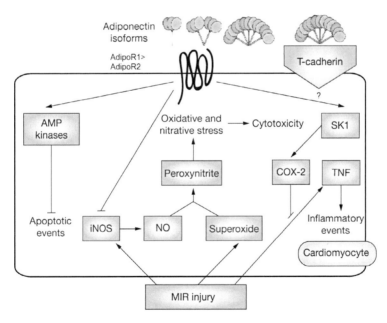

Fig. 4 Adiponectin signaling in cardiomyocytes. Adiponectin isoforms have been shown to exert multiple actions, such as activation of AMP kinases, suppression of TNF signaling via a COX-2-prostaglandin E2-linked cascade (Ikeda *et al.* 2008), and reduction of oxidative and nitrative stress after MIR injury that is associated with suppression of iNOS induction. AdipoR1 or AdipoR2, adiponectin receptor protein 1 or 2; AMP kinase, AMP-activated protein kinase; COX-2, cyclo-oxygenase 2; iNOS, inducible nitric oxide synthase; MI/R, myocardial ischemia-reperfusion; NO, nitric oxide; SK1, sphingosine kinase 1; TNF, tumor necrosis factor.

ADIPONECTIN DEFICIENCY AND OTHER DISEASES

We review here the relationship of adiponectin as it pertains to other disease pathologies beyond diabetes and the cardiovascular system (Table 2).

Table 2   Key features of adiponectin deficiency and non-cardiovascular diseases

- The lipodystrophic syndrome associated with highly active antiretroviral therapy administered to HIV patients is associated with adiponectin deficiency and is a potential risk factor for subclinical cardiac pathology.
- Associated with asymptomatic hepatic dysfunction, hypoadiponectinemia enhances hepatic steatosis and fibrosis and may accelerate tumor formation.
- Subjects with severe obstructive sleep apnea and smoking history have statistically significant attenuated circulating adiponectin levels (adjusted for obesity). Adiponectin knockout mice manifest increased allergic airway inflammation compared to wild type.
- In previously non-diabetic renal transplant patients, serum adiponectin levels were correlated with post-transplantation insulin sensitivity and glucose tolerance, after adjustment for age, steroidal dose, and family history of diabetes mellitus.
- Multiple studies have linked the association between hypoadiponectinemia and numerous obesity-related cancer types. Adiponectin's anti-tumor effects may have mechanistic basis in its stimulation of the AMP-activated protein kinase cell regulatory pathway and its anti-angiogenic properties. Further studies are required to elucidate clinical observations.

Hypoadiponectinemia and Infectious Disease

Highly active antiretroviral therapy (HAART) administered to HIV-infected patients is known to cause a lipodystrophic syndrome with characteristic metabolic abnormalities. Addy and colleagues' 2003 clinical study of 112 HIV-infected patients on HAART revealed an expected inverse relationship of adiponectin with visceral fat mass, serum triglycerides, and insulin resistance, and a direct correlation with HDL levels. In a smaller 2009 study of patients with HIV infection treated with HAART at least 4 yr, Bezante and colleagues demonstrated that hypoadiponectinemia in lipodystrophic individuals was a metabolic risk factor for subclinical cardiovascular damage.

Hypoadiponectinemia and Liver Disease

In Kamada's 2009 epidemiological study of nearly 4000 healthy Japanese people, both visceral obesity and hypoadiponectinemia were significantly associated with asymptomatic hepatic dysfunction by serum hepatic function assays. In 2004, Qi and colleagues demonstrated that adiponectin has attenuated both alcoholic and nonalcoholic steatotic liver diseases. Kamada and colleagues' subsequent 2007 study in a genetic NASH mouse model reaffirmed hypoadiponectinemia's enhancement of hepatic steatosis and fibrosis, and further discovered an acceleration of hepatic tumor formation.

Hypoadiponectinemia and Pulmonary Disease

As evidence accumulates suggesting obstructive sleep apnea may independently contribute to cardiovascular morbidity and mortality, Lam and colleagues' 2008 clinical study identified adiponectin suppression in subjects with severe obstructive sleep apnea, independent of obesity. Iwashima and colleagues' 2005 epidemiological study of Japanese men revealed that adiponectin levels decrease significantly 12 hr after cigarette smoking, and that plasma adiponectin concentration was significantly lower in smokers than in non-smokers. Using a murine model of chronic asthma, Medoff and colleagues (2009) demonstrated that adiponectin[-/-] animals developed increased allergic airway inflammation compared with wild-type mice, with elevated lung chemokine levels, progressing to severe pulmonary arterial hypertension and muscularization compared to wild type.

Hypoadiponectinemia and Renal Disease

In Hjelmesaeth and colleagues' 2006 study of 172 previously non-diabetic renal transplant patients, serum adiponectin levels were correlated with post-transplantation insulin sensitivity and glucose tolerance, after adjustment for age, steroidal dose, and family history of diabetes mellitus. The study also demonstrated divergent effects of glucocorticoids (increased) versus beta blockers (decreased) on serum adiponectin level.

Adiponectin and Cancer

Certain cancers have greater incidence in the obese population. The majority of studies have found an inverse relationship between adiponectin concentration and particular malignancy types, including breast, colorectal, endometrial, prostate, myelodysplastic syndromes and preleukemia, and gastric cancers. Renal cell carcinoma also has association with lower adiponectin levels, to greater degree in metastatic versus localized renal cancer states. Contrary to this pattern, Dalamaga's 2009 case control study investigating pancreatic cancer found a positive association between the disease and adiponectin level. In a subset of this study's samples, increased adiponectin receptor expression was also found in relation to pancreatic cancer, but further study is needed to ascertain whether the finding was primary or compensatory. Other case studies have shown the up-regulation of adiponectin receptors, particularly AdipoR1, in prostate, breast, and colorectal malignancies.

Adiponectin stimulates the AMPK pathway, which regulates cell proliferation, decreases transcriptional regulator expression, and positively regulates growth arrest and apoptosis-controlling proteins (such as p21 and p53). Adiponectin may exert antitumor effects via caspase-mediated endothelial cell apoptosis, and its anti-angiogenic effects by inhibiting tumor-derived basic fibroblast growth factor and interleukin 8 or endothelial-derived platelet-derived growth factor BB. Fujisawa and colleagues' 2008 study found that adiponectin suppressed colorectal carcinogenesis in mice fed a high-fat diet. More studies need to be conducted before the relationship and utility of adiponectin in obesity-related malignancies is fully realized.

Summary Points

- Adiponectin, an endogenous insulin-sensitizing hormone, is an abundantly produced adipocytokine of primarily adipose origin and circulates in plasma as various multimeric complexes.
- Deficiency and/or imbalance of adiponectin's oligomeric forms has well-described association with insulin resistance, obesity, type-2 diabetes, cardiovascular disease, atherosclerosis, metabolic syndrome, diseases of the liver, lung, and kidney, and even several types of cancers.
- The clinical significance of hypoadiponectinemia with obesity-related disease is well established.
- Adiponectin's roles in disease processes via mechanisms yet to be elucidated provide the potential for multiple therapeutic targets and ultimately clinical applications. Increasing attention to adiponectin in recent years is well warranted given the potential salutary benefits of the protein in numerous studies.
- Pharmacological agents (1) augmenting adiponectin's circulating concentrations (including novel or existing medications), (2) increasing adiponectin receptor expression, or (3) dispersing specific adiponectin receptor agonists capable of downstream signaling pathway induction are all potential therapeutic modalities for ameliorating the various disease pathologies discussed in this review.
- Further work will reveal the definitive effects and contributions of the multiple adiponectin oligomeric forms in the vasculature, cardiac, and metabolic systems.

Definition of Terms

Adiponectin (Ad): Protein hormone responsible for the modulation of numerous metabolic processes, whose deficiency is increasingly found to be linked with various human disease pathologies.

Adiponectin knockout (AdKO): Transgenetic animal model in which the adiponectin gene has been turned off.

Adiponectin receptor (either AdipoR1 of AdipoR2): One of the two identified transmembrane protein receptors mediating the downstream signaling mechanisms for adiponectin.

AMP-activated protein kinase (AMPK): Enzyme that plays a role in cellular energy homeostasis; the major downstream effector pathway of adiponectin.

Full-length adiponectin (fAd): Uncleaved, full length form of adiponectin that circulates in trimeric, hexameric, and multimeric high molecular weight forms.

Globular domain of adiponectin (gAd): Proteolytic cleavage product of full length adiponectin containing purely globular C-terminal domain.

Ischemia/reperfusion (MI/R): Series of metabolic events occuring in circulation or target end organ after the restoration of blood flow to ischemic tissues.

References

Addy, C.L. and A. Gavrila, S. Tsiodras, K. Brodovicz, A.W. Karchmer, and C.S. Mantzoros. 2003. Hypoadiponectinemia is associated with insulin resistance, hypertriglyceridemia, and fat redistribution in human immunodeficiency virus-infected patients treated with highly active antiretroviral therapy. J. Clin. Endocrinol. Metab. 88: 627-636.

Arita, Y. and S. Kihara, N. Ouchi, K. Maeda, H. Kuriyama, Y. Okamoto, *et al.* 2002. Adipocyte-derived plasma protein adiponectin acts as a platelet-derived growth factor-BB-binding protein and regulates growth factor-induced common postreceptor signal in vascular smooth muscle cell. Circulation 105: 2893-2898.

Arita, Y. and S. Kihara, N. Ouchi, M. Takahashi, K. Maeda, J. Miyagawa, *et al.* 1999. Paradoxical decrease of an adipose-specific protein, adiponectin, in obesity. Biochem. Biophys. Res. Comm. 257: 79-83.

Bezante, G.P. and L. Briatore, D. Rollando, D. Maggi, M. Setti, M. Ghio, S. Agosti, G. Murdaca, M. Balbi, A. Barsotti, and R. Cordera. 2009. Hypoadiponectinemia in lipodystrophic HIV individuals: a metabolic marker of subclinical cardiac damage. Nutr. Metab. Cardiovasc. Dis. 19: 277-282.

Cao, Y. and L. Tao, Y. Yuan, X. Jiao, W.B. Lau, Y. Wang, *et al.* 2009. Endothelial dysfunction in adiponectin deficiency and its mechanisms involved. J. Mol. Cell. Cardiol. 46: 413-419.

Civitarese, A.E. and C.P. Jenkinson, D. Richardson, M. Bajaj, K. Cusi, S. Kashyap, *et al.* 2004. Adiponectin receptors gene expression and insulin sensitivity in non-diabetic mexican americans with or without a family history of type 2 diabetes. Diabetologia 47: 816-820.

Dalamaga, M. and I. Migdalis, J.L. Fargnoli, E. Papadavid, E. Bloom, N. Mitsiades, K. Karmaniolas, N. Pelecanos, S. Tseleni-Balafouta, A. Dionyssiou-Asteriou, and C.S. Mantzoros. 2009. Pancreatic cancer expresses adiponectin receptors and is associated with hypoleptinemia and hyperadiponectinemia: a case-control study. Cancer Causes Control 20: 625-633.

Fujisawa, T. and H. Endo, A. Tomimoto, M. Sugiyama, H. Takahashi, S. Saito, M. Inamori, N. Nakajima, M. Watanabe, N. Kubota, T. Yamauchi, T. Kadowaki, K. Wada, H. Nakagama, and A. Nakajima. 2008. Adiponectin suppresses colorectal carcinogenesis under the high-fat diet condition. Gut. 57: 1531-1538.

Hjelmesaeth, J. and A. Flyvbjerg, T. Jenssen, J. Frystyk, T. Ueland, M. Hagen, and A. Hartmann. 2006. Hypoadiponectinemia is associated with insulin resistance and glucose intolerance after renal transplantation: impact of immunosuppressive and antihypertensive drug therapy. Clin J Am Soc Nephrol 1: 575-582.

Hotta, K. and T. Funahashi, N.L. Bodkin, H.K. Ortmeyer, Y. Arita, B.C. Hansen, *et al.* 2001. Circulating concentrations of the adipocyte protein adiponectin are decreased in parallel with reduced insulin sensitivity during the progression to type 2 diabetes in rhesus monkeys. Diabetes 50: 1126-1133.

Hu, E. and P. Liang, B.M. Spiegelman. 1996. AdipoQ is a novel adipose-specific gene dysregulated in obesity. J. Biol. Chem. 271: 10697-10703.

Hug, C. and J. Wang, N.S. Ahmad, J.S. Bogan, T.S. Tsao, and H.F. Lodish. 2004. T-cadherin is a receptor for hexameric and high-molecular-weight forms of Acrp30/adiponectin. Proc. Natl. Acad. Sci. USA 101: 10308-10313.

Hulver, M.W. and O. Saleh, K.G. MacDonald, W.J. Pories, and H.A. Barakat. 2004. Ethnic differences in adiponectin levels. Metab. Clin. Exp. 53: 1-3.

Ikeda, Y. and K. Ohashi, R. Shibata, D.R. Pimentel, S. Kihara, N. Ouchi, et al. 2008. Cyclooxygenase-2 induction by adiponectin is regulated by a sphingosine kinase-1 dependent mechanism in cardiac myocytes. FEBS Lett. 582: 1147-1150.

Iwashima, Y. and T. Katsuya, K. Ishikawa, I. Kida, M. Ohishi, T. Horio, N. Ouchi, K. Ohashi, S. Kihara, T. Funahashi, H. Rakugi, and T. Ogihara. 2005. Association of hypoadiponectinemia with smoking habit in men. Hypertension 45: 1094-1100.

Kamada, Y. and H. Matsumoto, S. Tamura, J. Fukushima, S. Kiso, K. Fukui, T. Igura, N. Maeda, S. Kihara, T. Funahashi, Y. Matsuzawa, I. Shimomura, and N. Hayashi. 2007. Hypoadiponectinemia accelerates hepatic tumor formation in a nonalcoholic steatohepatitis mouse model. J. Hepatol. 47: 556-564.

Kamada, Y. and T. Nakamura, T. Funahashi, M. Ryo, H. Nishizawa, Y. Okauchi, J. Fukushima, Y. Yoshida, S. Kiso, I. Shimomura, and N. Hayashi. 2009. Visceral obesity and hypoadiponectinemia are significant determinants of hepatic dysfunction: An epidemiologic study of 3827 Japanese subjects. J. Clin. Gastroenterol. 43: 995-1000.

Kubota, N. and Y. Terauchi, T. Yamauchi, T. Kubota, M. Moroi, J. Matsui, et al. 2002. Disruption of adiponectin causes insulin resistance and neointimal formation. J. Biol. Chem. 277: 25863-25866.

Kumada, M. and S. Kihara, S. Sumitsuji, T. Kawamoto, S. Matsumoto, N. Ouchi, et al. 2003. Association of hypoadiponectinemia with coronary artery disease in men. Arterioscler. Thromb. Vasc. Biol. 23: 85-89.

Kusminski, C.M. and P.E. Scherer. 2009. The road from discovery to clinic: Adiponectin as a biomarker of metabolic status. Clin. Pharmacol. Ther. 86: 592-595.

Lam, J.C. and A. Xu, S. Tam, P.I. Khong, T.J. Yao, D.C. Lam, A.Y. Lai, B. Lam, K.S. Lam, and S.M. Mary. 2008. Hypoadiponectinemia is related to sympathetic activation and severity of obstructive sleep apnea. Sleep 31: 1721-1727.

Li, R. and W.Q. Wang, H. Zhang, X. Yang, Q. Fan, T.A. Christopher, et al. 2007. Adiponectin improves endothelial function in hyperlipidemic rats by reducing oxidative/nitrative stress and differential regulation of eNOS/iNOS activity. Am. J. Physiol. Endocrinol. Metabol. 293: E1703-1708.

Li, S. and H.J. Shin, E.L. Ding, and R.M. van Dam. 2009. Adiponectin levels and risk of type 2 diabetes: A systematic review and meta-analysis. JAMA 302: 179-188.

Lindsay, R.S. and T. Funahashi, R.L. Hanson, Y. Matsuzawa, S. Tanaka, P.A. Tataranni, et al. 2002. Adiponectin and development of type 2 diabetes in the pima indian population. Lancet 360: 57-58.

Maeda, N. and I. Shimomura, K. Kishida, H. Nishizawa, M. Matsuda, H. Nagaretani, et al. 2002. Diet-induced insulin resistance in mice lacking adiponectin/ACRP30. Nat. Med. 8: 731-737.

Medoff, B.D. and Y. Okamoto, P. Leyton, M. Weng, B.P. Sandall, M.J. Raher, S. Kihara, K.D. Bloch, P. Libby, and A.D. Luster. 2009. Adiponectin deficiency increases allergic airway inflammation and pulmonary vascular remodeling. Am. J. Respir. Cell. Mol. Biol. 41: 397-406.

Nakano, Y. and T. Tobe, N.H. Choi-Miura, T. Mazda, and M. Tomita. 1996. Isolation and characterization of GBP28, a novel gelatin-binding protein purified from human plasma. J. Biochem. 120: 803-812.

Qi, Y. and N. Takahashi, S.M. Hileman, H.R. Patel, A.H. Berg, U.B. Pajvani, P.E. Scherer, and R.S. Ahima. 2004. Adiponectin acts in the brain to decrease body weight. Nat. Med. 10: 524-529.

Ouchi, N. and S. Kihara, Y. Arita, K. Maeda, H. Kuriyama, Y. Okamoto, *et al.* 1999. Novel modulator for endothelial adhesion molecules: Adipocyte-derived plasma protein adiponectin. Circulation 100: 2473-2476.

Pajvani, U.B. and M. Hawkins, T.P. Combs, M.W. Rajala, T. Doebber, J.P. Berger, *et al.* 2004. Complex distribution, not absolute amount of adiponectin, correlates with thiazolidinedione-mediated improvement in insulin sensitivity. J. Biol. Chem. 279: 12152-12162.

Retnakaran, R. and A.J. Hanley, N. Raif, P.W. Connelly, M. Sermer, and B. Zinman. 2004. Hypoadiponectinaemia in south asian women during pregnancy: Evidence of ethnic variation in adiponectin concentration. Diabetic Med.: J. Br. Diabetic Assoc. 21: 388-392.

Scherer, P.E. and S. Williams, M. Fogliano, G. Baldini, and H.F. Lodish. 1995. A novel serum protein similar to C1q, produced exclusively in adipocytes. J. Biol. Chem. 270: 26746-26749.

Shibata, R. and Y. Numaguchi, T. Matsushita, T. Sone, R. Kubota, T. Ohashi, *et al.* 2008. Usefulness of adiponectin to predict myocardial salvage following successful reperfusion in patients with acute myocardial infarction. Am J. Cardiol. 101: 1712-1715.

Shibata, R. and K. Sato, D.R. Pimentel, Y. Takemura, S. Kihara, K. Ohashi, *et al.* 2005. Adiponectin protects against myocardial ischemia-reperfusion injury through AMPK- and COX-2-dependent mechanisms. Nat. Med. 11: 1096-1103.

Tao, L. and E. Gao, X. Jiao, Y. Yuan, S. Li, T.A. Christopher, *et al.* 2007. Adiponectin cardioprotection after myocardial ischemia/reperfusion involves the reduction of oxidative/nitrative stress. Circulation 115: 1408-1416.

Tao, L. and Y. Wang, E. Gao, H. Zhang, Y, Yuan, W.B. Lau, *et al.* 2010. Adiponectin, an indispensable molecule in rosiglitazone cardioprotection following myocardial infarction. Circ. Res. 409: 409-417.

Tsuchida, A. and T. Yamauchi, Y. Ito, Y. Hada, T. Maki, S. Takekawa, *et al.* 2004. Insulin/Foxo1 pathway regulates expression levels of adiponectin receptors and adiponectin sensitivity. J. Biol. Chem. 279: 30817-30822.

Wang, Y. and E. Gao, L. Tao, W.B. Lau, Y. Yuan, B.J. Goldstein, *et al.* 2009a. AMP-activated protein kinase deficiency enhances myocardial ischemia/reperfusion injury but has minimal effect on the antioxidant/antinitrative protection of adiponectin. Circulation 119(6): 835-844.

Wang, Y. and L. Tao, Y. Yuan, W.B. Lau, R. Li, B.L. Lopez, *et al.* 2009b. Cardioprotective effect of adiponectin is partially mediated by its AMPK-independent antinitrative action. Am. J. Physiol. Endocrinol. Metabol. 297: E384-391.

Weyer, C. and T. Funahashi, S. Tanaka, K. Hotta, Y. Matsuzawa, R.E. Pratley, *et al.* 2001. Hypoadiponectinemia in obesity and type 2 diabetes: Close association with insulin resistance and hyperinsulinemia. J. Clin. Endocrinol. Metabol. 86: 1930-1935.

CHAPTER 34

Adiponectin Enhances Inflammation in Rheumatoid Synovial Fibroblasts and Chondrocytes

Natsuko Kusunoki,[1] **Kanako Kitahara**[1] **and Shinichi Kawai**[1,*]

[1]Division of Rheumatology, Department of Internal Medicine, Toho University School of Medicine, 6-11-1 Omori-Nishi, Ota-ku, Tokyo 143-8541, Japan

ABSTRACT

Adiponectin is an adipokine that modulates insulin sensitivity and has an anti-atherosclerotic effect. It has also been reported to show an association with rheumatoid arthritis (RA), which is a chronic systemic inflammatory disease. It was reported that the serum/plasma concentration of adiponectin is higher in patients with RA than in healthy controls or patients with osteoarthritis (OA). In the synovial fluid of RA patients, the adiponectin level is also higher than in fluid from OA patient. These reports suggest that adiponectin might contribute to the pathogenesis of arthritis. Because synovial fibroblasts are a major source of inflammation in RA, the possible effect of adiponectin on these cells has been investigated. It was reported that adiponectin stimulates production of interleukin (IL)-6, a very important pro-inflammatory cytokine for RA, by rheumatoid synovial fibroblasts (RSF). This action was related to activation of 5'-adenosine monophosphate-activated protein kinase (AMPK), p38 MAP kinase, and nuclear factor-κB (NF-κB). We found that adiponectin enhanced the production of a chemokine, IL-8, by RSF. These findings suggest that adiponectin may trigger the excessive immune response in RA. Furthermore, adiponectin induces the production of prostaglandin E_2 by RSF via up-regulation of the expression of cyclooxygenase-2 and membrane-associated prostaglandin E synthase-1. This effect of adiponectin might arise from activation of the nuclear translocation of

*Corresponding author

NF-κB. Moreover, increased production of vascular endothelial growth factor (VEGF) and matrix metalloproteinases (MMPs) by RSF and/or chondrocytes were reported. Therefore, it is possible that adiponectin regulates not only synovial inflammation, but also joint destruction in RA patients, suggesting that it may have important pathophysiological effects on RA joints.

INTRODUCTION

Adipose tissue has long been known as a structural component of many organs and a site for energy storage. Recently, some *in vitro* and *in vivo* studies have demonstrated that the major cellular component of adipose tissue, the adipocyte, has the ability to synthesize and release various physiologically active molecules, including adiponectin, leptin, and resistin, as well as cytokines like interleukin (IL)-6 and tumor necrosis factor α (TNFα) (Kadowaki and Yamauchi 2005). Adiponectin was discovered in Japan (Maeda *et al.* 1996), and initial reports were related to insulin resistance in diabetics and atherosclerosis. Some reports raised our interest with respect to the role of adiponectin in inflammation and immunity, including diseases such as rheumatoid arthritis (RA). Therefore, we investigated the role of adiponectin in the pathogenesis of RA.

PATHOPHYSIOLOGY OF RHEUMATOID ARTHRITIS

RA is a systemic chronic inflammatory disease that arises because of autoimmunity. Table 1 shows the key features of RA. RA is characterized by extensive inflammation and proliferation of the synovium in various joints. Synovial tissue is activated by inflammatory stimuli such as cytokines, after which it proliferates and becomes a mass of tissue that is called pannus. As shown in Fig. 1, some adipose tisssues were attached to pannus (proliferative synovial tissue). This raises the possibility that adipokines have a paracrine effect on pannus. Pannus also contains cells such as

Table I   Key features of rheumatoid arthritis

- RA is characterized by extensive inflammation and proliferation of synovial tissue in various joints.
- As progressive joint inflammation and joint destruction eventually occur, the patient's quality of life is significantly reduced.
- RA affects all races, with the prevalence being about 0.3-1.5% of the population.
- The etiology is not clear yet, but DMARDs, glucocorticoids, and NSAIDs have been used for treatment of RA.
- Recently, anti-cytokine agents (so-called biologics) have been developed to neutralize cytokines such as TNFα and IL-6, and have an excellent efficacy for RA so that their use is growing.
- Despite the development of these new drugs, some patients still become worse, so RA is a disease which cannot be explained entirely as due to known inflammatory cytokines.

RA, rheumatoid arthritis; DMARDs, disease modifying anti-rheumatic drugs; NSAIDs, nonsteroidal anti-inflammatory drugs; TNFα, tumor necrosis factor α; IL-6, interleukin-6.

RSF and macrophages that secrete pro-inflammatory cytokines. As pannus grows, it invades cartilage and bone, causing joint destruction. Thus, physical and/or biochemical inhibition of synovial hyperplasia may be an effective treatment for RA (Kusunoki *et al.* 2008).

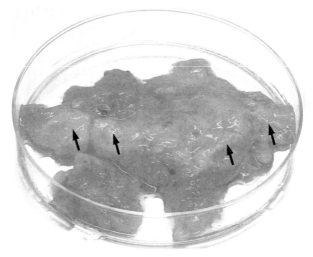

Fig. 1   Synovial tissue of a patient with rheumatoid arthritis. There are numerous adipose tissues in the synovial tissue (arrow).

Color image of this figure appears in the color plate section at the end of the book.

Since pro-inflammatory cytokines, including TNFα, IL-1β, and IL-6, have a central role in the pathophysiology of RA, novel strategies that neutralize these cytokines by using monoclonal antibodies or soluble receptors have recently been developed as new treatments for RA (Kawai 2003). Although blockade of these cytokines is beneficial, it is not curative and the effect is only partial, with failure to respond being common (Brennan and McInnes 2008). Therefore, it seems possible that other pro-inflammatory cytokines may also contribute to the pathophysiology of inflammation in RA.

ADIPONECTIN IN RA

Several reports promoted our interest in the role of adiponectin in the pathogenesis of arthritis. Schäffler *et al.* (2003) reported that the adiponectin level of RA synovial fluid was significantly higher than that of fluid from patients with osteoarthritis (OA). In their study, the adiponectin level in the synovial fluid of OA patients was correlated with the C-reactive protein level, which suggested that adiponectin might have a role in the progression of arthritis. Several other reports also appeared about the adiponectin concentration in serum/plasma and synovial fluid of RA patients (Table 2). In general, the serum/plasma and synovial fluid concentrations of adiponectin are higher in RA patients than in disease controls such as OA patients, although variations in patient background factors make direct comparison problematic. It is interesting that adiponectin might have a role in the aggravation of arthritis. For instance, it was reported that the serum concentration of adiponectin was marginally associated with a higher Larsen score (a useful index of radiographic joint damage) in RA patient (Rho *et al.* 2009).

Table 2 Adiponectin concentration in plasma/serum or synovial fluid

	Concentration of case/control (μg/mL)			
	RA	Normal	OA	Reference
plasma/serum	37.64	25.60	32.95	Senolt *et al.*
	13.56	7.6		Otero *et al.*
	4.116 (M), 6.017 (F)	2.352 (M), 3.487 (F)		Popa *et al.*
	4.543	3.369		Pemberton *et al.*
	22.8	16.7		Rho *et al.*
	8.9 (E), 11.6 (C)	4.8	14.1	Laurberg *et al.*
	17.9	10.9		Dermemezis *et al.*
	10.2	9.1	11.0	Tan *et al.*
synovial fluid	2.2		1.1	Schäffler *et al.*
	11.67		5.8	Senolt *et al.*
	1.2		0.6	Tan *et al.*

M, male; F, female; E, early RA; C, chronic RA; RA, rheumatoid arthritis; OA, osteoarthritis.

EFFECT OF ADIPONECTIN ON RSF AND CHONDROCYTES

There have been arguments about the role of adiponectin in the disease state of RA. However, Ehling *et al.* found the expression of adiponectin mRNA and protein in synovial tissue, not only in adipocytes but also in the synovial layer (Ehling *et al.* 2006). It has also been demonstrated that adiponectin receptors are expressed by RSF (Fig. 2) and chondrocytes (Lago *et al.* 2008). These reports indicate that adiponectin might regulate synovial inflammation and joint destruction in RA.

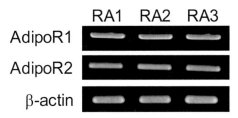

Fig. 2 Adiponectin receptor mRNA expression in synovial fibroblasts of patients with rheumatoid arthritis. Total RNA from rheumatoid synovial fibroblasts was subjected to the reverse transcription–polymerase chain reaction for adiponectin receptor 1 (AdipoR1), adiponectin receptor 2 (AdipoR2), and β-actin. Cells from three patients with RA (RA1-3) were used.

CYTOKINES

Pro-inflammatory Cytokines

IL-6 is a multifunctional cytokine that regulates the inflammatory response and immune reaction in RA patients. IL-6 induces not only mature B cell differentiation, but also the formation of osteoclasts, production of cytokines by macrophages, and Th17 cell differentiation. Ehling *et al.* (2006) reported that adiponectin enhances the production of IL-6 by RSF. Tang *et al.* (2007) showed that adiponectin receptor 1 (AdipoR1) was associated with the induction of IL-6 production from RSF by adiponectin through the activation of 5'-adenosine monophosphate-activated protein kinase (AMPK), p38 MAP kinase, and nuclear factor-κB (NF-κB). In addition, Lee *et al.* (2008) reported that adiponectin increases IL-6 expression in IL-1β–stimulated RSF. Therefore, adiponectin might regulate RA joint disease via up-regulation of IL-6 production by RSF. Moreover, Lago *et al.* (2008) showed that adiponectin stimulates IL-6 production by ATDC5 cells, a chondrogenic cell line. There is thus a possibility that adiponectin could regulate bone turnover by acting on chondrocytes.

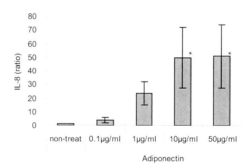

Fig. 3 Adiponectin induces Interleukin-8 (IL-8) production in synovial fibroblast of patient with rheumatoid arthritis. Cells were treated with each dose of adiponectin for 24 h and levels of IL-8 in culture supernatant were measured by ELISA. The IL-8 level of control (non-treat) was 0.11-10.03 ng/ml. *p < 0.05 vs. non-treat. Reprinted from Kitahara *et al.* (2009), with permission from Elsevier Inc.

Chemokines and Growth Factors

In the synovial fluid of RA patients, there is an increase of IL-8 and monocyte chemoattractant protein 1 (MCP-1). These chemokines induce the migration of neutrophils and monocytes that initiate the immune reaction and also promote overproduction of reactive oxygen species by these cells. Recently, we (Kitahara *et al.* 2009) reported that adiponectin induces IL-8 production by RSF (Fig. 3) and

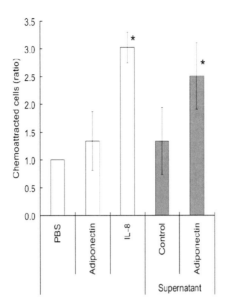

Fig. 4 Effect of adiponectin on chemotaxis of human polymorphonuclear cells. Rheumatoid synovial fibroblasts were treated with 10 μg/ml of adiponectin or phosphate-buffered saline (PBS); culture supernatants were collected after 24 h. Either 29 μl of vehicle, adiponectin (10 μg/ml), cell culture supernatants, or IL-8 dilutions (100 ng/ml) (positive control) was added to the bottom chamber, and 25 μl of polymorphonuclear cell suspension was placed onto a filter. After incubation for 1 h, the ratio of directed migration to random migration (D/R ratio) was calculated; D/R = total fluorescence of wells containing chemoattractant/total fluorescence of wells containing vehicle. *p < 0.05 vs. vehicle. Reprinted from Kitahara *et al.* (2009), with permission from Elsevier Inc.

enhances the chemotaxis of polymorphonuclear cells (Fig. 4). Tan *et al.* (2009) demonstrated that the adiponectin concentration was higher in the synovial fluid of RA patients than in OA patients, and they showed that adiponectin induces MCP-1 production by RSF. Lago *et al.* (2008) reported that adiponectin stimulates MCP-1 production by chondrocytes, with this effect being suppressed by addition of LY294002, an inhibitor of phosphatidylinositol 3 (PI3)-kinase. These studies have suggested that adiponectin may trigger the excessive immune response involved in the pathogenesis of RA.

Growth of new vessels is required for the development of rheumatoid pannus, which then leads to extensive synovial inflammation and joint destruction. VEGF is the best known mediator of angiogenesis. Choi *et al.* (2009) investigated whether adiponectin regulates VEGF expression by RSF and found that adiponectin enhances VEGF production from RSF, as it does for IL-1β. This suggested that adiponectin might regulate the development of pannus by modulating angiogenesis in RA synovial tissue.

ENZYMES

Matrix Metalloproteinases (MMPs)

MMPs are a family of zinc-dependent proteinases involved in the degradation and remodeling of extracellular matrix proteins (Table 3). Pro-inflammatory cytokines, such as IL-1β and TNFα, stimulate the production of MMPs. The collagenases MMP-1 and MMP-13 have an important role in RA because they are rate-limiting enzymes in the process of collagen degradation. MMP-13 also degrades the proteoglycan aggrecan, giving it a dual role in matrix destruction. MMP-3 and MMP-9 activity is also elevated in RA, and these enzymes degrade non-collagen matrix components in the joints. Adiponectin has been reported to induce the expression of MMP-1 and MMP-13 in RSF (Choi *et al.* 2009), and it also enhances MMP-3 and MMP-9 production by chondrocytes (Lago *et al.* 2008). Therefore, adiponectin might promote joint destruction through the induction of MMPs.

Table 3   Key features of matrix metalloproteinases

- MMPs are proteases that degrade extracellular matrix proteins, such as collagens, elastin, proteoglycans, and laminins.
- MMPs family includes over 20 proteins in mammals.
- MMPs are secreted or anchored to the cell surface.
- Some members of MMPs are expressed in normal tissues, implying a role in homeostasis.
- Most of MMPs are not expressed in resting tissue yet are induced in repair or remodeling processes and in diseased or inflammed tissues.
- Since MMPs can degrade components of cartilage, they play a central role in cartilage destruction of RA patients.

RA, rheumatoid arthritis.

Nitric Oxide Synthase (NOS)

Pro-inflammatory cytokines increase the synthesis of NO through the inducible enzyme iNOS. NO has various effects as a mediator of pro-inflammatory cytokine-induced responses, including inhibition of aggrecan and collagen synthesis, enhancement of MMP activity, and promotion of chondrocyte apoptosis. It has been reported that adiponectin induces iNOS expression in chondrocytes, resulting in the overproduction of NO (Lago *et al.* 2008).

Cyclooxygenase and Prostaglandin E Synthase

Prostaglandins (PGs) were identified and characterized in the 1960s. PGs and cyclooxygenase (COX), a key enzyme in their production, have been demonstrated to play a significant role in articular diseases. There are two isozymes of COX, designated COX-1 and COX-2. COX-1 is expressed constitutively by various cells and tissues, and it has an important role in maintaining homeostasis. In contrast, production of COX-2 by inflammatory cells and tissues is induced via various stimuli, suggesting that this isozyme has a key role in the process of inflammation. Among the PGs, PGE_2 is one of the major mediators of inflammation and high PGE_2 levels have been detected in the synovial fluid of RA patients.

Prostaglandin E synthase (PGES) is a key enzyme in the biosynthesis of PGE_2 that acts downstream of COX. To date, at least three forms of PGES (mPGES-1, mPGES-2, and cPGES) have been cloned and characterized. mPGES-1 has an important role in the production of PGE_2 during the development of chronic inflammation in patients with arthritic disorders such as RA or OA (Kojima *et al.* 2005). PGE_2 is a strong enhancer of IL-1β–induced mPGES-1 expression in RSF and chondrocyte (Kojima *et al.* 2003). Autoregulation of mPGES-1 expression by PGE_2 may play an important role in the vicious circle of inflammation associated with arthritis.

Recently, our study revealed that adiponectin induced the expression of COX-2 and mPGES-1 by RSF, and also enhanced PGE_2 production (Kusunoki *et al.* 2010). This PGE_2 production was inhibited by RNA interference for AdipoR1 and AdipoR2 gene, or co-incubation with the receptor signal inhibitors. We also confirmed nuclear translocation of NF-κB, similarly to IL-1β in RSF. These effects of adiponectin suggest that it may play an important role in inflammation associated with arthritis.

EFFECTS OF ADIPONECTIN ON OTHER CELLS

With regard to its role in inflammation, physiological concentrations of adiponectin have been shown to inhibit TNFα-induced adhesion of human monocytic THP-1 cells in a dose-dependent manner. Adiponectin also decreases TNFα-induced expression of vascular cell adhesion molecule 1, endothelial-leukocyte adhesion molecule 1, and intracellular adhesion molecule 1 by human aortic endothelial cells (Ouchi *et al.* 1999). In contrast, adiponectin activates NF-κB, a transcription factor that is essential for the expression of inflammatory proteins, in U937 cells in a time- and dose-dependent manner (Haugen and Drevon 2007). These findings suggest that adiponectin might have either anti-inflammatory and/or pro-inflammatory properties under different conditions.

IN THE FUTURE

Adiponectin might promote inflammation in RA joints by inducing the production of pro-inflammatory molecules in RSF and chondrocytes (Table 4). However, TNFα (production of which is increased in the insulin-resistant state) reduces the expression of adiponectin by adipocytes through suppression of promoter activity (Maeda *et al.* 2001). Because TNFα is a key mediator of the pathophysiology of RA, the correlation between adiponectin levels and disease activities of RA patients has

Table 4 Molecules reported to show increased/decreased expressions of mRNA and /or protein in response to adiponectin in rheumatoid synovial fibroblast or articular chondrocytes

| Molecule | Cells | | Reference |
	RSF	Chondrocytes	
IL-6	Increased	Increased	Ehling *et al.* 2006, Tang *et al.* 2007, Lee *et al.* 2006, Tan *et al.* 2009, Lago *et al.* 2008
IL-8	Increased		Kitahara *et al.* 2009
MCP-1	Increased	Increased	Tan *et al.* 2009, Lago *et al.* 2008
VEGF	Increased		Choi *et al.* 2009
MMP-1	Increased		Choi *et al.* 2009
MMP-3	Decreased	Increased	Lee *et al.* 2008, Lago *et al.* 2008
MMP-9		Increased	Lago *et al.* 2008
MMP-13	Increased		Choi *et al.* 2009
iNOS		Increased	Lago *et al.* 2008
COX-2	Increased		Kusunoki *et al.* 2010
mPGES-1	Increased		Kusunoki *et al.* 2010

COX-2, cyclooxygenase-2; IL, interleukin; iNOS, inducible nitric oxide synthase; MCP-1, monocyte chemoattractant protein 1; MMP, matrix metalloproteinase; mPGES-1, membrane-associated prostaglandin E synthase-1; VEGF, vascular endothelial growth factor.

been assessed during anti-TNFα therapy. There have been reports that indicate a positive, negative, or no correlation with disease amelioration. Radiographic joint damage is an important factor influencing the quality of life of RA patients. Analysis of the relation of adiponectin levels to radiographic damage has also been done by several researchers, but conflicting results have been obtained. Recently, Giles *et al.* (2009) reported that adiponectin may represent the link between low adiposity and increased radiographic damage in RA. They observed a strong cross-sectional association between increasing serum adiponectin concentration and both radiographic erosions and narrowing of joint space. Detailed basic and clinical investigation will be performed in the future.

Summary Points

- It seems possible that other pro-inflammatory cytokines, in addition to interleukin (IL)-1β, tumor necrosis factor α, and IL-6, may contribute to the pathophysiology of inflammation in patients with rheumatoid arthritis (RA).
- There are abundant adipocytes in the synovial tissue of RA patients.
- Adiponectin stimulates the production of IL-6, monocyte chemoattractant protein (MCP) 1, IL-8, vascular endothelial growth factor, and prostaglandin E_2 by rheumatoid synovial fibroblasts.

- Adiponectin stimulates the production of MCP-1 and enhances the expression of matrix metalloproteinase and inducible nitric oxide synthase by chondrocytes.
- In RA joints, adiponectin might have a pro-inflammatory effect.

Abbreviations

AdipoR	:	adiponectin receptor
AMPK	:	5'-adenosine monophosphate-activated protein kinase
COX	:	cyclooxygenase
IL	:	interleukin
MCP-1	:	monocyte chemoattractant protein-1
MMP	:	matrix metalloproteinase
mPGES-1	:	membrane-associated prostaglandin E synthase-1
NF-κB	:	nuclear factor-κB
NOS	:	nitric oxide synthase
OA	:	osteoarthritis
PG	:	prostaglandin
PI3	:	phosphatidylinositol 3
RA	:	rheumatoid arthritis
RSF	:	rheumatoid synovial fibroblasts
TNFα	:	tumor necrosis factor α
VEGF	:	vascular endothelial growth factor

Definition of Terms

Chondrocyte: A type of cell found in cartilage. It produces and maintains the cartilaginous matrix, which mainly consists of collagen and proteoglycans.

Cyclooxygenase-2 (COX-2): A key enzyme in the metabolism of arachidonic acid to prostaglandins such as prostaglandin E_2.

Interleukin-8 (IL-8): A chemokine secreted by several cell types, which is one of the major mediators of the inflammatory response and functions as a chemoattractant.

Rheumatoid arthritis (RA): A systemic chronic inflammatory condition of uncertain etiology with an autoimmune background, which is characterized by extensive inflammation and proliferation of the synovium in various joints.

Synovial fibroblast: A type of cell found in synovial tissue. When activated by pro-inflammatory cytokines, it proliferates and produces several bioactive molecules that influence the microenvironment of the joint.

Acknowledgements

This work was supported in part by research grants from The Japanese Ministry of Education, Culture, Sports, Science and Technology (no. 17590477 to SK), The Ministry of Health, Labour and Welfare of Japan (to SK), and a Research Promotion grant from Toho University Graduate School of Medicine (no. 07-04 to SK). We wish to express our gratitude to Professor Toru Suguro, who provided the synovial tissue of RA patients. We also thank Sonoko Sakurai for her secretarial assistance.

References

Brennan, F.M. and I.B. McInnes. 2008. Evidence that cytokines play a role in rheumatoid arthritis. J. Clin. Invest. 118: 3537-3545.

Choi, H.M. and Y.A. Lee, S.H. Lee, S.J. Hong, D.H. Hahm, S.Y. Choi, H.I. Yang, M.C. Yoo, and K.S. Kim. 2009. Adiponectin may contribute to synovitis and joint destruction in rheumatoid arthritis by stimulating vascular endothelial growth factor, matrix metalloproteinase-1, and matrix metalloproteinase-13 expression in fibroblast-like synoviocytes more than proinflammatory mediators. Arthritis Res. Ther. 11: R161.

Derdemezis, C.S. and T.D. Filippatos, P.V. Voulgari, A.D. Tselepis, A.A. Drosos, and D.N. Kiortsis. 2009. Effects of a 6-month infliximab treatment on plasma levels of leptin and adiponectin in patients with rheumatoid arthritis. Fundam. Clin. Pharmacol. 23: 595-600.

Ehling, A. and A. Schäffler, H. Herfarth, I.H. Tarner, S. Anders, O. Distler, G. Paul, J. Distler, S. Gay, J. Schölmerich, E. Neumann, and U. Müller-Ladner. 2006. The potential of adiponectin in driving arthritis. J. Immunol. 176: 4468-4478.

Giles, J.T. and M. Allison, C.O. Bingham 3rd, W.M. Scott Jr., and J.M. Bathon. 2009. Adiponectin is a mediator of the inverse association of adiposity with radiographic damage in rheumatoid arthritis. Arthritis Rheum. 61: 1248-1256.

Haugen, F. and C.A. Drevon. 2007. Activation of nuclear factor-kappaB by high molecular weight and globular adiponectin. Endocrinology 148: 5478-5486.

Kadowaki, T. and T. Yamauchi. 2005. Adiponectin and adiponectin receptors. Endocrine Rev. 26: 439-451.

Kawai, S. 2003. Current drug therapy for rheumatoid arthritis. J. Orthop. ScI. 8: 259-263.

Kitahara, K. and N. Kusunoki, T. Kakiuchi, T. Suguro, and S. Kawai. 2009. Adiponectin stimulates IL-8 production by rheumatoid synovial fibroblasts. Biochem. Biophys. Res. Commun. 378: 218-223.

Kojima, F. and H. Naraba, Y. Sasaki, M. Beppu, H. Aoki, and S. Kawai. 2003. Prostaglandin E2 is an enhancer of interleukin-1β–induced expression of membrane-associated prostaglandin E synthase in rheumatoid synovial fibroblasts. Arthritis Rheum. 48: 2819-2828.

Kojima, F. and S. Kato, and S. Kawai. 2005. Prostaglandin E synthase in the pathophysiology of arthritis. Fundam. Clin. Pharmacol. 19: 255-261.

Kusunoki, N. and R. Yamazaki, and S. Kawai. 2008. Pro-apoptotic effect of nonsteroidal anti-inflammatory drugs on synovial fibroblasts. Mod. Rheumatol. 18:542-551.

Kusunoki, N. and K. Kitahara, F. Kojima, N. Tanaka, K. Kaneko, H. Endo, T. Suguro, and S. Kawai. 2010. Adiponectin stimulates prostaglandin E_2 production in rheumatoid synovial fibroblasts. Arthritis Rheum. 62: 1641-1649.

Lago, R. and R. Gomez, M. Otero, F. Lago, R. Gallego, C. Dieguez, J.J. Gomez-Reino, and O. Gualillo. 2008. A new player in cartilage homeostasis: adiponectin induces nitric oxide synthase type II and pro-inflammatory cytokines in chondrocytes. Osteoarthritis Cartilage 16: 1101-1109.

Laurberg, T.B. and J. Frystyk, T. Ellingsen, I.T. Hansen, A. Jørgensen, U. Tarp, M.L. Hetland, K. Hørslev-Petersen, N. Hornung, J.H. Poulsen, A. Flyvbjerg, and K. Stengaard-Pedersen. 2009. Plasma adiponectin in patients with active, early, and chronic rheumatoid arthritis who are steroid- and disease-modifying antirheumatic drug-naive compared with patients with osteoarthritis and controls. J. Rheumatol. 36: 1885-1891.

Lee, S.W. and J.H. Kim, M.C. Park, Y.B. Park, and S.K Lee. 2008. Adiponectin mitigates the severity of arthritis in mice with collagen-induced arthritis. Scand. J. Rheumatol. 37: 260-268.

Maeda, K. and K. Okubo, I. Shimomura, T. Funahashi, Y. Matsuzawa, and K. Matsubara. 1996. cDNA cloning and expression of a novel adipose specific collagen-like factor, apM1 (AdiPose Most abundant Gene transcript 1). Biochem. Biophys. Res. Commun. 221: 286-289.

Maeda, N. and M. Takahashi, T. Funahashi, S. Kihara, H. Nishizawa, K. Kishida, H. Nagaretani, M. Matsuda, R. Komuro, N. Ouchi, H. Kuriyama, K. Hotta, T. Nakamura, I. Shimomura, and Y. Matsuzawa. 2001. PPARgamma ligands increase expression and plasma concentrations of adiponectin, an adipose-derived protein. Diabetes 50: 2094-2099.

Otero, M. and R. Lago, R. Gomez, F. Lago, C. Dieguez, J.J. Gómez-Reino, and O. Gualillo. 2006. Changes in plasma levels of fat-derived hormones adiponectin, leptin, resistin and visfatin in patients with rheumatoid arthritis. Ann. Rheum. Dis. 65: 1198-1201.

Ouchi, N. and S. Kihara, Y. Arita, K. Maeda, H. Kuriyama, Y. Okamoto, K. Hotta, M. Nishida, M. Takahashi, T. Nakamura, S. Yamashita, T. Funahashi, and Y. Matsuzawa. 1999. Novel modulator for endothelial adhesion molecules: adipocyte-derived plasma protein adiponectin. Circulation 100: 2473-2476.

Popa, C. and M.G. Netea, J. de Graaf, F.H. van den Hoogen, T.R. Radstake, H. Toenhake-Dijkstra, P.L. van Riel, J.W. van der Meer, A.F. Stalenhoef, and P. Barrera. 2009. Circulating leptin and adiponectin concentrations during tumor necrosis factor blockade in patients with active rheumatoid arthritis. J. Rheumatol. 36: 724-730.

Pemberton, P.W. and Y. Ahmad, H. Bodill, D. Lokko, S.L. Hider, A.P. Yates, M.G. Walker, I. Laing, and I.N. Bruce. 2009. Biomarkers of oxidant stress, insulin sensitivity and endothelial activation in rheumatoid arthritis: a cross-sectional study of their association with accelerated atherosclerosis. BMC Res. Notes 2: 83.

Rho, Y.H. and J. Solus, T. Sokka, A. Oeser, C.P. Chung, T. Gebretsadik, A. Shintani, T. Pincus, and C.M. Stein. 2009. Adipocytokines are associated with radiographic joint damage in rheumatoid arthritis. Arthritis Rheum. 60: 1906-1914.

Schäffler, A. and A. Ehling, E. Neumann, H. Herfarth, I. Tarner, J. Schölmerich, U. Müller-Ladner, and S. Gay. 2003. Adipocytokines in synovial fluid. JAMA 290: 1709-1710.

Senolt, L. and K. Pavelka, D Housa, and M. Haluzík. 2006. Increased adiponectin is negatively linked to the local inflammatory process in patients with rheumatoid arthritis. Cytokine 35: 247-252.

Tan, W. and F. Wang, M. Zhang, D. Guo, Q. Zhang, and S. He. 2009. High adiponectin and adiponectin receptor 1 expression in synovial fluids and synovial tissues of patients with rheumatoid arthritis. Semin. Arthritis Rheum. 38: 420-427.

Tang, C.H. and Y.C. Chiu, T.W. Tan, R.S. Yang, and W.M. Fu. 2007. Adiponectin enhances IL-6 production in human synovial fibroblast via an AdipoR1 receptor, AMPK, p38, and NF-kappa B pathway. J. Immunol. 179: 5483-5492.

CHAPTER 35

Resistin, Inflammation and Seminal Fluid[†]

Uwe Paasch,[2] Juergen Kratzsch,[1] Sonja Grunewald,[2] Michael Gerber[1] and Hans-Juergen Glander[2,*]

[1]Institute of Laboratory Medicine, Clinical Chemistry and Molecular Diagnostics, Paul-List-Str.13-15
[2]Training Centre of the European Academy of Andrology, Campus Leipzig, Department of Dermatology, University of Leipzig, Ph.-Rosenthal-Strasse 23, 04103 Leipzig, Germany

ABSTRACT

Resistin is supposed to play a role in the inflammatory process in humans under the stimulation of inflammatory cytokines such as interleukin 6 (IL-6). Therefore, it is of interest whether adipocytokine resistin also correlates with markers of inflammation such as elastase and IL-6 in human seminal fluid and with semen sample quality. Therefore, non-obese infertility patients without clinical symptoms of genital tract inflammation and post-vasectomy patients were investigated with regard to concentrations of resistin, elastase and IL-6 in the seminal fluid and classical spermiogram parameters (sperm concentration, motility and morphology). Resistin expression in human seminal fluid, in testicular tissue on Leydig cells and within tubuli seminiferi was detectable at a median concentration of 1.62 ng/ml together with IL-6 (12.0 pg/ml) and elastase (16.7 ng/ml). The concentrations of resistin correlated well with the inflammatory markers elastase (r = 0.86; P < 0.01) and IL-6 (r = 0.71; P < 0.01) in seminal fluid, but not with semen sample quality. There were no significant differences between the resistin concentration in the semen samples with normal spermiogram parameters and those with subnormal

---

[†]*Declaration: The authors report no financial or commercial conflicts of interest.*
*Corresponding author

spermiogram parameters. In addition, no significant relationship was observed between resistin in semen and the serum hormones FSH, inhibin B, testosterone and serum resistin or body mass index. Moreover, vasectomy did not significantly reduce the total amount of resistin in the semen samples.

INTRODUCTION

Being overweight or obese is suspected to impair semen sample quality (Jensen *et al.* 2004). Men who are overweight show elevated numbers of sperm with activated apoptosis signaling, DNA fragmentation (Grunewald *et al.* 2009) and proteomic changes (Kriegel *et al.* 2009) compared with men of normal weight. These observations drew attention to the adipocytokines. One of the adipocytokines is resistin, a 12.5 kDa cysteine-rich adipocyte-secreted peptide, which is known as resistin-like molecule (RELM) or FIZZ (found in the inflammatory zones) (Holcomb *et al.* 2000; Steppan *et al.* 2001). In rodents resistin released from adipocytes was initially postulated to contribute to insulin resistance with subsequent suppression of insulin-stimulated glucose uptake (Steppan *et al.* 2001). The expression of resistin is induced in rodents by several endocrine factors, e.g., testosterone, growth hormone and prolactin (Banerjee and Lazar 2003). Other studies in humans have indicated that resistin might also play a role in the inflammatory process (Steppan and Lazar 2004). In humans resistin is mainly expressed in mononuclear leukocytes, macrophages (Yang *et al.* 2003) and endothelial cells (Kawanami *et al.* 2004) under the stimulation of inflammatory cytokines such as interleukin 6, IL-6 (Reilly *et al.* 2005). Discrepancies in the biological role of resistin between the species are likely based on the lack of similarity between rodent and human resistin sequence and function (Banerjee and Lazar 2003), as well as its site of production (Nogueiras *et al.* 2004, Bokarewa *et al.* 2005). Moreover, on one hand resistin appears to control testosterone secretion, and on the other hand resistin expression is modulated by hormones (Nogueiras *et al.* 2004, Gerber *et al.* 2005). These associations suggest the need for examination of the role of resistin in the male genital tract. Therefore, resistin in human seminal fluid and testicular tissue was investigated in relation to elastase and IL-6, serum male sexual hormones, spermiogram variables and body mass index (BMI).

Table I Characteristics of the 72 examined patients

Characteristics	Value
Age (yr)	36.0; 28.0-43.0
Body mass index	24.5; 22-28
FSH (IU/L)	3.7; 1.8-8.4
Inhibin B (ng/L)	105.0; 49.2-159.0
Testosterone (nmol/L)	16.3; 10.6-22.8

Values are given as median; 10th to 90th percentiles.

Table 2   Characteristics of the semen sample groups

	Normospermia (N = 21)	Pathozoospermia (N = 33)
Volume (ml)	3.5; 2.0-5.0	3.5; 2.0-5.0
Total sperm count (Mio/ejaculate)	308.0; 50.0-529.0 [a]	116.0; 27.0-345.0 [a]
Sperm with normal morphology (%)	16.0; 15-21.0 [b]	3.0; 1.0-8.0 [b]
Progressively motile sperm (%)	63.3; 50.8-80.3 [c]	36.8; 11.4-72.5 [c]
VSL (µm/s)	43.9; 22.8-53.4	38.4; 22.5-58.1
VAP (µm/s)	59.8; 27.9-66.8	50.6; 32.6-78.5
Sperm with intact acrosome (%)	28.0; 11.0-62.0	21.0; 9.0-35.0
Seminal plasma fructose (µmol/ejaculate)	43.5; 16.8-114.0	38.6; 17.3-76.8

Mann Whitney U-Test; VAP= velocity average path; VSL= velocity straight line; values with an identical character are significantly different. Values are given as median; 10th to 90th percentiles.

ANALYSIS OF THE ROLE OF RESISTIN, INFLAMMATION SIGNALING AND ITS ORIGINS IN MEN

Concentrations of Resistin, IL-6 and Elastase in Seminal Plasma of Infertility Patients

Resistin, IL-6, and elastase were present in the human seminal plasma of all infertility patients without clinical signs of inflammation at concentrations described in Table 3. The resistin concentration in the serum of patients amounted to 4.03 (3.06 5.84) ng/ml and differed significantly from that in the seminal plasma, which was determined at 1.62 (0.6-23.6) ng/ml (median, 10th–90th percentiles; $p = 0.03$). The semen samples were divided into 'normal' and 'non-normal' or 'pathological' samples with respect to the cut-off point and lower limit of reference values according to World Health Organization (1999) guidelines. There were no significant differences ($P > 0.05$) between the concentrations of resistin, elastase, and IL-6 in the 'pathological' semen sample group and the 'normal' group (Table 4).

Table 3   Concentration and quantity of resistin, interleukin 6 and elastase in seminal plasma from 54 infertility patients

Parameter	Seminal plasma concentration	Total amount per ejaculate
Resistin	1.62; 0.6-23.6 (ng/ml)	5.61; 1.8-82.0 (ng)
Interleukin 6	12.0; 3.0-52.0 (pg/ml)	47.6; 10.5-180.6 (pg)
Elastase	16.7; 2.0-432.0 (ng/ml)	72.8; 6.0-1446.0 (ng)

Values are given as median; 10th to 90th percentiles.

Table 4 Concentration and quantity of resistin, interleukin 6 (IL-6) and elastase in seminal plasma from infertility patients classified as either normozoospermic or pathozoospermic

Parameter	Normozoospermia (N = 21)	Pathozoospermia (N = 33)
Resistin concentration (ng/ml)	1.76; 0.4-34.3	1.2; 0.7-23.4
Total resistin (ng/ejaculate)	6.23; 1.8-70.6	4.6; 2.0-82.0
IL-6 concentration (pg/ml)	11.9; 3.0-54.4	12.0; 4.0-51.6
Total IL-6 (pg/ejaculate)	47.6; 7.5-227.2	46.4; 13.2-158.4
Elastase concentration (ng/ml)	30.4; 2.0-265.0	15.7; 2.0-467.0
Total elastase (ng/ejaculate)	103.6; 5.0-1105.0	63.0; 6.0-1512.0

IL-6 = interleukin 6. Values are given as median; 10th to 90th percentiles. No significant difference was found between any of the values for the two groups.

Potential of Resistin and Macrophages in Testicular Tissue

Resistin was found by immunohistochemistry within the tubuli seminiferi of human testis of our patients with severe pathozoospermia and Sertoli cell only syndrome. Besides the Sertoli cells, the perinuclear region of spermatogonia localized at lamina propria were labeled by anti-resistin antibodies. In patients with Sertoli cell only syndrome, resistin was preferentially found within the tubuli seminiferi. Outside of the tubuli, resistin was weakly detectable on Leydig cells. Macrophages were seen in all the tissues sections. Figure 1 shows an example of localization of resistin and labeling of macrophages by anti-CD68 antibodies in human testicular tissue.

Relationship between Resistin and Elastase and Il-6 in Seminal Plasma to Routine Spermiogram Parameters and Hormones

Resistin concentrations in the seminal plasma were significantly correlated with the concentration of elastase ($r = 0.86$; $P = 0.00001$; Fig. 2a) and IL-6 ($r = 0.71$; $P < 0.00001$; Fig. 2b). Elastase and IL-6 expression significantly correlated in the seminal plasma ($r = 0.63$; $P < 0.00001$). Apart from these correlations, no significant associations of seminal plasma resistin were observed, either with spermiogram parameters or with hormones FSH ($r = 0.14$), inhibin B ($r = 0.04$), testosterone ($r = 0.04$) and BMI ($r = 0.05$). The resistin concentration in serum did not significantly correlate with either its concentration in the seminal plasma ($r = 0.07$) or BMI ($r = 0.04$).

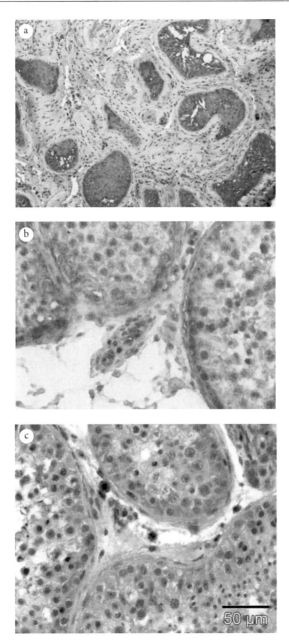

Figure 1. Examples of immunohistochemistry of resistin staining in a testis (a) without germ cells (Sertoli cell only syndrome), (b) with germ cells (impaired spermatogenesis) and (c) CD68-positive macrophages in human testis

Color image of this figure appears in the color plate section at the end of the book.

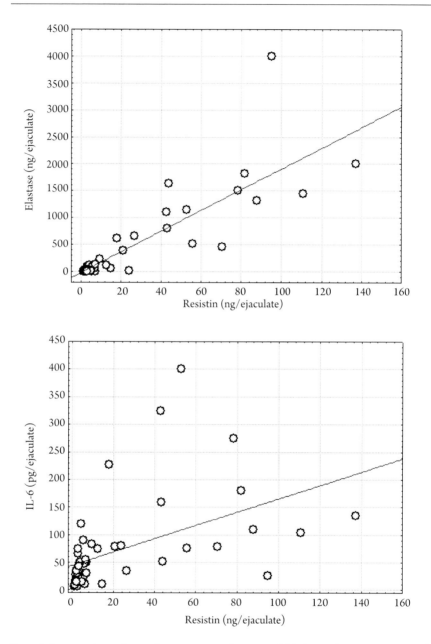

Figure 2. Spearman's correlation of concentration of resistin and (a) elastase (r = 0.86; P< 0.00001) and (b) interleukin-6 (IL-6) (r = 0.71; P< 0.00001) in the human seminal plasma of infertility patients.

Effect of vasectomy on concentration of resistin in seminal plasma

Vasectomy did not cause a significant alteration (P > 0.05) of total content of resistin in seminal fluid (4.11; 0.7-24.2 ng/ejaculate in the vasectomy group vs. 5.61; 1.8-82.0 ng/ejaculate in the non-vasectomized patients), although patients showed a significantly lower volume of semen samples of 2.6 ± 0.2 ml (mean ± SEM) after vasectomy when compared with semen volume from non-vasectomized patients (3.8 ± 0.3 ml, P = 0.002). The age of both groups differed significantly (P < 0.0001), but there was no significant correlation between age and seminal plasma concentration of resistin (r = 0.03).

Significance of Adipocytokines in Testis and Seminal Fluids and Its Correlation with Their Serum Equivalents

Resistin was detected in human seminal plasma, weakly on Leydig cells and within the tubuli seminiferi of human testis. A highly significant association between resistin and elastase and IL-6 was evident in the seminal plasma but no significant relationships between seminal plasma resistin and spermiogram parameters, FSH, inhibin B, testosterone and serum resistin were observed. In addition to the known localization of resistin in human, e.g., bone marrow, lung, spleen (Patel *et al.* 2003), adipose tissue (Fain *et al.* 2003), skeletal muscle cells (Nagaev and Smith 2001), pancreatic islet cells (Minn *et al.* 2003), serum (Yannakoulia *et al.* 2003), and synovial fluid (Bokarewa *et al.* 2005), we found resistin in the human seminal plasma, on Leydig cells and within the tubuli seminiferi of human testes with an association between resistin and markers of inflammation in human seminal plasma.

The physiological role of resistin and its regulation is still under debate. Increasingly, it has been recognized that in addition to metabolic effects resistin is also involved in inflammatory process and is expressed primarily in inflammatory cells (Kunnari *et al.* 2006). Inflammatory cells, such as macrophages, are found in testicular tissue (Frungieri *et al.* 2002) as we confirmed in our patients. Resistin expression in these cells is induced by pro-inflammatory cytokine (Kaser *et al.* 2003). Additionally, resistin up-regulates the expression of pro-inflammatory cytokines, such as IL-6 in leukocytes (Bokarewa *et al.* 2005). Thus, it is plausible that resistin correlates with markers of inflammation such as IL-6 and elastase.

The localization of resistin in human seminal plasma and in testicular tissue suggests three main questions: (1) What is the role of resistin in seminal plasma? (2) What are the target cells for resistin? (3) Which cells produce and secrete seminal plasma resistin? Because resistin is involved in the regulation of numerous substances, understanding its function and biology in the male genital tract is complicated. However, the results of highly significant (P < 0.00001) correlations between resistin, elastase and IL-6 in seminal plasma support the concept of a function of resistin as a pro-inflammatory cytokine. Thus, human seminal fluid resistin shows an analogous relationship to pro-inflammatory cytokines similar to what has already been described in the synovial fluid of rheumatoid arthritis patients (Bokarewa *et al.* 2005). The target tissue for resistin in seminal fluid is unknown and the source of its production is speculative. A possible source of resistin in seminal fluid could be mononuclear leukocytes (Kunnari *et al.* 2006) and the testicular Leydig cells (Nogueiras *et al.* 2004).

In rat testes, regulated resistin gene expression was demonstrated in Leydig cells and Sertoli cells (Nogueiras *et al.* 2004). In human testes, as shown in our patients, resistin was present in human tubuli seminiferi, but in contrast to the rat expression was not restricted to the Sertoli cells but was also found in the basal levels of spermatogenesis. Moreover, Leydig cells bound our anti-resistin antibodies very weakly. This might explain the lack of a relationship between resistin and testosterone in our patient group and by Ling *et al.* (2001) in rodents. Vasectomy did not significantly alter the total amount of resistin in seminal fluid, suggesting that resistin is multilocularly produced in the male genital tract. Seminal plasma resistin showed a significantly lower concentration ($P < 0.05$) than serum resistin, whereas the lack of a correlation between resistin concentration in serum and seminal plasma argues against a simple diffusion and distribution process in the male genital tract.

The concentration of serum resistin in the patient group corresponded to the concentrations determined by other authors (Reilly *et al.* 2005). Additionally, the concentrations of IL-6 (Eggert-Kruse *et al.* 2001) and elastase in the seminal plasma from these patients (Ludwig *et al.* 1998) are comparable with data from control groups of studies published earlier. However, most of the IL-6 and elastase concentrations were not in the pathological range, as defined by > 30 pg/ml for IL-6 (Eggert-Kruse *et al.* 2001) and by > 600 ng/ml for elastase (Ludwig *et al.* 1998). Moreover, all patients were free of genital tract clinical symptoms. This situation may explain the missing correlation between IL-6 and elastase with respect to spermiogram variables. The important question as to whether clinical inflammation of the genital tract can be mediated by resistin in seminal plasma must be addressed in a subsequent study. Nevertheless, in the male genital tract, resistin seems to be associated with inflammatory markers irrespective of semen sample quality.

SUmmary Points

- Adipocytokine resistin shows no correlation with its serum levels.
- Adipocytokine resistin in human seminal plasma significantly correlates with the inflammatory markers elastase and IL-6.
- Adipocytokine resistin in human seminal plasma does not significantly correlate with semen quality.
- These results suggest an inflammation-related role of resistin in the seminal fluid.
- As vasectomy did not significantly reduce the total amount of resistin in the semen samples, a multilocular production of resistin is postulated.

Key terms

- Conventional spermiogram parameters are used to determine male fertility.
- In infertility patients there are often subnormal spermiogram parameters.
- Ongoing inflammation may not only be detected by the presence of leukocyte within the seminal plasma.
- Elastase is a good indicator of ongoing adnexal inflammation like IL-6.
- Resistin is secreted under the stimulation of inflammatory cytokines such as interleukin 6 in humans.
- Adipocytokine resistin, a 12.5 kDa cysteine-rich adipocyte-secreted peptide, is known as resistin-like molecule (RELM) or FIZZ (found in the inflammatory zones).
- There is a lack of similarity between rodent and human resistin sequence.
- Resistin correlates with markers of inflammation such as IL-6 and elastase.

Acknowledgements

The authors gratefully acknowledge Gabriele Kersten and Annette Drechsler for their excellent technical assistance and Heidrun Janus for her care of the patients.

References

Banerjee, R.R. and M.A. Lazar. 2003. Resistin: molecular history and prognosis. J. Mol. Med. 81: 218-226.

Bokarewa, M. and I. Nagaev, L. Dahlberg, U. Smith, and A. Tarkowski. 2005. Resistin, an adipokine with potent proinflammatory properties. J. Immunol. 174: 5789-5795.

Cooper, T.G. and L. Björndahl, J. Vreeburg, and E. Nieschlag. 2002. Semen analysis and external quality control schemes for semen analysis need global standardization. Int. J. Androl. 25: 306-311.

Corea, M. and J. Campagnone, and M. Sigman. 2005. The diagnosis azoospermia depends on the force of centrifugation. Fertil. Steril. 83: 920-922.

Eggert-Kruse, W. and R. Boit, G. Rohr, J. Aufenanger, M. Hund, and T. Strowitzki. 2001. Relationship of seminal plasma interleukin (IL) -8 and IL-6 with semen quality. Hum. Reprod. 16: 517-528.

Fain, J.N. and P.S. Cheema, S.W. Bahouth, and M. Lloyd Hiler. 2003. Resistin release by human adipose tissue explants in primary culture. Biochem. Biophys. Res. Commun. 300: 674-678.

Frungieri, M.B. and R.S. Calandra, L. Lustig, V. Meineke, F.M. Köhn, H.J. Vogt, and A. Mayerhofer. 2002. Number, distribution pattern, and identification of macrophages in the testes of infertile men. Fertil. Steril. 78: 298-306.

Gerber, M. and A. Boettner, B. Seidel, A. Lammert, J. Bär, E. Schuster, J. Thiery, W. Kiess, and J. Kratzsch. 2005. Serum resistin levels of obese and lean children and adolescents: biochemical analysis and clinical relevance. J. Clin. Endocrinol. Metab. 90: 4503-4509.

Glander, H.-J. and L.-C. Horn, W. Dorschner, U. Paasch, and J. Kratzsch. 2000. Probability to retrieve testicular spermatozoa in azoospermic patients. Asian J. Androl. 2: 199-205.

Grunewald, S. and U. Paasch, T. Kriegel, F. Heidenreich, C. Roessner, and H.J. Glander. 2009. Obesity-related impairment of human sperm quality. Hum. Reprod. 24Suppl.: i33.

Haugen, T.B. and T. Egeland, and O. Magnus. 2006. Semen parameters in Norwegian fertile men. J. Androl. 27: 66–71.-

Holcomb, I.N. and R.C. Kabakoff, B. Chan, T.W. Baker, A. Gurney, W. Henzel, C. Nelson, H.B. Lowman, B.D. Wright, N.J. Skelton, G.D. Frantz, D.B. Tumas, F.V. Jr. Peale, D.L. Shelton, and C.C. Hébert. 2000. FIZZ1, a novel cysteine-rich secreted protein associated with pulmonary inflammation, defines a new gene family. EMBO J. 19: 4046-4055.

Jensen, T.K. and A.M. Andersson, N. Jorgensen, A.G. Andersen, E. Carlsen, J.H. Petersen, and N.E. Skakkebaek. 2004. Body mass index in relation to semen quality and reproductive hormones among 1,558 Danish men. Fertil. Steril. 82: 863-870.

Kaser, S. and A. Kaser, A. Sandhofer, C.F. Ebenbichler, H. Tilg, and J.R. Patsch. 2003. Resistin messenger-RNA expression is increased by proinflammatory cytokines in vitro. Biochem. Biophys. Res. Commun. 309: 286-290.

Kawanami, D. and K. Maemura, N. Takeda, T. Harada, T. Nojiri, Y. Imai, I. Manabe, K. Utsunomiya, and R. Nagai. 2004. Direct reciprocal effects of resistin and adiponectin on vascular endothelial cells: a new insight into adipocytokine–endothelial cell interactions. Biochem. Biophys. Res. Commun. 314: 415-419.

Kriegel, T.M. and F. Heidenreich, K. Kettner, T. Pursche, B. Hoflack, S. Grunewald, K. Poenicke, H.J. Glander, and U. Paasch. 2009. Identification of diabetes- and obesityassociated proteomic changes in human spermatozoa by difference gel electrophoresis. Reprod. BioMed. Online 19: 660-670.

Kunnari, A. and O. Ukkola, M. Paivansalo, and Y.A. Kesaniemi. 2006. High plasma resistin level is associated with enhanced highly sensitive C-reactive protein and leukocytes. J. Clin. Endocrinol. Metab. 91: 2755-2760.

Ling, C. and J. Kindblom, H. Wennbo, and H. Billig. 2001. Increased resistin expression in the adipose tissue of male prolactin transgenic mice and in male mice with elevated androgen levels. FEBS Lett. 507: 147-150.

Ludwig, M. and C. Kummel, I. Schroeder-Printzen, R.H. Ringert, and W. Weidner. 1998. Evaluation of seminal plasma parameters in patients with chronic prostatitis or leukocytospermia. Andrologia 30 (Suppl.1): 41-47.

Menkveld, R. and T.F. Kruger. 1995. Advantages of strict (Tygerberg) criteria for evaluation of sperm morphology. Int. J. Androl. 18 (Suppl.): 36-42.

Minn, A.H. and N.B. Patterson, S. Pack, S.C. Hoffmann, O. Gavrilova, C. Vinson, D.M. Harlan, and A. Shalev. 2003. Resistin is expressed in pancreatic islets. Biochem. Biophys. Res. Commun. 310: 641-645.

Nagaev, I. and U. Smith. 2001. Insulin resistance and type 2 diabetes are not related to resistin expression in human fat cells or skeletal muscle. Biochem. Biophys. Res. Commun. 285:561-564.

Nogueiras, R. and M.L. Barreiro, J.E. Caminos, F. Gaytan, J.S. Suominen, V.M. Navarro, F.F. Casanueva, E. Aguilar, J. Toppari, C. Dieguez, and M. Tena-Sempere. 2004. Novel expression of resistin in rat testis: functional role and regulation by nutritional status and hormonal factors. J. Cell Sci. 117: 3247-3257.

Patel, L. and A.C. Buckels, I.J. Kinghorn, P.R. Murdock, J.D. Holbrook, C. Plumpton, C.H. Macphee, and S.A. Smith. 2003. Resistin is expressed in human macrophages and directly regulated by PPAR gamma activators. Biochem. Biophys. Res. Commun. 300: 472-476.

Reilly, M.P. and M. Lehrke, M.L. Wolfe, A. Rohatgi, M.A. Lazar, and D.J. Rader. 2005. Resistin is an inflammatory marker of atherosclerosis in humans. Circulation 111: 932-939.

Steppan, C.M. and M.A. Lazar. 2004. The current biology of resistin. J. Int. Med. 255: 439-447.

Steppan, C.M. and S.T. Bailey, S. Bhat, E.J. Brown, R.R. Banerjee, C.M. Wright, H.R. Patel, R.S. Ahima, and M.A. Lazar. 2001. The hormone resistin links obesity to diabetes. Nature 409: 307-312.

World Health Organization. 1999. Laboratory manual for the examination of human semen and sperm–cervical mucus interaction, 4th ed. Cambridge University Press, Cambridge, UK.

Yang, R.Z. and Q. Huang, A. Xu, J.C. McLenithan, J.A. Eisen, A.R. Shuldiner, S. Alkan, and D.W. Gong. 2003. Comparative studies of resistin expression and phylogenomics in human and mouse. Biochem. Biophys. Res. Commun. 310: 927-935.

Yannakoulia, M. and N. Yiannakouris, S. Bluher, A.L. Matalas, D. Klimis-Zacas, and C.S. Mantzoros. 2003. Body fat mass and macronutrient intake in relation to circulating soluble leptin receptor, free leptin index, adiponectin, and resistin concentrations in healthy humans. J. Clin. Endocrinol. Metab. 88: 1730-1736.

CHAPTER 36

Resistin in Amniotic Fluid

Shali Mazaki–Tovi, Juan Pedro Kusanovic, Edi Vaisbuch and
Roberto Romero[*]
Perinatology Research Branch, Intramural Division, NICHD/NIH/DHHS
Department of Obstetrics and Gynecology, Wayne State University/Hutzel
Women's Hospital, 3990 John R, Detroit, Michigan 48201, USA

ABSTRACT

Resistin, a novel adipocytokine, has been implicated in the regulation of the innate immune response. Specifically, resistin has pro-inflammatory properties that are abrogated by NF-κB inhibitor, indicating the importance of NF-κB signaling pathway for resistin-induced inflammation. Consistent with this view, circulating plasma resistin concentrations are increased in patients with sepsis, and it has been proposed that plasma concentrations of this adipocytokine can serve as a marker of sepsis severity. Resistin has also been detected in several body fluids, including saliva, urine and synovial fluid. Indeed, resistin accumulates in the inflamed joints of patients with rheumatoid arthritis and its synovial concentrations correlate with other markers of inflammation. Recently, resistin was found to be a physiological constituent of amniotic fluid. In addition, high amniotic fluid resistin concentrations are associated with intra-amniotic infection and/or inflammation, and amniotic fluid resistin concentrations are associated with well-established indices of intra-amniotic inflammation such as amniotic fluid interleukin (IL) 6 concentration and white blood cell count, as well as with amniocentesis-to-delivery interval. The strong association with intra-amniotic infection/inflammation has been corroborated in an unbiased, high-throughput proteomics analysis of amniotic fluid. The aim of this chapter is to present the

---

[*]Corresponding author

available evidence regarding amniotic fluid resistin and to discuss their importance and possible implications.

INTRODUCTION

Resistin is a novel adipocytokine with a molecular weight of 12.5 kDa. Resistin belongs to the FIZZ (found in inflammatory zone) family, also know as RELM (resistin-like-molecules) (Kim *et al.* 2001, Steppan *et al.* 2001). The latter is a family of cysteine-rich proteins that include RELM-α, RELM-β and RELM-γ. In serum, resistin circulates in two different isoforms: a high-molecular-weight hexamer and a low-molecular-weight form with a significantly increased bioactivity (Patel *et al.* 2004). Resistin was originally identified as FIZZ3 in mice by Holcomb *et al.* (2000). Subsequently, two independent groups (Kim *et al.* 2001, Steppan *et al.* 2001) identified resistin as a potential link between obesity and insulin resistance. Steppan *et al.* (2001) discovered a protein that was called resistin after screening for genes induced during adipocyte differentiation but down-regulated in mature adipocytes exposed to thiazolidinediones (insulin-sensitizing agents). In mice, resistin is secreted almost exclusively by adipocytes, whereas studies in humans have shown that this protein is not tissue-specific. Indeed, resistin expression and/ or secretion has been determined in muscle, pancreatic islets, mononuclear cells, macrophages, neutrophils and placenta.

PHYSIOLOGICAL ROLE OF RESISTIN

The physiological role of resistin has not yet been fully clarified. Nevertheless, it has been proposed that this adipocytokine plays a role in glucose homeostasis, as well as in the regulation of the innate immune response. Studies in mice have provided compelling evidence for the metabolic effects of resistin. These findings include the following: (1) *in vitro* neutralization of resistin results in enhanced insulin-stimulated glucose uptake by adipocytes; (2) resistin mRNA expression in adipocytes is low during fasting but increases significantly when fasting mice are re-fed with a high-carbohydrate diet (25-fold) or treated with insulin (23-fold); (3) administration of resistin impairs glucose tolerance and insulin action in normal mice; (4) administration of anti-resistin IgG improves blood glucose and insulin action; and (5) resistin concentration is increased in both diet-induced and genetic obesity.

In contrast to murine models, the importance of resistin as a significant metabolic mediator in humans has been a subject of controversy (Arner 2005). Plasma resistin concentration correlates with insulin resistance indices and obesity (Silha *et al.* 2003) and was found to be higher in subjects with insulin resistance (Pagano *et al.* 2005). In addition, polymorphisms in the promoter region of the resistin gene were associated with insulin resistance in obese patients (Wang *et al.* 2002). In contrast to the above-mentioned studies, lack of correlation between circulating resistin, insulin resistance and obesity has been reported by other investigators (Heilbronn *et al.* 2004).

A compelling body of evidence supports a role for resistin in the regulation of the innate immune system and inflammation (Tilg and Moschen 2006). The following observations support this view: (1) resistin induces nuclear factor (NF) κB activity and its translocation from the cytoplasm to the nucleus (Bokarewa *et al.* 2005); (2) resistin mRNA expression by human peripheral blood mononuclear cells (PBMC) increases after exposure to interleukin (IL) 1 beta, IL-6, tumor necrosis factor (TNF) α and lipopolysaccharide (LPS) (Kaser *et al.* 2003); (3) stimulation of human macrophages with LPS results in an increase in resistin mRNA expression (Lehrke *et al.* 2004); (4) resistin up-regulates the expression of TNF-α and IL-6 by human PBMC (Bokarewa *et al.* 2005); (5) administration of LPS to human subjects results in a significant increase in circulating resistin concentrations (Lehrke *et al.* 2004); (6) patients with sepsis or septic shock have a higher circulating resistin concentration than controls (Sunden-Cullberg *et al.* 2007); and (7) serum resistin concentrations are higher in patients with chronic inflammatory processes such as alcoholic liver disease, hepatitis C virus–induced chronic hepatitis (Sunden-Cullberg *et al.* 2007), rheumatoid arthritis and osteoarthritis (Schaffler *et al.* 2003), and chronic kidney disease (Schaffler *et al.* 2003).

ROLE OF MATERNAL CIRCULATING RESISTIN IN HUMAN GESTATION

Consistent with its suggested role in the non-pregnant state, resistin has been implicated in the metabolic adaptations to normal gestation, as well as in complications of pregnancy. In normal pregnant women, circulating resistin concentrations increased as a function of gestational age (Chen *et al.* 2005, Nien *et al.* 2007) and alterations in maternal resistin concentrations have been reported in preeclampsia (Cortelazzi *et al.* 2007) and gestational diabetes mellitus. These observations are consistent with reports in which other cytokines and adipocytokines (Kusanovic *et al.* 2008, Nien *et al.* 2007, Vaisbuch *et al.* 2009, Mazaki-Tovi *et al.* 2008, 2009a, b, c, d, e, f) have been associated with pathological conditions in human pregnancy. In a study that included 441 women (Nien *et al.* 2007), we reported the following observations: (1) pregnant women of normal weight had a significantly higher median plasma resistin concentration than non-pregnant women; (2) maternal plasma concentrations of resistin were significantly higher at term than in the first, second, or early third trimester; and (3) during normal gestation there were no significant differences in maternal plasma resistin concentrations between normal weight and overweight/obese women. In addition, we reported a nomogram for maternal plasma resistin during normal pregnancy (Nien *et al.* 2007).

RATIONALE TO DETERMINE RESISTIN CONCENTRATIONS IN AMNIOTIC FLUID

The rationale to investigate amniotic fluid resistin in normal and abnormal pregnancy rests on the following findings: (1) resistin has been identified in several other body fluids including synovial fluid, saliva, and urine. Furthermore, alteration in its concentrations in these compartment is associated with pathological conditions; (2) resistin is produced and secreted by innate immune cells such as neutrophils and macrophages that are found in amniotic fluid in pathological conditions such as intra-amniotic infection/inflammation (IAI); (3) other adipocytokines such as leptin, adiponectin (Mazaki-Tovi *et al.* 2009f), visfatin (Mazaki-Tovi *et al.* 2008), and retinol binding protein-4 (Vaisbuch *et al.* 2009) have been determined in amniotic fluid and alterations in their concentrations have been associated with adverse pregnancy outcome; (4) resistin can be secreted by amnion cells (Lappas *et al.* 2005) and its release is greatly affected by a variety of inflammatory stimuli and mediators such as LPS, TNF-α and IL-6.

To date, only three studies have reported resistin concentrations in amniotic fluid. We were the first to report the presence of this adipocytokine in amniotic fluid (Kusanovic *et al.* 2008). This observation was made using commercially available sensitive and specific ELISA kits. Subsequently, using Isobaric Tag for Relative and Absolute Quantitation (iTRAQ^{TM}) method, we determined that resistin is differentially expressed in the amniotic fluid of patients with IAI (Romero *et al.* 2009). Bugatto *et al.* (2010) have reported a significant correlation between amniotic fluid resistin concentrations and maternal body mass index.

AMNIOTIC FLUID RESISTIN CONCENTRATIONS IN MID-TRIMESTER

Our group has conducted a cross-sectional study in which amniotic fluid resistin concentrations were determined in women in the mid-trimester of pregnancy (14-18 wk) who underwent amniocentesis for genetic indications and delivered a normal neonate at term (n = 61), as well as normal pregnant women at term with (n = 49) and without (n = 50) spontaneous labor. Resistin was detected in all amniotic fluid samples. Women with a normal pregnancy at term not in labor had a significantly higher median resistin concentration in amniotic fluid than women in the mid-trimester (23.6 ng/mL vs. 10 ng/mL, p < 0.001) (Fig. 1). The results of this study characterized resistin as a physiological constituent of amniotic fluid.

Bugatto *et al.* (2010) addressed the question of whether amniotic fluid resistin concentrations in the mid-trimester (15-20 wk, n = 70) are associated with maternal body mass index. The authors reported a positive correlation between maternal body mass index and amniotic fluid concentrations of this adipocytokine (r = 0.396; p = 0.01). In addition, a comparison between overweight/obese (n = 35) and normal-weight pregnant women (n = 35) revealed that the former have higher amniotic fluid resistin concentrations (20.68 ng/mL vs. 15.26 ng/mL, p = 0.01). The authors propose that exposure of the fetus to altered amniotic fluid resistin concentrations, as well as to other adipokines and cytokines, may have long-term deleterious effects on the fetus' health.

SPONTANEOUS LABOR AT TERM AND AMNIOTIC FLUID RESISTIN CONCENTRATIONS

The association between term labor and amniotic fluid concentrations of resistin was determined in one study that included normal pregnant women at term with (n = 49) and without (n = 50) spontaneous labor (Kusanovic *et al.* 2008). No significant differences were observed in the median amniotic fluid resistin concentration between patients with spontaneous labor at term and those at term not in labor (28.7 ng/mL vs. 23.6 ng/mL, p = 0.07) (Fig. 1), suggesting that this protein does not play a major role in term parturition.

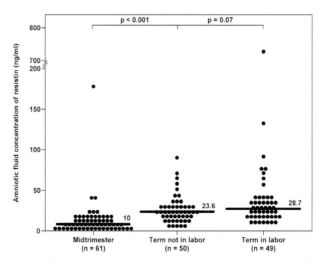

Fig. 1 Amniotic fluid concentrations of resistin in normal pregnancies at mid-trimester and in those at term with and without labor. The median amniotic fluid concentration of resistin was significantly higher in pregnancies at term not in labor than in those in the mid-trimester (23.6 ng/mL, IQR 16.3-31.1 vs. 10 ng/mL, IQR 6.2-15.3; p < 0.001). In contrast, no significant differences were observed in the median amniotic fluid resistin concentration between women with spontaneous labor at term and those at term not in labor (28.7 ng/mL, IQR 19.3-39.4 vs. 23.6 ng/mL, IQR 16.3-31.1; p = 0.07). Reproduced from Kusanovic *et al.* 2008, with permission.

In contrast, intra-amniotic infection at term is associated with high resistin concentrations in the presence of clinical chorioamnionitis (Kusanovic *et al.* 2008). We reported that among patients with clinical chorioamnionitis at term, those with microbial invasion of the amniotic cavity (MIAC) had a significantly higher median amniotic fluid concentration of resistin than those with a negative amniotic fluid culture (110.5 ng/mL vs. 40.3 ng/mL, p = 0.008) (Kusanovic *et al.* 2008) (Fig. 2).

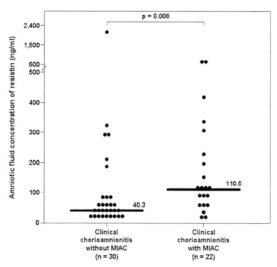

Fig. 2 Amniotic fluid concentrations of resistin in patients with clinical chorioamnionitis at term. The median amniotic fluid concentration of resistin was significantly higher in patients with microbial invasion of the amniotic cavity (MIAC) than in those without MIAC (110.5 ng/mL, IQR 52.4–243.4 vs. 40.3 ng/mL, IQR 26.2-85.9; p = 0.008). Reproduced from Kusanovic *et al.* 2008, with permission.

Table 1   Key features of resistin in amniotic fluid

- Resistin is a 12.5 kDa protein that belongs to the FIZZ (found in inflammatory zone) family, also known as RELM (resistin-like-molecules).
- While in mice resistin is produced almost exclusively by adipocytes, in humans it is not a tissue-specific protein. Several cells and tissues can express this protein, including neutrophils, macrophages and muscle.
- While the role of resistin as a significant metabolic mediator in humans is still controversial, emerging evidence implicates resistin in the regulation of innate immunity by virtue of its pro-inflammatory effect.
- Hyperresistinemia is a feature of sepsis and septic shock. Circulating plasma resistin concentrations are a marker of sepsis severity and have a predictive value for survival of critically ill patients.
- Resistin is a physiological constituent of amniotic fluid.
- The presence of intra-amniotic infection/inflammation is associated with elevated amniotic fluid resistin concentrations.
- The source of amniotic fluid resistin, as well as the mechanism(s) underlying its differential expression in amniotic fluid in the context of intra-amniotic infection and/or inflammation, is unclear. It is conceivable that amniotic fluid white blood cells secrete resistin in response to pro-inflammatory cytokines such as TNF-α.

AMNIOTIC FLUID RESITIN IN PATIENTS WITH SPONTANEOUS PRETERM LABOR WITH AND WITHOUT INTRA-AMNIOTIC INFECTION/INFLAMMATION

Consistent with the findings described above, among patients with MIAC at term, emerging evidence supports a role for resistin in the host response to intra-amniotic infection in patients with preterm labor (PTL). These observations include the following: (1) among patients with spontaneous PTL and intact membranes, those with IAI had a significantly higher median amniotic fluid concentration of resistin than those who delivered preterm without IAI (144.9 ng/mL vs. 18.7 ng/mL, p < 0.001) and those who subsequently delivered at term (16.3 ng/mL, p < 0.001) (Fig. 3). There were no differences in the median amniotic fluid resistin concentration between patients with PTL without IAI who delivered preterm and those who delivered at term (Fig. 3) (Kusanovic *et al.* 2008); (2) when the analysis was restricted to patients with PTL without IAI who delivered within 48 h from

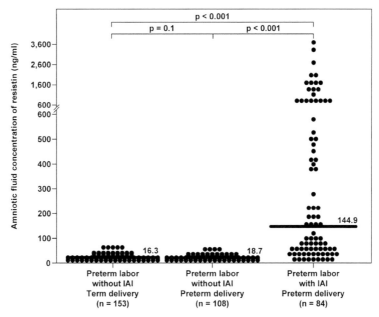

Fig. 3 Amniotic fluid concentrations of resistin among women with spontaneous preterm labor (PTL) and intact membranes. The median amniotic fluid concentration of resistin was significantly higher in patients with intra-amniotic infection/inflammation (IAI) than in women without IAI (144.9 ng/mL, IQR 44.6-623.2 vs. 18.7 ng/mL, IQR 12.1-25.8; p < 0.001) as well as in those who delivered at term (144.9 ng/mL, IQR 44.6-623.2 vs. 16.3 ng/mL, IQR 12.5-22.3; p < 0.001). There was no significant difference in the median amniotic fluid concentration of resistin between those who delivered preterm and those who delivered at term. Reproduced from Kusanovic *et al.* 2008, with permission.

amniocentesis, this subgroup had a significantly higher median amniotic fluid concentration of resistin than those with PTL and intact membranes who delivered at term (23.2 ng/mL vs. 16.3 ng/mL p = 0.02) (Kusanovic *et al.* 2008) (Fig. 4). Similar results were found in patients with PTL and intact membranes who delivered within 7 d (p = 0.007); (3) patients with histological chorioamnionitis and/or funisitis who delivered within 72 h from amniocentesis had a significantly higher median resistin concentration in amniotic fluid than those without histological inflammation (495.1 ng/mL vs. 29.1 ng/mL, p < 0.001) (Kusanovic *et al.* 2008) (Fig. 5); (4) the median amniocentesis-to-delivery interval was significantly shorter in patients with an amniotic fluid resistin concentration ≥ 37 ng/mL (derived from an ROC curve) compared to those with a concentration < 37 ng/mL [3 d (95% CI 2-4) vs. 43 d (95% CI 39-47); p < 0.001] (Kusanovic *et al.* 2008) (Fig. 6).

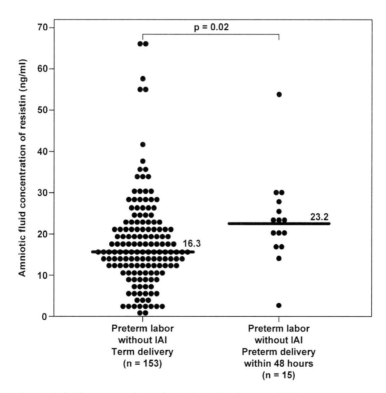

Fig. 4   Amniotic fluid concentrations of resistin in the absence of IAI: spontaneous preterm labor (PTL) vs. term delivery. Patients with spontaneous preterm labor without IAI who delivered within 48 h of amniocentesis had a significantly higher median amniotic fluid concentration of resistin than those who delivered at term (23.2 ng/mL, IQR 18.4-28.3 vs. 16.3 ng/mL, IQR 12.5-22.3; p = 0.02). Reproduced from Kusanovic *et al.* 2008, with permission.

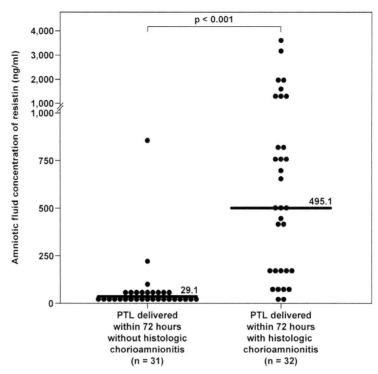

Fig. 5 Amniotic fluid concentrations of resistin in patients with spontaneous preterm labor with and without histologic chorioamnionitis who delivered within 72 h from amniocentesis. Patients with histologic chorioamnionitis and/or funisitis had a significantly higher median resistin concentration in amniotic fluid than those without histologic inflammation (495.1 ng/mL, IQR 15.7-3554.1 vs. 29.1 ng/mL, IQR 3-846.4, respectively; p < 0.001). Reproduced from Kusanovic *et al.* 2008, with permission.

Recently, we used iTRAQ to identify proteins differentially regulated in amniotic fluid samples of women with spontaneous PTL and intact membranes with and without IAI. Resistin was one of the amniotic fluid proteins that were up-regulated in the presence of IAI and in patients who delivered preterm in the absence of IAI (Romero *et al.* 2009). The results of this unbiased approach using a high-throughput technique strongly support the aforementioned observations. Taken together, these findings suggest a strong association between elevated amniotic fluid resistin concentrations and IAI in patients with spontaneous PTL and intact membranes.

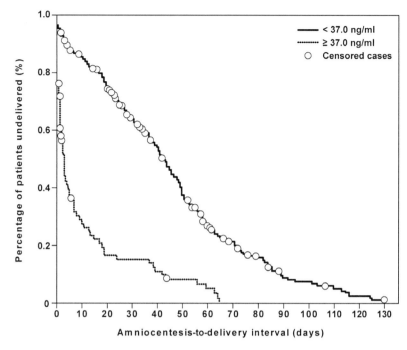

Fig. 6 Survival analysis of the amniocentesis-to-delivery interval (days) according to amniotic fluid resistin concentration cutoff ≥ 37 ng/mL in patients with spontaneous preterm labor and intact membranes. Patients with an amniotic fluid resistin concentration ≥ 37 ng/mL (dotted line) had a shorter median amniocentesis-to-delivery interval than those with an amniotic fluid resistin concentration < 37 ng/mL (solid line) (3 d, 95% CI 2-4 vs. 43 d, 95% CI 39-47, respectively; p < 0.001, log rank test). Reproduced from Kusanovic *et al.* 2008, with permission.

AMNIOTIC FLUID RESISTIN CONCENTRATIONS IN PATIENTS WITH PRETERM PRELABOR RUPTURE OF MEMBRANES WITH AND WITHOUT INTRA-AMNIOTIC INFECTION/INFLAMMATION

Similarly to pregnant women with PTL and intact membranes, patients with preterm prelabor rupture of membranes (pPROM) and IAI had a significantly higher median amniotic fluid resistin concentration than women with pPROM without IAI (132.6 ng/mL vs. 13 ng/mL, p < 0.001) (Fig. 7). From these findings, it can be concluded that amniotic fluid resistin concentrations are elevated in the presence of IAI regardless of the chorioamniotic membrane status.

Additional observations lend support to the association between amniotic fluid resistin concentrations and infection/inflammation. Amniotic fluid concentrations of resistin are correlated with amniotic fluid indices of infection/inflammation

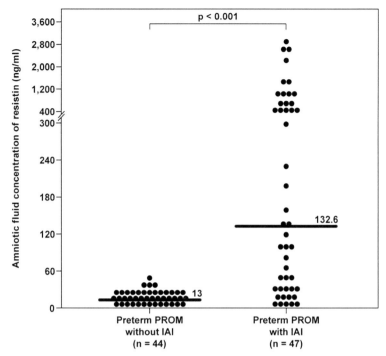

Fig. 7 Amniotic fluid concentrations of resistin in women with preterm prelabor rupture of the membranes (pPROM). The median amniotic fluid concentration of resistin was significantly higher in patients with intra-amniotic infection/inflammation (IAI) than in those without IAI (132.6 ng/mL, IQR 32.3-869.7 vs. 13 ng/mL, IQR 6.9-19.4; p = 0.02). Reproduced from Kusanovic *et al.* 2008, with permission.

including IL-6 and glucose concentrations, and white blood cell count, in patients with spontaneous PTL and those with pPROM (IL-6: r = 0.64, p < 0.001. Glucose: r = -0.24, p < 0.001. White blood cell count: r = 0.4, p < 0.001) (Kusanovic *et al.* 2008).

Collectively, the available evidence strongly suggests that resistin in amniotic fluid participates in the host response against intra-amniotic infection. These findings corroborate reports in the non-pregnant state in which resistin has been proposed to be a mediator of the innate immune response. It is important to mention that septic patients had a higher circulating resistin concentration than controls (Sunden-Cullberg *et al.* 2007). Moreover, hyperresistinemia characterizes human experimental endotoxemia (Lehrke *et al.* 2004). Nevertheless, the specific physiological role of resistin in amniotic fluid has not been fully clarified. Likewise, it needs to be determined whether resistin plays a role in the pathophysiology of IAI or rather alterations in its amniotic fluid concentration are secondary to the disease.

A possible explanation for the higher amniotic fluid resistin concentration in patients with IAI can be its secretion by neutrophils, which can be found in the amniotic cavity in the presence of infection and/or inflammation. This view is supported by the observations that resistin is produced and secreted by neutrophils (Johansson *et al.* 2009) and macrophages (Lehrke *et al.* 2004). Moreover, a positive correlation has been demonstrated between white blood cell count and circulating resistin concentration (Sunden-Cullberg *et al.* 2007, Johansson *et al.* 2009). It has been proposed that the secretion of resistin by neutrophils is mediated through TNF-α (Lehrke *et al.* 2004). Of note, IAI is associated with increased amniotic fluid concentrations of TNF-α (Romero *et al.* 1989, 1992), thus providing a putative molecular mechanism for the association between amniotic fluid resistin concentrations and IAI.

In conclusion, resistin is a physiological constituent of amniotic fluid. Independent lines of evidence strongly suggest that alterations in amniotic fluid concentrations of this adipocytokine are associated with intra-amniotic infection. The specific role of amniotic fluid resistin in normal gestation and complications of pregnancy merits further investigation.

Summary Points

- Resistin is a physiological constituent of the amniotic fluid.
- Amniotic fluid concentration of resistin increases as a function of gestational age.
- Spontaneous labor at term is not associated with alteration in amniotic fluid resistin concentrations.
- Intra-amniotic infection/inflammation is associated with increased amniotic fluid resistin concentrations regardless of chorioamniotic membrane status.
- High amniotic fluid resistin concentrations are associated with shorter amniocentesis-to-delivery interval in patients with preterm labor with intact membranes.
- High-throughput protein analysis of amniotic fluid reveals that resistin is one of the proteins up-regulated in patients with intra-amniotic infection/inflammation.

Abbreviations

IAI	:	intra-amniotic infection and/or inflammation
iTRAQ^{TM}	:	isobaric tag for relative and absolute quantitation
MIAC	:	microbial invasion of the amniotic cavity
pPROM	:	preterm prelabor rupture of membranes
PTL	:	preterm labor

Definition of Terms

Adipocytokines: Adipocyte-specific or enriched proteins that engage in extensive cross-talk within adipose tissue and with other tissues.

Clinical chorioamnionitis: Diagnosed in the presence of maternal fever ($\geq$ 37.8°C) and two or more of the following criteria: uterine tenderness, malodorous vaginal discharge, maternal leukocytosis ($\geq$ 15,000 cells/mm^3), maternal tachycardia (> 100 beats/min) and fetal tachycardia (> 160 beats/min).

Intra-amniotic infection: A positive amniotic fluid culture for micro-organisms. This condition is most often sub-clinical.

Intra-amniotic inflammation: An amniotic fluid interleukin 6 concentration $\geq$ 2.6 ng/mL or an amniotic fluid white blood cell count of more than 100 cells/mm^3.

Prelabor rupture of membranes: Rupture of the chorioamniotic membranes before the onset of labor.

Spontaneous preterm labor: The presence of regular uterine contractions occurring at a frequency of at least two every 10 min associated with uterine cervical change before 37 completed weeks of gestation.

References

Arner, P. 2005. Resistin: yet another adipokine tells us that men are not mice. Diabetologia 48: 2203-2205.

Bokarewa, M. and I. Nagaev, L. Dahlberg, *et al.* 2005. Resistin, an adipokine with potent proinflammatory properties. J. Immunol. 174: 5789-5795.

Bugatto, F. and A. Fernandez-Deudero, A. Bailen, *et al.* 2010. Second-trimester amniotic fluid proinflammatory cytokine levels in normal and overweight women. Obstet. Gynecol. 115: 127-133.

Chen, D. and M. Dong, Q. Fang, *et al.* 2005. Alterations of serum resistin in normal pregnancy and pre-eclampsia. Clin. Sci. (Lond.) 108: 81-84.

Cortelazzi, D. and S. Corbetta, S. Ronzoni, *et al.* 2007. Maternal and foetal resistin and adiponectin concentrations in normal and complicated pregnancies. Clin. Endocrinol. (Oxf.) 66: 447-453.

Heilbronn, L.K. and J. Rood, L. Janderova, *et al.* 2004. Relationship between serum resistin concentrations and insulin resistance in nonobese, obese, and obese diabetic subjects. J. Clin. Endocrinol. Metab. 89: 1844-1848.

Holcomb, I.N. and R.C. Kabakoff, B. Chan, *et al.* 2000. FIZZ1, a novel cysteine-rich secreted protein associated with pulmonary inflammation, defines a new gene family. EMBO J. 19: 4046-4055.

Johansson, L. and A. Linner, J. Sunden-Cullberg, *et al.* 2009. Neutrophil-derived hyperresistinemia in severe acute streptococcal infections. J. Immunol. 183: 4047-4054.

Kaser, S. and A. Kaser, A. Sandhofer, *et al.* 2003. Resistin messenger-RNA expression is increased by proinflammatory cytokines *in vitro*. Biochem. Biophys. Res. Commun. 309: 286-290.

Kim, K.H. and K. Lee, Y.S. Moon, *et al.* 2001. A cysteine-rich adipose tissue-specific secretory factor inhibits adipocyte differentiation. J. Biol. Chem. 276: 11252-11256.

Kusanovic, J.P. and R. Romero, S. Mazaki-Tovi, *et al.* 2008. Resistin in amniotic fluid and its association with intra-amniotic infection and inflammation. J. Matern. Fetal Neonatal Med. 21: 902-916.

Lappas, M. and K. Yee, M. Permezel, *et al.* 2005. Release and regulation of leptin, resistin and adiponectin from human placenta, fetal membranes, and maternal adipose tissue and skeletal muscle from normal and gestational diabetes mellitus-complicated pregnancies. J. Endocrinol. 186: 457-465.

Lehrke, M. and M. P. Reilly, S. C. Millington *et al.*, "An inflammatory cascade leading to hyperresistinemia in humans. PLoS. Med. 1: e45.

Mazaki-Tovi, S. and R. Romero, J.P. Kusanovic, *et al.* 2008. Visfatin/Pre-B cell colony-enhancing factor in amniotic fluid in normal pregnancy, spontaneous labor at term, preterm labor and prelabor rupture of membranes: an association with subclinical intrauterine infection in preterm parturition. J. Perinat. Med. 36: 485-496.

Mazaki-Tovi, S. and R. Romero, J.P. Kusanovic, *et al.* 2009a. Maternal visfatin concentration in normal pregnancy. J. Perinat. Med. 37: 206-217.

Mazaki-Tovi, S. and R. Romero, J.P. Kusanovic, *et al.* 2009b. Visfatin in human pregnancy: maternal gestational diabetes vis-a-vis neonatal birthweight. J. Perinat. Med. 37: 218-231.

Mazaki-Tovi, S. and R. Romero, E. Vaisbuch, *et al.* 2009c. Maternal serum adiponectin multimers in preeclampsia. J. Perinat. Med. 37: 349-363.

Mazaki-Tovi S. and R. Romero, E. Vaisbuch, *et al.* 2009d. Maternal plasma visfatin in preterm labor. J. Matern. Fetal Neonatal Med. 22: 693-704.

Mazaki-Tovi S. and R. Romero, E. Vaisbuch, *et al.* 2009e. Maternal serum adiponectin multimers in gestational diabetes. J. Perinatal Med. 37: 637-650.

Mazaki-Tovi S. and R. Romero, E. Vaisbuch, *et al.* 2009f. Adiponectin in amniotic fluid in normal pregnancy, spontaneous labor at term, and preterm labor: A novel association with subclinical intrauterine infection/inflammation. J. Matern. Fetal Neonatal Med. 23: 120-130.

Nien, J.K. and S. Mazaki-Tovi, R. Romero, *et al.* 2007. Resistin: a hormone which induces insulin resistance is increased in normal pregnancy. J. Perinat. Med. 35: 513-521.

Pagano, C. and O. Marin, A. Calcagno, *et al.* 2005. Increased serum resistin in adults with prader-willi syndrome is related to obesity and not to insulin resistance. J. Clin. Endocrinol. Metab. 90: 4335-4340.

Patel, S.D. and M.W. Rajala, L. Rossetti, *et al.* 2004. Disulfide-dependent multimeric assembly of resistin family hormones. Science 304: 1154-1158.

Romero, R. and J.P. Kusanovic, F. Gotsch, *et al.* 2010. Isobaric labeling and tandem mass spectrometry: A novel approach for profiling and quantifying proteins differentially expressed in amniotic fluid in preterm labor with and without intra-amniotic infection/inflammation. J. Matern. Fetal Neonatal Med. 23: 261-280.

Romero, R. and K.R. Manogue, M.D. Mitchell, *et al.* 1989. Infection and labor. IV. Cachectin-tumor necrosis factor in the amniotic fluid of women with intraamniotic infection and preterm labor. Am. J. Obstet. Gynecol. 161: 336-341.

Romero, R. and M. Mazor, W. Sepulveda, *et al.* 1992. Tumor necrosis factor in preterm and term labor. Am. J. Obstet. Gynecol. 166: 1576-1587.

Schaffler, A. and A. Ehling, E. Neumann, *et al.* 2003. Adipocytokines in synovial fluid. JAMA 290: 1709-1710.

Silha, J.V. and M. Krsek, J.V. Skrha, *et al.* 2003. Plasma resistin, adiponectin and leptin levels in lean and obese subjects: correlations with insulin resistance. Eur. J. Endocrinol. 149: 331-335.

Steppan, C.M. and S.T. Bailey, S. Bhat, *et al.* 2001. The hormone resistin links obesity to diabetes. Nature 409: 307-312.

Sunden-Cullberg, J. and T. Nystrom, M.L. Lee, *et al.* 2007. Pronounced elevation of resistin correlates with severity of disease in severe sepsis and septic shock. Crit. Care Med. 35: 1536-1542.

Tilg, H. and A.R. Moschen. 2006. Adipocytokines: mediators linking adipose tissue, inflammation and immunity. Nat. Rev. Immunol. 6: 772-783.

Vaisbuch, E. and S. Mazaki-Tovi, J.P. Kusanovic, *et al.* 2009. Retinol binding protein 4: an adipokine associated with intra-amniotic infection/inflammation. J. Matern. Fetal Neonatal Med.: PMID: 19900011.

Wang, H. and W.S. Chu, C. Hemphill, *et al.* 2002. Human resistin gene: molecular scanning and evaluation of association with insulin sensitivity and type 2 diabetes in Caucasians. J. Clin. Endocrinol. Metab. 87: 2520-25.

Index

About the Editors

Victor R. Preedy BSc, PhD, DSc, FIBiol, FRCPath, FRSPH is Professor of Nutritional Biochemistry, King's College London, Professor of Clinical Biochemistry, Kings College Hospital and Director of the Genomics Centre, King's College London. Presently he is a member of the Kings College London School of Medicine. Professor Preedy graduated in 1974 with an Honours Degree in Biology and Physiology with Pharmacology. He gained his University of London PhD in 1981 when he was based at the Hospital for Tropical Disease and The London School of Hygiene and Tropical Medicine. In 1992, he received his Membership of the Royal College of Pathologists and in 1993 he gained his second doctoral degree, i.e. DSc, for his outstanding contribution to protein metabolism in health and disease. Professor Preedy was elected as a Fellow to the Institute of Biology in 1995 and to the Royal College of Pathologists in 2000. Since then he has been elected as a Fellow to the Royal Society for the Promotion of Health (2004) and The Royal Institute of Public Health (2004). In 2009, Professor Preedy became a Fellow of the Royal Society for Public Health. In his career Professor Preedy has carried out research at the National Heart Hospital (part of Imperial College London) and the MRC Centre at Northwick Park Hospital. He has collaborated with research groups in Finland, Japan, Australia, USA and Germany. He is a leading expert on the pathology of disease and has lectured nationally and internationally. He has published over 570 articles, which include over 165 peer-reviewed manuscripts based on original research, 90 reviews and 20 books.

Ross J. Hunter MD BSc MRCPath trained in medical sciences at King's College London (Times University ranking 11th in UK). He spent a further year at Imperial College London (Times University ranking 3rd in UK) and was awarded his BSc in Cardiovascular medicine in 1998. Since returning to his medical training at King's College School of Medicine, he has remained an honorary research fellow at The Department of Nutritional Sciences, researching the effect of different nutritional states and alcoholism on the cardiovascular system. He was awarded his Bachelor of Medicine & Surgery (MBBS) with distinction in 2001. He trained in general medicine in London and Brighton and was made a member of the Royal College of Physicians (UK) in 2005. He trained as a Registrar in the London Deanery from 2005-2008. Since 2008 he has been a research fellow at the Department of

Cardiology & Electrophysiology at St Bartholomew's Hospital London, conducting clinical research and clinical trials in cardiology and electrophysiology. He has published over 60 scientific articles of various kinds.

Color Plate Section

Chapter 2

Fig. 2   Structure of adiponectin (Shapiro and Scherer 1998). The homotrimeric crystal structure of Acrp30/adiponectin was downloaded from the PDB protein databank (structure 1C3H).

Chapter 3

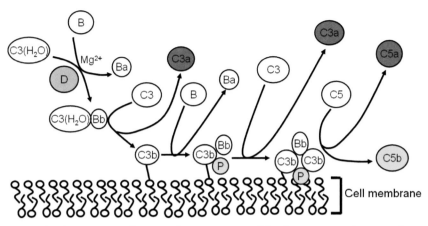

Fig. 1 The alternative complement pathway. Activation of the alternative complement pathway is initiated by the spontaneous hydrolysis of C3. Binding of factor B to $C3(H_2O)$ allows factor D/ adipsin to cleave factor B into Ba and Bb. Bb and $C3(H_2O)$ forms the complex, also known as C3 convertase. This convertase, although only produced in small amounts, can cleave multiple C3 proteins into C3a and C3b. After the creation of C3 convertase, the complement system follows the same path to the classical pathway. Binding of another C3b-fragment to the C3-convertase of the alternative pathway creates a C5-convertase analogous to the classical pathway. (This figure is modified from a file in Wikimedia Commons. Wikimedia Commons declares that permission is granted to copy, distribute and/or modify this document under the terms of the GNU Free Documentation License.)

Chapter 8

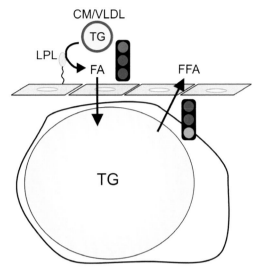

Fig. 2 Role of Angptl4 in lipid metabolism. Angptl4 inhibits LPL-mediated lipolytic processing of TG-rich lipoprotein particles. CM, chylomicron; VLDL, very low density lipoprotein; LPL, lipoprotein lipase; FA, fatty acid. Traffic lights indicate whether Angptl4 has a stimulatory (green) or inhibitory (red) effect.

Chapter 10

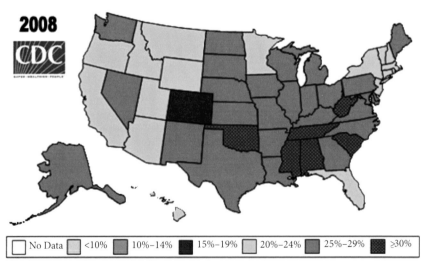

Fig. 2 Obesity rates, by state, for the USA in 2008. This illustration is made available for use by the CDC website.

Chapter 12

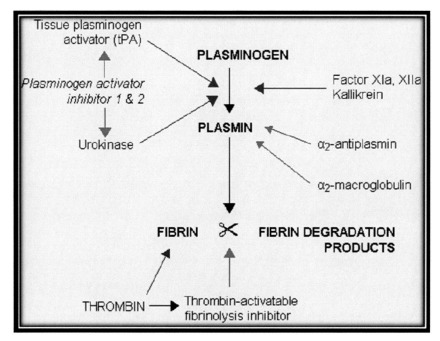

Fig. 3 Fibrinolytic pathway. Blue arrows denote stimulation and red arrows denote inhibition (*Source:* Wikipedia, the Free Encyclopedia). Fibrinolysis is normally mediated by plasmin, which circulates in blood as its pro-enzyme, plasminogen. Conversion and activation of plasminogen into plasmin is effected by two plasminogen activators: tissue-type plasminogen activator (t-PA) and urokinase-type plasminogen activator (u-PA). The process of fibrinolysis can thus be inhibited by inhibition plasmin itself, by α-2 anti-plasmin, or by inhibition of the plasminogen activators by PAI-1.

Chapter 13

Fig. 2 Crystal structure of mouse resistin. The homotrimeric crystal structure of resistin was downloaded from the PDB protein databank (structure 1RGX) (Patel *et al.* 2004).

Chapter 15

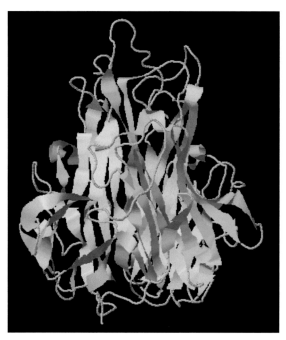

Fig. 1 Structure of TNF-α. Note the homotrimeric structure and jelly-roll β-sandwich fold characteristic of the C1q/TNF superfamily (Eck and Sprang 1989).

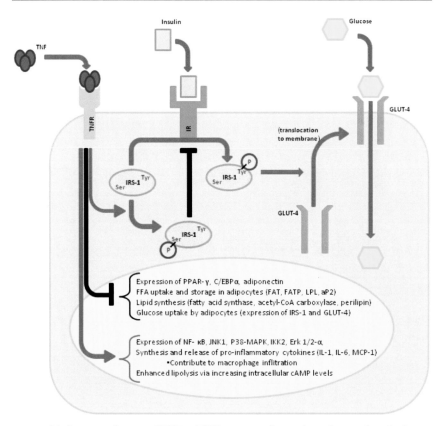

Fig. 3 Mechanisms of action of TNF-α. TNF-α can interfere with insulin signaling, leading to decreased translocation of GLUT4 to the plasma membrane. It also influences transcription of many metabolic genes.

Chapter 20

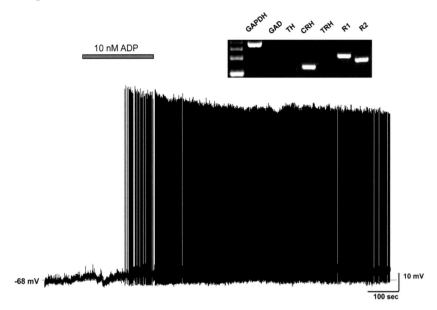

Fig. 2   Adiponectin depolarizes CRH neurons in the paraventricular nucleus of the hypothalamus. This electrophysiological trace shows a current clamp recording from an electrophysiologically identified neuroendocrine neuron identified post hoc using single cell RT-PCR (gel inset) as a CRH neuron that expresses both AdipoR1 (R1) and AdipoR2 (R2). This cell was negative for GAD (GABA marker), tyrosine dehydroxylase (TH, dopamine marker), and TRH. The red bar shows the time of bath application of 10 nM adiponectin and this CRH neuron responded with a small depolarization (8 mV) and a robust increase in action potential frequency (adapted from Hoyda *et al.* 2009a, with permission).

PVN OT Neurons

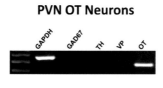

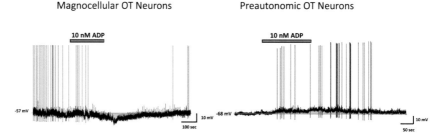

Fig. 3 Adiponectin has opposite effects on different subpopulations of paraventricular nucleus oxytocin neurons. This figure shows examples of recordings from two PVN neurons identified as OT cells using post hoc single cell RT-PCR techniques as shown in the agarose gel presented at the top of the figure. Using electrophysiological fingerprinting we can further subdivide these cells into magnocellular (expression of a dominant transient potassium current, projecting to posterior pituitary), or preautonomic (expression of a low threshold calcium current, projecting to caudal autonomic centers). Intriguingly, while magnocellular OT neurons respond to bath administration of 10 nmol adiponectin with hyperpolarizations (as illustrated in the current clamp recording on the left side), preautonomic OT neurons are depolarized by adiponectin (adapted from Hoyda *et al.* 2007, 2009a, with permission).

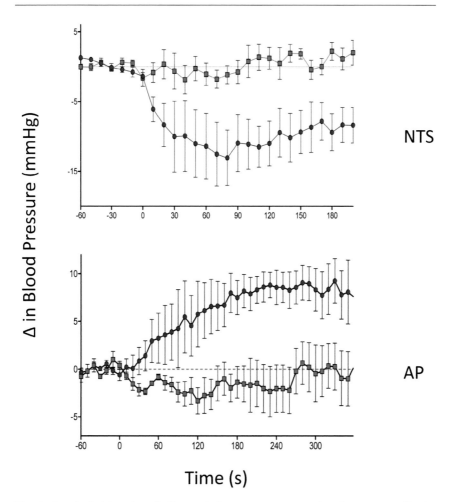

Fig. 4 Local administration of adiponectin into area postrema and nucleus tractus solitarius of anaesthetized rats influences blood pressure. Normalized mean blood pressure (upper) traces showing the different cardiovascular responses to adiponectin microinjection (0.5 µl at time = 0) into the medial nucleus tractus solitarius (NTS - 50 fmol, upper graph blue circles) and the area postrema (AP - 1 pmol, lower graph blue circles). Importantly, despite the close anatomical proximity of these two regions in the medulla, microinjections of adiponectin exert opposite effects on blood pressure. In both graphs the red square indicates the effects of control microinjections (vehicle or inactive peptide) into the same region (adapted from Fry *et al.* 2006, Hoyda *et al.* 2009b, with permission).

Chapter 21

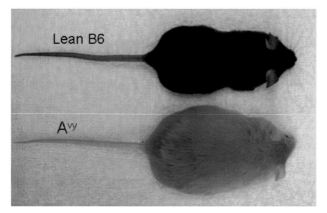

Fig. 1 Two-month-old adult-onset obese A^{vy} mouse and its lean 2-mo-old B6 control.

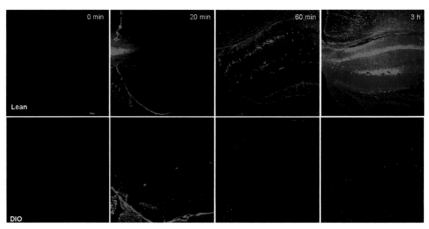

Fig. 4 Distribution of fluorescently labeled leptin in the hippocampus at various times after icv administration. In the lean control (top), Alexa568-leptin reached neurons in the hippocampus by 60 min. In the obese mouse (bottom), prominent fluorescence was seen in the choroid plexus and part of the hilus at 20 min, but was no longer present in the hippocampus at 60 min and 3 h.

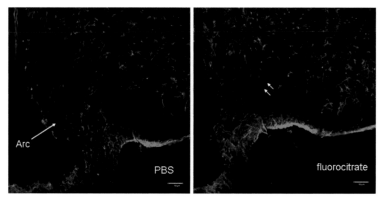

Fig. 5 Distribution of fluorescently labeled leptin after fluorocitrate pretreatment preceding icv injection of Alexa568-leptin. In the presence of this metabolic inhibitor of astrocytic activity, there was an increased amount of Alexa568-leptin (+) cells in the arcuate nucleus (arrows) of the hypothalamus as compared with the diluent-treated controls.

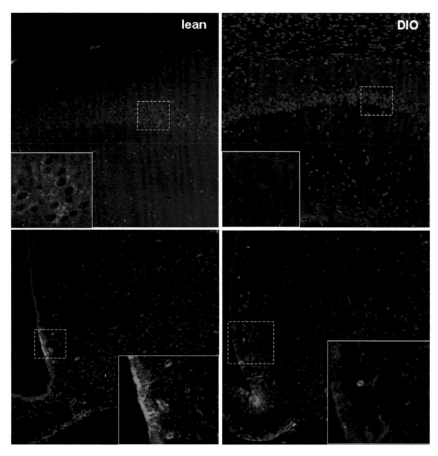

Fig. 7 pSTAT3 signaling in lean (left) and DIO (right) mice 20 min after leptin administered icv. The activation of pSTAT3 in the lean B6 mice, seen in focal areas in the CA1 of the hippocampus, in the α2 tanycytes lining the third ventricle, and in the adjacent arcuate nucleus (boxes) was much less in the corresponding regions of the DIO mouse.

Chapter 22

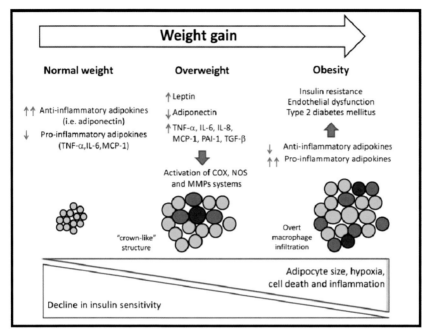

Fig. 1   Patterns of adipokine secretion in the natural history of obesity. Schematic representation of the change in the pattern of secretion of adipokines and insulin sensitivity from the healthy lean state (left) to frank obesity (right). Hypoxia and cell death and adipokines have been shown to trigger macrophage infiltration in adipose tissue with the resultant increase in adipokines and the resultant vicious circle that activates alternate pathways of inflammation and eventually progresses to type 2 diabetes mellitus and cardiovascular disease. COX = cycloxygenase, NOS = nitric oxide synthase, MMP = matrix metalloproteinase; TGF-beta = transforming growth factor-beta. Dark circle = dead adipocyte, red circles = macrophages (M1 stage).

Chapter 24

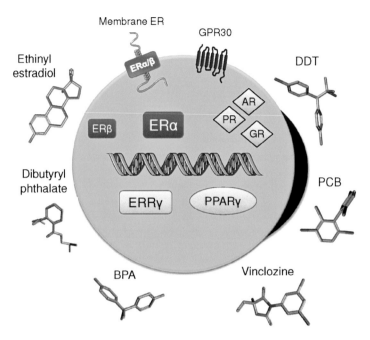

Fig. 2 A model of the adipocyte illustrating the various receptors, located in the membrane, cytoplasm or nucleus, which have been reported to mediate the action of endocrine-disrupting compounds (EDC). The structure of selected EDCs is shown outside of the adipocyte. BPA, bisphenol A; DDT, dichlorodiphenyltrichloroethane; ER, estrogen receptor; ERR, estrogen-related receptor; GPR30, G protein-coupled receptor 30; GR, glucocorticoid receptor; PCB, polychlorinated biphenyl; PPAR, peroxisome proliferator-activated receptor; PR, progesterone receptor.

Chapter 25

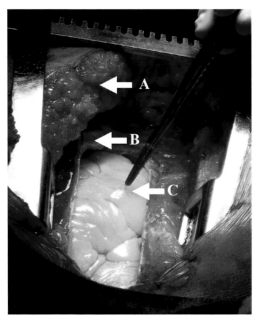

Fig. 1   Adipose tissue of the heart. Large cardiac adipose tissue in a patient with coronary artery disease during a triple coronary artery bypass. The adipose tissue abnormally covers the entire heart in this patient. (A) Adipose tissue deposited in the chest under the skin (subcutaneous). (B) Pericardial adipose tissue. (C) Epicardial adipose tissue pulled up by forceps. (This figure is unpublished in this current version and comes from the author's own collection. Patient's informed consent had been obtained.)

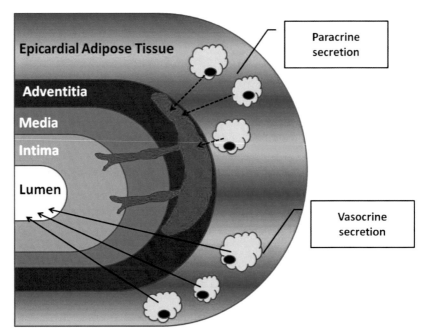

Fig. 3 Paracrine or vasocrine secretion of adipokines from the epicardial fat. A diagram of two possible mechanisms by which adipokines might reach the coronary artery lumen from the epicardial fat. Adipokines from periadventitial epicardial fat could traverse the coronary wall by diffusion from outside to inside by a paracrine mechanism. Adipokines might be released from epicardial tissue directly into vasa vasorum and be transported downstream into the arterial wall by a vasocrine mechanism. (This artwork is unpublished and comes from the author's own collection.)

Chapter 26

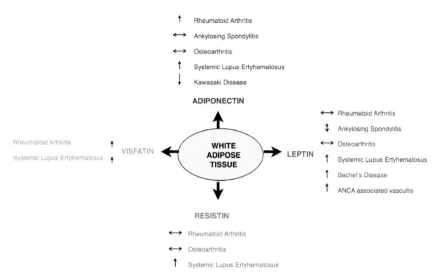

Fig. 1 Levels of adiponectin, leptin, resistin and visfatin in rheumatic diseases. The upward arrow indicates increased levels as compared to controls while the downward arrow indicates decreased levels as compared to controls.

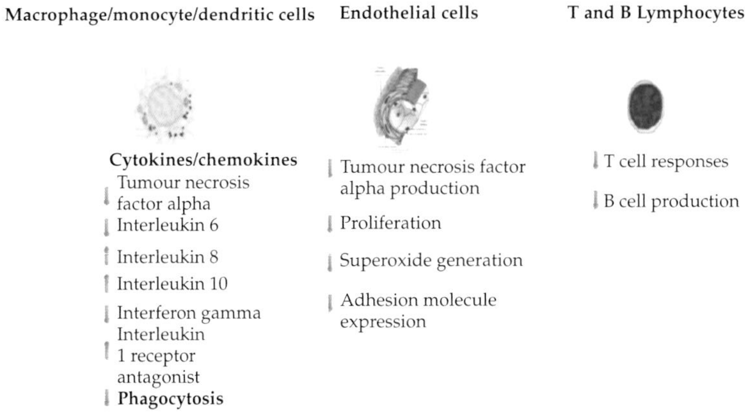

Macrophage/monocyte/dendritic cells	Endothelial cells	T and B Lymphocytes

Cytokines/chemokines
↓ Tumour necrosis factor alpha

↓ Interleukin 6

↑ Interleukin 8

↑ Interleukin 10

↓ Interferon gamma

↑ Interleukin 1 receptor antagonist

↓ **Phagocytosis**

↓ Tumour necrosis factor alpha production

↓ Proliferation

↓ Superoxide generation

↓ Adhesion molecule expression

↓ T cell responses

↓ B cell production

Fig. 2 Known immunological effects of adiponectin on macrophages/ monocytes/dendritic cells, endothelial cells and T and B cells as the effects pertain to rheumatic diseases. The upward arrow indicates enhancement and the downward arrow indicates down-regulation. Although macrophages, monocytes and dendritic cells each serve different function, the effects of adiponectin on both cytokine/chemokine production and phagocytosis are similar in all three cells and therefore they are put under one column for simplicity. The overall effect of adiponectin on these cells is anti-inflammatory. Similarly, adiponectin decreases endothelial cell proliferation, superoxide generation, and the production of the pro-inflammatory cytokine tumour necrosis factor alpha with an overall anti-inflammatory effect. It also decreases the up-regulation of adhesion molecules and thus will not all allow inflammatory cells to migrate through the vessel wall. The last part of the figure demonstrates that although the mechanism by which adiponectin affects T and B cells is not well studied, the overall effect is to decrease the response of these cells. However, each of these effects alter individual rheumatic diseases as outlined in the text.

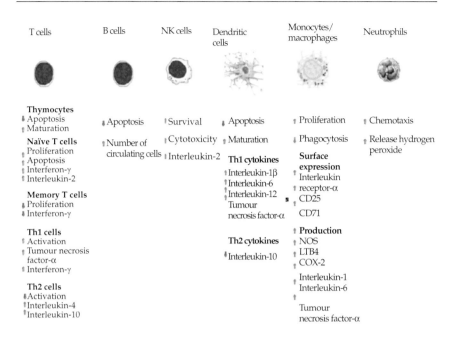

Fig. 3 Immunological effects of leptin in rheumatic diseases. The details for each disease are outlined in the text. The upward arrow indicates enhancement and the downward arrow indicates down-regulation. As can be seen the immunological effects of leptin have been well studied for many cell lines. Leptin decreases apoptosis of thymocytes, naïve T cells and B cells and increases the number of these cells. It promotes the production of the pro-inflammatory cytokines interleukin 2, tumour necrosis factor alpha, and interferon-gamma leading to the activation Th1 cells (pro-inflammatory cells) while also decreasing the activation of Th2 cells although it increases interleukin 10 and interleukin 4 for an overall net pro-inflammatory response from T cells. Many of these cytokines have important roles in the pathogenesis of rheumatic diseases. It also activates natural killer cells and increases their cytotoxicity and survival, which further activates the immune system. The effect of leptin on the antigen-presenting cells, dendritic cells and monocytes/macrophages is pro-inflammatory as it increases the maturation of these cells to become more effective and decreases apoptosis. Furthermore, it results in the production of multiple pro-inflammatory cytokines by these cells as shown in the illustration as well as enhancing the production of the pro-inflammatory molecules/enzymes nitric oxide (NOS), leutriene B4 (LBT4) and cylcooxygenase enzyme 2 (COX-2). The overall effect again is pro-inflammatory. Lastly, it activates neutrophils, enhances phagocytosis, and releases hydrogen peroxide to allow this non-specific effector to perform its killing function better (similar to the effect on natural killer cells).

Chapter 32

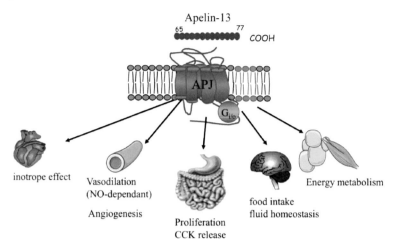

Fig. 1 Principal central and peripheral effects of apelin. Apelin peptides (such as apelin-13) bind to membrane APJ receptor and exert different effects in the cardiovascular and the gastrointestinal systems but also in the central nervous system. Recently, apelin was shown to play a role in energy metabolism.

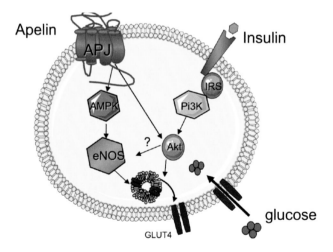

Fig. 2 Apelin signaling pathway involved in muscle glucose transport. Apelin stimulates AMPK, eNOS and Akt activation and then GLUT4 transporters are translocated to the membrane allowing glucose uptake in muscle. It is not known whether apelin can activate Akt independently of AMPK activation. Apelin pathway is partly independent of the first steps of insulin signaling requiring PI3K-associated IRS proteins.

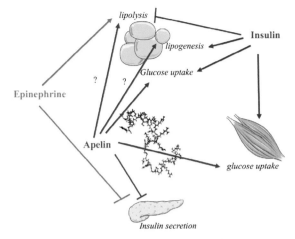

Fig. 3 Apelin and insulin effects on energy metabolism. Insulin increases energy storage by stimulating lipogenesis and inhibiting lipolysis (hydrolysis of triglycerides) in adipose tissue and by stimulating glucose uptake mainly in skeletal muscle. Insulin effects are opposite to those of epinephrine, which known to stimulate lipolysis and to inhibit insulin secretion. Apelin shares effects with both insulin and epinephrine, since apelin stimulates glucose uptake (like insulin) and inhibits insulin secretion (like epinephrine). However, the effects of apelin on lipolysis and lipogenesis are not known but apelin seems rather to stimulate lipid utilization.

Chapter 34

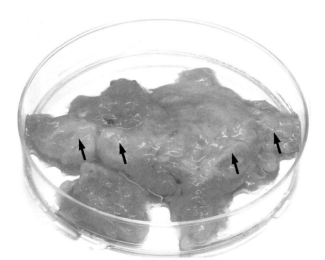

Fig. 1 Synovial tissue of a patient with rheumatoid arthritis. There are numerous adipocytes in the synovial tissue (arrow).

Chapter 35

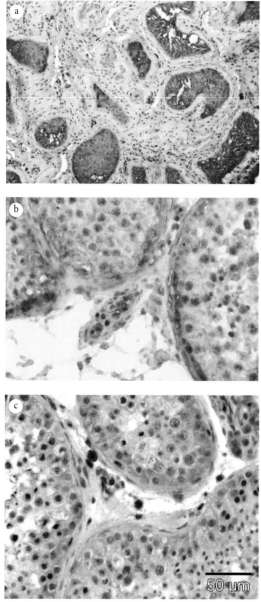

Figure 1. Examples of immunohistochemistry of resistin staining in a testis (a) without germ cells (Sertoli cell only syndrome), (b) with germ cells (impaired spermatogenesis) and (c) CD68-positive macrophages in human testis

Printed and bound by CPI Group (UK) Ltd, Croydon, CR0 4YY

18/10/2024

01776270-0010